Perspectival Realism

OXFORD STUDIES IN PHILOSOPHY OF SCIENCE

RECENTLY PUBLISHED IN THE SERIES:

Perspectival Realism

MICHELA MASSIMI

OXFORD
UNIVERSITY PRESS

OXFORD
UNIVERSITY PRESS

Oxford University Press is a department of the University of Oxford. It furthers
the University's objective of excellence in research, scholarship, and education
by publishing worldwide. Oxford is a registered trade mark of Oxford University
Press in the UK and certain other countries.

Published in the United States of America by Oxford University Press
198 Madison Avenue, New York, NY 10016, United States of America.

Library of Congress Cataloging-in-Publication Data
Names: Massimi, Michela, author.
Title: Perspectival realism / Michela Massimi.
Description: New York : Oxford University Press, [2022] |
Series: Oxford studies in philos science series |
Includes bibliographical references and index.
Identifiers: LCCN 2021039979 (print) | LCCN 2021039980 (ebook) |
ISBN 9780197555620 (hardback) | ISBN 9780197555644 (epub) | ISBN 9780197555651
Subjects: LCSH: Science—Philosophy.
Classification: LCC Q175 .M388 2022 (print) |
LCC Q175 (ebook) | DDC 501—dc23/eng/20211027
LC record available at https://lccn.loc.gov/2021039979
LC ebook record available at https://lccn.loc.gov/2021039980

DOI: 10.1093/oso/9780197555620.001.0001

3 5 7 9 8 6 4

Printed by Integrated Books International, United States of America

To Mark and Edward

Contents

I. PERSPECTIVAL MODELLING

THREE CASE STUDIES

II. THE WORLD AS WE PERSPECTIVALLY MODEL IT

PART I
PERSPECTIVAL MODELLING

1

The short tale of a long journey

1.1. Why I wrote a book on realism about science

'Realism is dead', intimated Arthur Fine back in 1984. Its death was announced by 'the neopositivists, who realized that they could accept all the results of science, including all the members of the scientific zoo, and still declare that the questions raised by the existence claims of realism were mere pseudo-questions' (Fine 1984/1991, p. 261). I share some of Fine's sentiment.[1] But I ultimately disagree with him about the demise of realism.

My original motivations for writing this book were fairly simple and, in a way, pre-philosophical. I have always been of the view that a realist stance on science offered a safeguard to a society where trust in science was being eroded before our eyes. I watched with apprehension TV news about measles and COVID-19 outbreaks due to anti-vaccine movements gaining traction among the public; international talks on climate change breaking down under the pressure of powerful political lobbies; and scientists forced to take to the streets and march for science.

This is in no way to suggest that *philosophical* anti-realism underwrites an anti-science stance. Scientific anti-realism has *never* been a bedfellow of anti-vaccine movements or climate change denials. On the contrary, it has encouraged an appropriately critical stance towards science as a way of reflecting on its empirical foundations and resisting metaphysical grand vistas. I have always found empiricists' arguments irresistible. And I learned from philosophical anti-realists like Bas van Fraassen (1980) never to underestimate the empirical origins of our scientific knowledge, its situated nature, or particular point of view. The story I am going to tell is thoroughly empiricist and perspectival from beginning to end.

---

[1] In a later paper, Fine (2018, pp. 42 and 45) returns to the topic and argues that there is no 'reliable "best practice" guide that links generic scientific tasks (build theories, measure parameters, look for novel phenomena, etc.) with meta-attitudes like realism, or instrumentalism, or empiricism. . . . Parity between truth and reliability marks a permanent impasse in arguments between realists and instrumentalists'. I broadly agree. Where I differ is that I see reliability no less than truth as key to the kind of realism I want to defend. And I believe that a closer look at a variety of situated scientific practices makes a strong case for a perspectival kind of realism that focuses on the reliability of the claims of knowledge advanced.

Perspectival Realism. Michela Massimi, Oxford University Press. © Oxford University Press 2022.
DOI: 10.1093/oso/9780197555620.003.0001

Thus, my attraction to realism has never been a 'call to arms', a defence of Realism with a capital R or Truth with a capital T (I find them almost empty notions.) Indeed, ever since I was a graduate student, I have had mixed feelings about realism. Like Arthur Fine, I too was concerned that the debate on realism/anti-realism in science had obsessed over questions about 'existence'. Do electrons exist? Do DNA strands exist? Does caloric exist? What about more complex or elusive entities such as dyslexia or dark matter?

I see now that this might not be the right question. Or necessarily the defining question about what it's like to be a realist about science. (Hats off to the neopositivists for their foreknowledge.) Realism often felt to me like a caricature for some non-better-qualified claim about what exists. But, surely (I thought), a realist about science cannot just be someone who believes in the existence of X, Y, and Z but rejects the existence of Q, P, and R.

The recent trend in philosophy of science to go *local* about realism has been a salutary response to this. Realism does not have to be all-or-nothing. Realist commitments may come in degrees and shades, depending on the context of inquiry and the entities at stake (on this see Hoefer and Martí 2020). Of course, we all want to know whether what exists is the electromagnetic field rather than the ether, or viruses as opposed to miasma, or an expanding universe instead of the ancient Greek crystalline spheres.

Yet the *philosophical* question about realism in science does not boil down to whether X (rather than Y) *exists*. Questions of existence—I follow W. V. O. Quine—are best left to scientists. Realism does not have to be some kind of philosophical running commentary on what scientists have discovered. I am not suggesting that getting clear on the *metaphysics* of science serves no cause. It has served the cause of realism well over decades. But to my mind it has done so at the cost of neglecting other important questions.

I'd like to think of realism as an answer to first and foremost an *epistemological question*. To be a realist about science is to be able to answer the question of how historically and culturally situated epistemic communities have over time come to *reliably advance* claims of scientific knowledge.[2] The explanandum for realism is not *what there is* but *how human beings come to reliably know* the natural world. If one reorients the realist commitment to science along these lines—*from within* scientific history, rather than from a scientific view from nowhere—not only is realism alive and well: a different flavour of realism about science is within our grasp.

---

[2] In what follows and throughout the book, I use the term 'scientific knowledge claims' as a shorthand for 'claims of scientific knowledge', namely, claims put forward by particular epistemic communities at particular historical times and whose truth and justification has to be established (rather than being already established as implicit in the notion of 'knowledge').

My starting point is, then, radically different from the concerns that over the past thirty years or so have shaped the debate on realism and anti-realism in science. The latter has revolved around the issue of whether or not our best theories in mature science tell us the truth about what *there is* in nature.

I won't have anything to say about scientific theories and their being approximately true or their terms being referential. Often in this debate scientific theories were portrayed as if they had a life of their own, living in Popper's (1978) 'World Three' of abstract objects. And questions of truth and reference were asked of scientific theories as we may ask of Beethoven's Sixth Symphony whether it is 'true' of bucolic life, or whether the fourth movement in F minor 'refers to' a thunderstorm fast approaching.

My focus will be instead on the reliability of the scientific knowledge claims. That reliability matters in science is not news. Think of Philip Kitcher's (1993, 2001) so-called Galilean strategy in arguing for the reliability of the telescope to observe ships coming to the Venetian shores no less than the craters of the Moon. Or of Naomi Oreskes's (2019) observation that the reliability of scientific claims and the ability of scientific communities to determine it are key to the trustworthiness of science.

What is different in my treatment is the role played by scientific perspectives in assessing reliability. I see reliability not as the sort of thing that individual epistemic communities can sanction or ratify on their own. My case for perspectival realism rests ultimately on the ability to assess reliable scientific knowledge claims *across a plurality of scientific perspectives*. These perspectives have often spanned long stretches of time and have historically 'interlaced'. The realism I articulate in this book arises from the deeply *social* and *cooperative* nature of scientific inquiry, with perspectival pluralism as its driving force.

1.2. What I mean by 'scientific perspective'

The term 'scientific perspective' has become common in philosophy of science with Giere's influential book (2006a) and a flurry of recent work on the topic of scientific perspectivism (see, e.g., the collection of essays in Massimi and McCoy 2019). No single generally agreed upon definition has emerged. The term lends itself to a variety of uses (metaphorical or not). For this book, I adopt the following working definition (expanding on Massimi 2018f and 2019a):

Scientific perspective (sp): A scientific perspective *sp* is the actual—historically and culturally situated—scientific practice of a real scientific community at a given historical time. Scientific practice should here be understood to include: (i) the body of *scientific knowledge claims*[3] advanced; (ii) the experimental, theoretical, and technological resources available to *reliably* make those scientific knowledge claims; and (iii) second-order (methodological-epistemic) principles that can *justify* the *reliability* of the scientific knowledge claims so advanced.

Metaphysical, philosophical, or religious beliefs may also have been influential in making the community endorse some claims of knowledge but do not count as part of a 'scientific perspective' as I am going to use the term. They are intended to explain how communities come to accept some knowledge claims but not how the community comes to *reliably* make, or *justify* the reliable procedures for advancing them.[4]

A few aspects in this working definition are worth highlighting. This notion of scientific perspective extends well beyond a class of scientific models (e.g. what might be called the Newtonian perspective, the Maxwellian perspective, etc.—for this terminology, see Giere 2006a). It encompasses the whole body of claims of knowledge advanced by specific epistemic communities at any particular historical time. This is a general enough characterization to encompass claims of knowledge generated via modelling practices such as those adopted by the Intergovernmental Panel on Climate Change Assessment Report 5 that I discuss in Chapter 4.b. But it is also meant to include a much larger body of claims elicited by other experimental and technological resources. I will discuss

---

[3] As already clarified, by 'scientific knowledge claims' I mean *claims of* scientific knowledge—the kind of claims that communities of epistemic agents advance at a particular historical time and using specific theoretical, experimental, and technological resources. Not all of them amount to genuine scientific knowledge (for some may prove wrong over time). Still, we would not want to deny the title of 'scientific perspective' to, say, Ptolemaic astronomy or similar just because some claims of knowledge proved false over time.

[4] Epistemic communities often come to accept and endorse some claims of knowledge on the basis of metaphysical, philosophical, and religious beliefs. But I have reasons for not including them in my definition of a 'scientific perspective'. I do not want scientific perspectives to be subject to the vagaries of Kuhnian paradigms (see Kuhn 1957) where, say, Renaissance Neoplatonism might be regarded as contributing to Copernican knowledge claims about the Earth orbiting the Sun. Neoplatonism, though of course influential at the time, did not play a direct role in establishing either the *truth* of or the *justification* for the *reliability* of Copernican knowledge claims (e.g. that the Earth orbits the Sun). Renaissance Neoplatonism might have been a contributing factor for the epistemic community to *accept and endorse* Copernicanism as an attractive view at the time. However, reasons for accepting and endorsing Copernicanism are not the same as reasons for reliably and justifiably coming to *know* Copernicanism. It would be odd to say, for example, either that Galileo's knowledge that the Earth orbits the Sun was *reliably* delivered by whatever metaphysical beliefs he might have held; or that his *reliably-forming* methods for such knowledge claims (e.g. the use of telescopic observations) were *justified* by metaphysical beliefs in Neoplatonism in the community at the time. Instead, I think we should say that Tycho Brahe's observational data, conjoined with Kepler's own laws (plus all the experimental, theoretical, and technological resources available at the time, including Galileo's own telescope), played an *evidential role* in reliably and justifiably coming to *know* that the Earth orbits the Sun. And that Neoplatonism played a key role in ensuring that Copernicanism got traction in the community as an acceptable view at the time.

some examples: from synthesizing *hachimoji* DNA in Chapter 7 to hydraulic techniques of the Alhambra engineers in Chapter 9; from local knowledge of honey-producing flora in the Yucatán peninsula in Chapter 8 to knowledge claims about Earth's magnetic field emerging from the use of Chinese 'wet' and 'dry' compasses and their use in medieval navigational practices in Chapter 11.

From an epistemic point of view, the *situatedness* of scientific knowledge runs deeper than just endorsing whichever fashionable scientific theory at any particular time. Scientific perspectives, as I'd like to think of them, are effectively proxies for scientific practices, broadly understood along the lines of (i)–(iii). For it is impossible to detach the body of scientific knowledge claims from the varieties of experimental, technological, and theoretical procedures employed in advancing them *reliably*; *and* from the methodological and epistemic principles that can in turn *justify* those *reliable procedures*.

It should be clear from these remarks that my definition of 'scientific perspective' owes a great deal to perspectival knowledge as described by Ernest Sosa in epistemology, who charts a middle ground beyond foundationalism and coherentism.[5] Sosa draws an important distinction between what he calls apt beliefs and justified beliefs. The former are *reliably* obtained at a first-order 'animal knowledge' level that we share with non-human animals. For example, my reliable belief that milk is in the fridge is something I share with my cat. Justified beliefs, by contrast, belong to a second-order 'reflective knowledge' level.[6] Here a perspectival ascent to an epistemic perspective is required in order to *reflect* on the sources of the *reliably forming* methods and procedures behind apt beliefs.

The epistemic perspective includes, then, first-order reliable claims of knowledge about the objects under investigation and second-order methodological and epistemic principles that *justify the reliability* of the experimental, theoretical, and technological resources used to make the first-order claims. There are a number of attractive features in this way of thinking. The first is that a clear distinction between *truth* and *justification* for claims of knowledge becomes immediately available. The truth of knowledge claims endorsed by particular epistemic communities is ultimately a matter of correspondence with the way the world is and depends on having *reliable* experimental, technological, and theoretical

---

[5] See in particular the essays 'The raft and the pyramid: coherence versus foundations in the theory of knowledge', 'The coherence of virtue and the virtue of coherence', and 'Intellectual virtue in perspective' (all in Sosa 1991). For a discussion of Sosa's work, see Greco (2004), especially Goldman (2004) and Sosa's reply (2004, pp. 312–313); and Carter (2020). My remarks here build on Massimi (2012a).

[6] For example, my reliable belief that milk is in the fridge may be justified by being part of an epistemic perspective which includes a coherent bunch of interrelated beliefs about, say, today being a Sunday, the grocery down the road being shut on a Sunday, and my husband having the foresight to buy additional milk bottles on a Friday.

procedures for arriving at these claims. How those reliably formed claims are in turn justified is, however, perspectival.[7]

What changes when historically shifting from one scientific perspective to another are not the reliably formed claims of knowledge (if they are indeed reliably formed), but the epistemic-methodological justificatory principles. The reliability and ultimately the truth of those knowledge claims is not fixed by the scientific perspective in which they might have originated. Scientific perspectives do not offer perspectival facts. Nor should truth be understood in terms of perspectival truthmakers (as I point out in Chapter 3), or as indexed to a perspective or relative to a perspective. As I explain in Chapter 5 (building on Massimi 2018e), while I see scientific perspectives as offering both justificatory principles and assertability conditions for specific claims of knowledge, I do not see truth conditions as something to be delegated to any specific scientific perspective.

As new scientific perspectives come to the fore, existing scientific knowledge claims can be cross-perspectivally assessed and retained or withdrawn accordingly. While truth as correspondence with the way the world is (loosely speaking) a cross-perspectival affair, scientific perspectives offer a second-order set of epistemic-methodological principles that can shed light on whether or not someone has justifiably come to reliably formed claims of knowledge. Therefore I see the plurality of scientific perspectives not as a disjoint set but as *intersecting with one another* to fulfil this crucial epistemic role for scientific knowledge.

It could be, for example, that while reliably formed, some scientific knowledge claims are suffering from justificatory principles that might be defective, or insufficient by themselves to ground the reliability of the procedure.[8]

---

[7] There has been a long tradition in epistemology of understanding justification in perspectival terms. Susan Haack (1993, p. 208), for example, defined perspectivalism as 'the thesis that judgments of justification are inherently perspectival, in that what evidence one takes to be relevant to the degree of justification of a belief unavoidably depends on other beliefs one has'. Building on Haack, Jay Rosenberg (2002, pp. 148–149, emphasis in original) argued that 'On this reading, the reason that we correctly judge that S does not know that *p* is that, given our richer informational state, we recognize that what *we* are (stipulatively) entitled to take to be S's epistemic circumstances demand a higher level of scrutiny than we are supposing S himself to have exercised. S, therefore, has not satisfied what, from our perspective, are the standards of *performance*-adequacy appropriate to his epistemic circumstances, and, hence, from our epistemic perspective, we judge that, despite his not having acted irresponsibly given the information available to him (judged from his own legitimate perspective on his epistemic circumstances), he has *not* justifiably come to believe that *p*. What shifts from one epistemic perspective to the other, on this interpretation, is not the relationship between S's *de facto* grounds of belief and the truth of what he believes, but rather the specific procedural norms relevant to the assessment of his epistemic conduct'. Rosenberg's observation provides the basis for the kind of perspectival truth I articulated in Massimi (2018e) and to which I return in Chapter 5, Section 5.7.

[8] By separating issues about reliability from those of justification, the aforementioned notion of scientific perspective does not fall prey to classical problems affecting, for example, Kuhn's view about scientific paradigms. For instance, there is no equivalent 'living in a new scientific world' scenario, under my definition of 'scientific perspective'. Scientific perspectives do not mould ontology. There is more. They might reliably identify modally robust phenomena but have the wrong

Cross-perspectival assessment of scientific knowledge claims is key to deliver scientific knowledge. Hence the heavy lifting done by the pluralistic, diverse, and fluid interplay of scientific perspectives.

I come back to the distinctively *philosophical* question at the heart of perspectival realism: how do *we human beings* come to know the natural world as being a certain way from a number of historically and culturally situated perspectives? When posed this way, the philosophical question about realism is less about mapping the existence of the 'scientific zoo', and more about exploring what makes *us wonderfully diverse human beings capable of reliable scientific knowledge over time*. Perspectival realism, as I articulate it in this book, is not a metaphysics-first view. It is a project in the epistemology of science.

1.3. Two main motivations for this book

There are two main and interrelated motivations behind perspectival realism as a project in the epistemology of science. The first is historical. I have always thought that epistemic stances about science (realism, instrumentalism, empiricism, or similar) should be able to speak to the history of science. This is not a rehash of 'philosophy of science without history of science is empty'. It is more the need to take seriously the historically situated nature of scientific knowledge when discussing its epistemic foundations. My overarching question demands engaging with the historical plurality of practices in the sciences.

Understanding how we have come to know the world as teeming with atomic nuclei, melliferous flowers, DNA strands, and so forth, requires understanding how particular epistemic communities at particular historical times have produced reliable knowledge claims about them. Moreover—and the most important part of my view—it requires understanding why the associated realist commitments could be retained across different epistemic communities over time (or cross-perspectivally, as I call it).

The second and related motivation has to do with what I am going to refer to loosely as 'multiculturalism' (knowing all too well that the term has acquired a very specific meaning in political theory which is not necessarily how I am going to use the term here). Debates on realism in science and scientific knowledge more broadly have a tendency to proceed in some weirdly engineered cultural vacuum. There are obviously methodological reasons why philosophers of science cash out narratives in terms of 'claims of knowledge', 'inferences', 'suppositional antecedents', 'indicative conditionals', and so forth. These are some of the

---

justificatory principles; or, lack well-defined truth conditions, despite having clearly defined assertability conditions.

epistemic-semantic tools in our profession, as much as quadrupole moments, spin, and so on, are tools for nuclear physicists. But there is a problem arising from the uncritical use of such philosophical tools.

They hide a presumption that scientific knowledge production proceeds on some kind of idealized frictionless plane rather than in well-defined historical and cultural contexts that affect the nature of the claims of knowledge advanced. One of the motivations for perspectival realism is to counteract this presumption. The realism I shall be articulating is realism within the bounds of a plurality of intersecting scientific perspectives. Therefore, I understand the notion of scientific perspective rather broadly to include any scientific practice that has resulted in reliable knowledge claims that have been cross-perspectivally retained. This implies re-assessing the role played by a great number of historically and culturally situated epistemic communities in knowledge production, especially those communities that are often severed by epistemological narratives and frictionless accounts of scientific knowledge. I am thinking, for example, of the local knowledge about the melliferous flora among the beekeepers of the Yucatán peninsula (Chapter 8); or about the rosy periwinkle in the communities of South Madagascar (Chapter 11); or about 'kelp-making' (i.e. producing ashes of seaweed used in glass production) by the Scottish Hebridean communities of the eighteenth and early nineteenth century (see Chapter 10). Similarly, my working notion of scientific perspective extends to the engineering practices of studying ground-water motion to build magnificent fountains in parks and public spaces, such as the fountains of Alhambra and Villa d'Este (which I discuss in Chapter 9).

In Chapter 11, I clarify the multicultural dimension in the notion of 'scientific perspectives' and the implications for how to think of scientific knowledge production and associated epistemic injustices. In particular, I urge us not to think of scientific perspectives in terms of 'shared membership', or 'shared scientific homeland', or disjoint scientific 'silos', but instead as historically 'interlacing' and stretching beyond specific geopolitical and national boundaries. While 'intersecting' is a *methodological* feature of how scientific perspectives can be brought to bear on one another to refine the *reliability* of particular claims of knowledge, 'interlacing' is a *historical* feature. It refers to how situated perspectives have encountered and traded with one another some of their tools, instruments and techniques over time. 'Interlaced' perspectives track the evolution of knowledge concerning particular phenomena in what I call a 'historical lineage'.

When seen through the lenses of perspectival realism, scientific knowledge becomes knowledge whose reliable production is not the prerogative of one single community at one historical moment. It is social and collective in a distinctively multicultural way. This view has far-reaching implications for two historical kinds of epistemic injustices about scientific knowledge: what in Chapter 11

I call *epistemic severing* and *epistemic trademarking*. Epistemic severing is the almost surgical excision of the contribution of particular communities (either within the same scientific perspective or across culturally diverse perspectives) from narratives about scientific knowledge production. Epistemic trademarking is the subsequent fencing and ultimately often merchandising of portions of scientific knowledge as a 'trademark' of one epistemic community at the expense of others who have historically contributed to such production.

I see perspectival realism as an antidote to these epistemic injustices and a prelude to what at the very end of the book I call *non-classist scientific cosmopolitanism*, taking my cue from a large literature in cultural studies, sociology, and anthropology: scientific knowledge at the genuine service of a diverse and multicultural 'world citizenship'.

1.4. Structure and highlights

The book is in two parts. In Part I (through Chapter 5), I delve into the epistemology of science. I analyse scientific practices which I cluster under the heading of 'Perspectival Modelling'. In Part II (Chapters 6 through 11), entitled 'The World as We Perspectivally Model It', I clarify the kind of realism that emerges. Perspectival realism comes down to what the perspectivally modelled world is going to look like.

I see perspectival modelling as a specific variety of model pluralism in science. The plurality of models here is 'exploratory' in enabling a particular kind of inferential reasoning that proves fruitful when we want to explore what is possible. This exploration, I argue, is an important guide to what is actual in science. Indeed, it is often the only way to find out what is actual given that we do not have a 'view from nowhere' on nature.

In this respect, perspectival models, as I understand them and use the term in this book, are not necessarily autonomous entities mediating between the theory and the experimental data (along the lines of the influential view of models as mediators articulated by Mary Morgan and Margaret Morrison [1999]). Perspectival models are not necessarily downstream from higher-level theories, even if theories are always at work in the modelling. Nor is their role to map onto or be isomorphic to particular empirical data, or patterns thereof.

Instead of taking scientific theories in mature science as my starting point, I take perspectival modelling—with its exploratory role—as the very starting point of scientific inquiry. I understand perspectival modelling as an integral part of scientific perspective in the broad sense in which I have defined it in Section 1.2. It is important to distinguish 'perspectival modelling' from 'perspectival models' as I do in Chapter 4.

The latter are a variety of scientific models. Perspectival modelling (as I use the term) is not restricted to scientific models *exclusively*. It refers more generally to the situated modelling practices of epistemic communities, including the way they use particular experimental and technological resources to advance claims of knowledge and make inferences from data to the phenomena of interest. It captures how situated epistemic communities across a number of intersecting scientific perspectives come to know the world as being a certain way, as I illustrate with my three case studies in Ch. 4.a, 4.b, and 4.c. Perspectival modelling is a general expression under which I include the situated practice of dendroclimatologists no less than that of CMIP5 climate modellers; the respective practices of educationalists and of developmental psychologists working with models for dyslexia; the practices of petrologists, cosmochemists, and physicists working with nuclear models. This is the distinctive way in which I use the term 'perspectival modelling' in Part I of this book.

Given this terminological distinction between 'perspectival modelling' and 'perspectival models', one should tread carefully here. What makes some modelling 'perspectival' is *not* that each model involved in the wider modelling practice offers a different perspective on a given target system (as one might be tempted to think of it in a colloquial sense). What makes some modelling 'perspectival' is instead its being embedded into historically and culturally situated scientific practices of particular communities and its fulfilling a distinctively exploratory role in delivering scientific knowledge over time.

Zooming into 'perspectival models', I characterise them *as inferential blueprints*: they enable a variety of situated epistemic communities over time to make inferences from a range of datasets to what I call *modally robust phenomena*. Thus, I see perspectival models as representing, and in turn, I understand their representational function in a broadly inferentialist way (following on Suárez's very influential view—see Suárez 2004, 2009, 2015a, 2015b).

The book begins with a discussion of how to understand the notion of perspectival representation, a theme that has intrigued me since the time of Bas van Fraassen's (2008) book. The metaphor of perspective has featured in the works of philosophers of science—from Ron Giere to Paul Teller, from Bas van Fraassen to Sandra Mitchell, among many others—who have appealed to the importance of perspectivism. Yet the metaphor must be handled with care.[9] What makes our representations—in science as well in art—perspectival in some philosophically

---

[9] The metaphor is not at all meant to relay the old image of us as passive 'spectators' of nature. If anything, it is the opposite. Our encounter with a natural world teeming with modally robust phenomena and Natural Kinds with a Human Face (Chapters 6 and 7) begins with fairly mundane considerations about how we devise and craft modelling practices—broadly construed—to explore what is possible (Chapters 4 and 5), and how we go about reliably identifying the modally robust phenomena from data (Chapter 6).

interesting sense? In Chapter 2, I draw attention to two (complementary) ways in which a representation (in art or in science) can be said to be 'perspectival'.

Think of your favourite perspectival drawing in art. What do you see? You are likely to see the scene being represented from a particular angle. If the painting concerns an interior, you might see the objects lined up, with the ones in the foreground being bigger than those in the background. If it is a landscape, trees and figures are displayed along coplanar lines, giving the impression that you are observing the scene from below, or from above; from the centre or from left or right. You might even see the landscape opening up in different directions if more than one vanishing point is used. In any case, what you are likely to see is something that looks very much like a 'window on reality', a space where objects are situated, rather than a chaotic aggregate of objects of all sizes piled up on each other. The represented scene follows coplanar lines towards vanishing points and situates the objects along it, the bigger in the foreground and the smaller in the background, creating a sense of space and depth in what is effectively only a two-dimensional canvas.

Now think of what it means for a representation to be 'perspectival'. You might reply that a representation can be perspectival$_1$ because it is drawn *from a particular vantage point*: the interior or the landscape is represented *as drawn from a particular angle* where you can see some objects more prominently than others. But a representation can also be said to be perspectival$_2$ because it is drawn *towards one or more vanishing points*. These two ways of thinking about what makes a representation perspectival are two sides of the same coin. It is because the representation has one or more vanishing points that it appears to be drawn from a particular point of view.

There does not have to be a tension between these two ways of thinking about what makes a representation perspectival. They are clearly compatible: any representation can be said to be perspectival *both* because it is drawn from a particular vantage point *and* because it is directed towards one or more vanishing points. Yet the emphasis that *we place* on one of these two ways of thinking has philosophically far-reaching consequences. The first notion stresses the situatedness of the representation. The second its directionality.

In Chapter 3, I disentangle a host of issues that are often found in association with the first notion in the literature on scientific modelling and scientific representation. I return to it in Chapter 11, where I unpack what I take to be the most important insight in the idea of *situated* representation. How should one understand the idea of historically and culturally situated scientific perspectives? How not to conflate it with the deceptively similar idea that scientific perspectives are somehow well insulated from one another, or defined by some kind of shared membership, maybe one that is restricted to specific temporal and geopolitical boundaries? In Chapter 11, I give my reasons for seeing scientific perspectives as historically 'interlacing' over long periods of time. This is what makes scientific knowledge production possible at all, as knowledge that belongs to all human beings (or cosmopolitan scientific knowledge, as I call it).

Before then, Chapters 4 through 10 focus on the second notion of perspectival representation. This is closely patterned on the analogy with perspectival drawings in art. A representation is perspectival$_2$ in having one or more vanishing points that transform a two-dimensional canvas into a three-dimensional 'window on reality', to echo art historian Erwin Panofsky (1991). Perspectival models, I contend in Chapters 3–5, offer a 'window on reality' because they make it possible for human beings to make model-based inferences about phenomena and ultimately natural kinds. They act as 'inferential blueprints'. The central idea of inferential blueprints is meant to capture a set of salient features of this modelling exercise:

(1) Perspectival models, like architectural blueprints, offer perspectival representations of the target system.
(2) Like architectural blueprints, the representation is often distorted.
(3) Perspectival models are collaborative efforts of several epistemic agents/communities and evolve over time with new additions and tweaks.
(4) As blueprints make it possible for carpenters, joiners, masons, and so forth, to make relevant and appropriate inferences about, say, the house to be built, perspectival models allow different epistemic communities to come together and make the relevant and appropriate inferences about the target system.

These inferences are often couched in terms of indicative conditionals with the following form: 'if x is the case, y will be the case'. I see these indicative conditionals as having antecedents (if x is the case) that invite us to imagine or 'physically conceive' some scenarios suggested by the model. And I see their consequents (y will be the case) as hiding an epistemic modal such as 'can' or 'may' or 'might' (following Angelika Kratzer's 2012 view of epistemic conditionals). In this way, I maintain, perspectival models enable inferential reasoning that over time and often through a plurality of models employed by different epistemic communities allows us to explore what is possible and gain a 'window on reality'.

But what can one see through this 'window'? What is the perspectivally modelled world going to look like? Part II of the book delivers my take on realism in science, a version of realism that is downstream from the epistemological framework of perspectival modelling broadly understood. I characterize the realism that emerges from perspectival modelling as 'bottom up': from data to phenomena to natural kinds. This is in contrast with familiar realist views that start with scientific theories, their main theoretical posits and theoretical terms, and ask which elements of reality correspond (or not) to those.[10] It is also in contrast with metaphysics-first approaches that engage in reading essential properties, dispositions, categorical properties off our current best theories. Reading realism through the kaleidoscope of scientific perspectives forces upon us a shift from metaphysics-first to epistemology-first. A move away from searching for what Gilbert Ryle (1949/2000) aptly called the 'hidden goings-on', towards the scientific practices that are the inferential sources of modally robust phenomena.

Thus, the realism emerging from perspectival modelling is not realism about unobservable entities or similar. It is realism about phenomena that do not just occur, but *could occur* under a range of different experimental, theoretical and modelling circumstances and across a variety of perspectival data-to-phenomena inferences. These are not the phenomena of scientific realists or anti-realists. They are neither faint copies of the real unobservable entities nor sheer empirical appearances. Their modal robustness is reminiscent instead of the Kantian 'objects of experience' as the only objects one can claim knowledge of.

Therefore my phenomena-first ontology departs from traditional empiricist views of phenomena as mere appearances of the unobservable entities beneath them. It follows instead a long-standing neo-Kantian tradition in packing enough modal strength into the notion of phenomena for them to do the heavy lifting when it comes to realism in science.

I present some of these modally robust phenomena in my three case studies in Part I: nuclear stability in Chapter 4.a; global warming in Chapter 4.b; and children's difficulties with reading in Chapter 4.c. I describe how one comes to know each of these phenomena through a plurality of intersecting scientific perspectives and associated perspectival data-to-phenomena inferences. We do not encounter in our scientific travels a realm of unobservable entities. Nor do we stumble upon a sparse Humean mosaic of natural properties with no causal glue in between them. The perspectival realist landscape that opens up is not inhabited by categorical properties, dispositional essences, potencies, tropes, or universals.

It is instead populated by modally robust phenomena such as the bending of cathode rays, the decay of the Higgs boson, the electrolysis of water, germline APC mutations, the pollination of melliferous flora, the growth of a mycelium, among countless other examples.

---

[10] Thus, in what follows, I shall not give any global argument for the realism I defend. I will not seek any counterpart of the 'no miracles argument' or 'inference to the best explanation' or similar. Nor shall I challenge the realist wisdom with some new version of traditional anti-realist arguments such as 'pessimistic meta-induction' or 'the problem of unconceived alternatives'. I will instead make a series of localized moves to the effect of motivating and articulating a realist view of science that takes seriously our situated nature and celebrates its diversity and multicultural roots.

In Chapter 6, I tease out the realist commitment to modally robust phenomena via what I call the *evidential inference problem*: on what distinctively epistemic grounds do data provide evidence that a particular phenomenon is real? I argue that answers to this problem point towards phenomena understood as *stable events* that are *modally robust* across a variety of perspectival data-to-phenomena inferences. I understand the *stability* of the event in terms of what I call 'lawlike dependencies' among relevant features. And I see modal robustness as a secondary quality that has to do with how a plurality of historically and culturally situated communities are able to tease out the network of inferences from a variety of datasets to the stable event in question. Stable events, in other words, act as the realist tether in guiding epistemic communities through a maze of alternatives in being able to identify and re-identify over time the phenomena in question as modally robust.

But obviously, it is not enough to say that perspectival realism is realism about modally robust phenomena. Traditionally, realism in science is associated with natural kinds. And realism about kinds has often come with a defence of metaphysical posits, be they categorical properties or dispositional essences. Scientific realists have often argued that a commitment to these metaphysical posits is compatible with scientific pluralism, maybe in the way in which these properties are clustered together (think of Anjan Chakravartty's [2007] so-called sociable properties coupled with dispositionalism). In Chapters 7 to 10, I delve into the topic of natural kinds and offer a view which I call 'Natural Kinds with a Human Face' (NKHF). In brief, I define natural kinds as the

(i) historically identified and open-ended groupings of modally robust phenomena,
(ii) each displaying lawlike dependencies among relevant features,
(iii) that enable truth-conducive conditionals-supporting inferences over time.

A few aspects are worth highlighting in this definition. First, not any grouping of phenomena counts as a NKHF. In Chapter 7, I distinguish among what I call *empty kinds* (e.g. caloric), *in-the-making kinds* (e.g. dark matter), and *evolving kinds* (e.g. electron) depending on their respective 'nomological resilience' across a number of perspectival data-to-phenomena inferences. Laws of nature, or, better, lawlike dependencies, are key in transforming over time any *in-the-making kind* into an *evolving kind*. As they are also key in revealing if an *in-the-making kind* may in fact be an *empty kind* (in the absence of suitable lawlike dependencies). I argue that all natural kinds we know and love are ultimately (and fallibly) evolving kinds as the groupings are always open-ended. Lawlikeness, under this view, is not supervenient on dispositional essences or categorical properties but is inherently a primitive relation among features of events that are candidates for modally robust phenomena (as I spell out in Chapters 5 and 6).

The other aspect worth highlighting in NKHF is its underlying *historical naturalism*. In Chapter 8, I argue that the naturalness of natural kinds is not necessarily the expression of natural joints through which nature comes pre-carved. It is rooted instead in stable events and the lawlike dependencies inherent in them. But it also depends on our perspectival scientific history—how historically situated communities have come to identify and group relevant phenomena together over time. Historical naturalism as such is the first step in my broadly inferentialist and anti-foundational view of natural kinds, which takes inspiration from Neurath's Boat. It is also a necessary component for both the historical and multicultural motivations I mentioned earlier. Shifting attention from 'scientific theories in mature science' to 'modally robust phenomena' (and how to perspectivally find them) can realign the debate on realism away from Western-centric narratives.

Since the notion of scientific perspective pays attention to the modelling, experimental and technological resources available to any epistemic community to reliably advance scientific knowledge claims, the ability to identify and re-identify modally robust phenomena becomes a collective enterprise. One to which, for example, ethnobotany contributes alongside cytogenetics, when it comes to botanic taxa, as I discuss in Chapter 8. Or think of how medieval hydraulic engineering in Alhambra was instrumental to gaining knowledge of the phenomenon of ground-water motion, which alongside other phenomena (from water droplet formation to chemical bonds) is associated with the natural kind 'water'. In a genuinely anti-foundational Neurathian spirit, each NKHF is an open-ended grouping of historically identified phenomena that different situated epistemic communities have robustly identified over time. NKHF do not take any particular phenomenon (and its associated lawlikeness) as more essential or foundational than any other.

However, NKHF are not another name for conventionalism about kinds. In Chapter 9, I clarify the specific brand of contingentism at play in the view and why it is not to be confused with conventionalism. I will cash out the Spinoza-inspired view of natural kinds as sortal concepts and show that what holds open-ended groupings of phenomena together is a sort-relative sameness relation.

The last element of my account of NKHF has to do with their inferential nature and in particular what I call truth-conducive conditionals-supporting inferences. In Chapter 5, I discuss the semantic nature of these inferences, and in particular the role of indicative and subjunctive conditionals. In Chapter 10, I develop this aspect by delving into how scientists between 1897 and 1906 came to identify the electric charge as a fundamental unit and property of the natural kind 'electron' qua an evolving kind. Once again, I illustrate the role played by a number of historically and culturally situated scientific perspectives in knowledge production. Some of these were historically 'minority scientific perspectives', perspectives *within Western science* that have been forgotten, or erased in the grand historical narrative of scientific realism.

I share the sentiments of many colleagues in history of science in thinking that such minority perspectives do not belong to the dustbin of history. Our philosophical tendency to regard them as intellectual curiosities usable at best for case studies shows how pervasive the narrative of winners vs. losers has been in this debate. My goal is to show that they played a vital part in enabling the chain of inferences that eventually led epistemic communities to reliably advance scientific knowledge claims over time. They were an integral part of how *culturally situated scientific knowers* reliably and justifiably have come to know the natural world as being a certain way.

We remember today J. J. Thomson as the discoverer of the electron. But he called it a 'corpuscle'. He believed corpuscles to be end points of what he called a 'Faraday tube' as a field-theoretical *fin de siècle* scientific perspective on the nature of electric charge. This was soon to be overtaken by the scientific perspective coming out of Max Planck's work on black body radiation. We have similarly forgotten the complex experimental practices that made Thomson's work on cathode rays possible in the first instance. These included mastering the glass-blowing techniques for making bespoke cathode ray tubes for his researches. But also practices such as 'kelp-making' (collecting ashes of burnt seaweeds) among local communities of the Scottish Islands (e.g. the Hebrides, Orkney, among others), which enabled the British glassware industry to flourish at the start of the nineteenth century and produce the lead-free glass required for electrical researches.

Along the same lines, the short-lived engineering practice of the ice calorimeter devised by Lavoisier and Laplace at the end of the eighteenth century proved important for identifying the relevant phenomena that eventually led to transitions of states (from solid ice to liquid water) to be regarded as physical (rather than chemical) in nature, as I discuss in Chapter 7. These examples show how modelling and experimental practices that were an integral part of what turned out to be two 'minority perspectives' from a historical point of view played in fact a pivotal role for knowledge production at two key historical junctures. They also show the crucial role that artisans, glass-blowers, hydraulic engineers, beekeepers, and kelp-makers, among others, play in making reliable scientific knowledge possible within and across situated scientific perspectives.

I return to this aspect at the end of the book, in Chapter 11, where I revisit the situatedness of knowledge, and in particular the first notion of perspectival representation introduced in Chapter 2. Scientific perspectives offer standpoints afforded by particular geographical, socioeconomic, political, historical, and cultural locations. Yet collectively over time they intersect with one another and make scientific knowledge production possible. This becomes an opportunity for briefly engaging with the important topic of multiculturalism and cosmopolitanism in science. The way in which situated scientific perspectives span over time and historically 'interlaced' deserves more attention to tease out when this has been virtuous (as opposed to exploitative) in producing knowledge. I give the broad normative contours of this process in Chapter 11, where I discuss two varieties of epistemic injustice I have mentioned already: *epistemic severing* and *epistemic trademarking*. I argue that they demand more than mere 'recognition remedies'. They call for the 'reinstatement' of epistemic communities that have been unjustly excised from narratives of scientific knowledge production as a stepping-stone to a non-classist and non-elitist variety of what I call 'cosmopolitan science'.

1.5. What flavour of realism in perspectival realism?

Perspectival realism tells a story about realism in science with no winners or losers, no dominant 'paradigm' taking over an old one 'in crisis', or 'progressive research programme' gaining ground over a 'degenerating' one. The kind of realism that I offer in this book does not treat the history of science as the base for writing the epistemic hagiography of the winners. It embraces pluralism in methodologies and models, and a commitment to a plurality of situated scientific practices. It sees the many varieties of local knowledge as an integral part of scientific knowledge production. It takes our engineering practices and synthetic kinds as continuous with our natural kinds.

I agree with van Fraassen that on matters of realism, instrumentalism, and so on, there are ultimately 'voluntaristic stances'. Thus, I do not harbour hopes to convert the die-hard metaphysicians to swap their ontology of dispositional essences or categorical properties for a phenomena-first ontology. I will rest content if I manage to put forward—as carefully as I can and with as many examples as I can muster—an alternative way of thinking about realism, a different lens through which one can see the ways in which realism and perspectival pluralism in science could fruitfully come together. I ultimately want to celebrate scientific knowledge as a distinctive kind of *social and cooperative knowledge*—knowledge that pertains to *us wonderfully diverse human beings* occupying a kaleidoscope of historically and culturally situated perspectives.

If ever there was anything miraculous about the success of science, it is not the success of an individual scientific theory T in latching onto a mind-independent world. It is the success of *our human species* in building over time an extraordinarily varied and reliable knowledge of the world we live in, despite having no privileged standpoint to occupy. What needs to be explained therefore is how historically and culturally diverse epistemic communities have achieved this epistemic feat. This book offers a possible story about it.

I believe there is a lot of work for philosophers of science still to do in this respect. Historically, science has too often been for the benefit of Western elites. Philosophers of science, as I see it, have a duty not to be self-complacent in writing narratives about science and scientific knowledge. How one *ought* to think about science is different from how science *has historically happened*. And paying attention to the historical practices should not become an automatic way of *epistemically ratifying* the underlying mechanisms of knowledge production (especially whenever exclusionary mechanisms might have been at play).

The tension between the normative and the descriptive components in how we think about science is tangible in the discipline of history and philosophy of science (HPS). But philosophers have a duty to reorient those narratives—and realism is a key one—in a way that does not end up ratifying templates of epistemic dominance and exclusion. I see our job as that of reflecting on the practice of science and laying out ways in which one ought to be thinking about scientific knowledge.

To conclude, let me return to the metaphor of perspective. The vanishing point(s) towards which the lines of perspective converge in a drawing should not be understood as a proxy for a metaphysical reality to which we all converge at the end of inquiry. After all, they are only *vanishing points*. Perspectival realism *is not* a form of convergent realism. It is not forced upon us by any metaphysics-first approach to the natural world.

The absence of a scientific 'view from nowhere' means there exists no ideal atlas, and no privileged catalogue of ontological units. There is no directionality to serve in worn-out discussions about convergence towards a final theory or a final metaphysical reality or end of inquiry. Nor one that is synonymous with consensus building as a way of homogenizing lines of inquiry and smoothing out pluralism. There is, in other words, no conclusion to the story of our scientific endeavours. Scientific knowledge is ongoing and open-ended.

I am often asked what perspectival realism is. It has occasionally been suggested to me that we can think of this asymptotically, in the same metaphysically innocent way in which the American pragmatist Charles Sanders Peirce envisaged the 'end of inquiry' as a regulative ideal rather than a factual state of affairs realized at any point in time. But my view of the directionality of perspectival$_2$ representations (their having one or more vanishing points) in no way resembles *convergence to something*, even in the mild regulative Peircean sense.

The directionality I describe is more like that of Marco Polo, leaving Venice to explore uncharted territories in the East. His journey was not guided by any universal atlas. Nor, similarly, does science provide any universal atlas to nature and all that it contains. We are all fellow travellers, together with our predecessors and successors, across countries and myriad cultures in the world.

Being a perspectival realist about science is therefore akin to walking through the garden of inferential forking paths (to echo Jorge Luis Borges's 1941/2000 tale). I'd like to think of the directionality of perspectival representations as the directionality of fellow travellers zig-zagging through the garden, with no ideal atlas, but with plenty of inferential resources for reliably tracing and retracing our journeys, and not getting lost or permanently stuck along the way. That's how we encounter nature as teeming with modally robust phenomena and Natural Kinds with a Human Face.

1.6. Short autobiographical note and acknowledgements

I grew up in a small village in the Sabine hills near Rome: one of those unassuming places of a few thousand inhabitants where not much goes on and the days, weeks, and months all resemble one another. The long summer holidays were the toughest to live through. The place came to a standstill from June to September in the unbearable heat. Only the relentless sounds of the cicadas kept me company during this season as a teenager. And my books, of course, mostly novels that kept me amused while the rest of the world outside seemed to fade away under the scorching temperatures.

I began to read Italo Calvino's books. I loved his short stories and novels permeated by realism sprinkled with a good pinch of imagination. I read *Invisible Cities* and thought of cities I'd like to see, new places I'd like to explore as in Calvino's book the Chinese ruler Kublai dreamt of imaginary cities he had never visited and asked the Venetian Marco Polo about them.

This book too is about journeys of exploration—though not into imaginary cities but the natural world. It is a philosophy of science book that looks at scientific knowledge as an exploratory journey and tells a story of *our travels* through it. A bit like in Calvino's stories, this too is a story of *realism* about science combined with a good pinch of imagination. In this case, the imagination concerns how we *model* the way the natural world *might be* at every step and junction along the route and how we come to encounter a natural world teeming with phenomena and natural kinds as a result of it. In Chapter 5, I call this kind of imagination 'physical conceivability' in the context of a particular type of scientific modelling—perspectival modelling—to which Part I of the book is dedicated.

By contrast with Calvino's Marco Polo, whose fervent imagination could conceive of a myriad of possible cities, our encounter with the natural world requires hard work on our part. We have to find our ways through swarms of data and empirical regularities with the help of cleverly designed modelling practices, and always within the boundaries afforded by the technological, experimental, and theoretical resources available in any culturally located perspective at any one time. I owe my other literary debt to Jorge Luis Borges's *Fictions*, and in particular the aforementioned short story 'The Garden of Forking Paths', which provides the background reference for my inferentialist account of natural kinds from Chapter 8 onwards.

My autobiographical journey of exploration led me out of the Sabine hills of Lazio in the late 1990s, almost catapulting me into thriving London and the buzzing London School of Economics, and taking me through lands and places surpassing my wildest teenage dreams. I would not have made it thus far without the help and support of countless fellow travellers along the way—far too many to recount and thank here.

But to some I owe a special debt of gratitude for having made this book possible. Sadly, some of these people have passed away and did not get to live long enough to see the end of this particular journey. But it is to them and their enthusiastic encouragement of my work at times when I was myself in full-swing self-doubt that I owe very special thanks.

My interest into perspectivism began unexpectedly in 2007 with an email I received from Peter Lipton. I was in Cambridge at the time as a Junior Research Fellow at Girton College and I regularly attended Peter Lipton's Epistemology Reading Group on Thursday afternoons. We read Langton's *Kantian Humility* but also Putnam's *Reason, Truth and History* and Nickles's edited collection on Thomas Kuhn.

One day, Peter sent me a copy of his book review of Ron Giere's *Scientific Perspectivism*, knowing about my ongoing fascination with Kant and Kuhn and Putnam's internal realism. In that book review, he located Giere's book in a tradition that he saw as beginning with Kant and continuing with Kuhn: a tradition he called 'constructivism'. Taking his cue from Giere's analysis of colours as irreducibly perspectival, Peter Lipton concluded that, 'Like Kant, Giere wants to extend his picture of colours to all science. Scientific descriptions capture only selected aspects of reality, and those aspects are not bits of the world seen as they are in themselves, but bits of the world seen from a distinctive human perspective' (Lipton 2007, p. 834).

Peter Lipton himself was never persuaded by what he saw as the Kantian line behind Giere's perspectivism. His review concluded: 'Maybe in the end constructivism is true, or as true as a constructivist can consistently allow. Nevertheless, the thought that the world has determinate objective structures is almost irresistible, and Giere has not ruled out the optimistic view that science is telling us something about them' (Lipton 2007, p. 834). His scientific realist reservations notwithstanding, he did send me the review with the accompanying note that he thought I was going to like this book.

He was right. I did like Ron Giere's book. I more than liked it. Before I knew it, that book set me on a path that I have been on for the past fifteen years. My dialogue with Peter Lipton continued as we were planning a joint Philosophy of Science Association 2008 symposium where he was going to give a talk aptly entitled 'Kantian Kinds and Natural Kinds'. But Peter's untimely death in November 2007 marked an abrupt end to our exchange.

The exchange continued indirectly when in 2010 Anjan Chakravartty edited a special issue of *Studies in History and Philosophy of Science* in Peter Lipton's memory in which he himself published a seminal article on perspectivism (Chakravartty 2010). To Anjan I owe many thought-provoking conversations over the years on the topic of realism and perspectivism that often made me pause and think hard about some of the issues I cover in this book.

In 2008, Bas van Fraassen's book *Scientific Representation: Paradoxes of Perspective* came out and it provided a further opportunity for me to get more familiar with the literature on perspectivism. I loved the subtle interplay between art and science in the book, between Alberti's and Dürer's pictorial perspective and the use of machines and engines to offer representations of phenomena *not as they are*—van Fraassen says with a subtle Kantian undertone—but *as they appear from the particular vantage point of an observer*.

I wrote a review of Bas van Fraassen's book where I briefly touched on the Kantian theme of the distinction between phenomena and appearances (Massimi 2009, p. 326). I had been following his constructive empiricism since my graduate days. He too, like Peter Lipton, showed great generosity in replying to my philosophical questions and engaging with my often tentative (and mostly half-baked) ideas. I was thrilled to see the Kantian line becoming more tangible in his 2008 book. That was the time when I was embarking on my philosophical detour into Kant's philosophy of nature. My interest in Kant goes back to my undergraduate times and to my then dissertation supervisor Silvano Tagliagambe at the University of Rome. His Kantian-inspired epistemology provided the fertile breeding ground for my research interests in this area in the decades that followed.

I have never thought of Kant as a 'constructivist', despite Peter Lipton's assessment, unless one understands the word in a rather broad sense—in the same sense in which, for example, van Fraassen is a 'constructivist' in taking scientific models as human constructions that save the observable phenomena. Hence the self-declared 'constructive empiricism' of his view. By the same token, Ron Giere's *Scientific Perspectivism* has never struck me as a 'constructivist' piece of work. I never asked Ron, but I suspect he would have rejected the label for himself, unless again one understands it in a broad sense. But understood thus, the term 'constructivism' loses most of its philosophical force. In a way, we have all been constructivists all along so long as one takes scientific models as human constructions designed to explore the natural world. (After all, what are scientific models if not human constructions?)

To put it differently, insisting on the 'constructivist' label (for Kant, or Kuhn, or van Fraassen, or Giere) seems to be missing a crucial point here. Namely, that scientific realism qua 'the optimistic view that science is telling us something' about the world's 'determinate objective structures' (Lipton 2007, p. 834) cannot evade the following puzzle: how do scientific models qua human constructions match or fit or accurately represent 'determinate objective structures'? If they are human constructions, how can scientific models ever claim to even rise to the challenge of solving this puzzle between objective reality, on the one hand, and the scientific image, on the other hand?

This epistemological question cannot be eschewed by off-loading the explanatory task to metaphysics, in my view. Or by packaging the natural world with causal properties, dispositional essences, causal powers, and so forth. The answer to the epistemological question *How is scientific knowledge of the natural world possible?* cannot be outsourced and delegated to an arsenal of presumed metaphysical entities, whose very reason for existence is precisely to take off our shoulders the uncomfortable burden of the question.

This same question, of course, led Kant to give his own answer without taking refuge in any metaphysical realm. In my view, the particular answer Kant gave to it does not make him a 'constructivist' any more than Giere's answer to the same question in terms of scientific perspectivism, or van Fraassen's answer to it in terms of empirical adequacy, makes either of them 'constructivist' in some genuine sense.

If anything, Kant described his view as a kind of 'empirical realism', but of a very nuanced kind. He was someone for whom reality consisted of phenomena, but phenomena are not mere copies of things, or faint images of some underlying unobservable realm. Kant's empirical realism—I thought—sat very well with van Fraassen's take on scientific representation and perspectivism, much as van Fraassen would of course resist the label 'realism' altogether.

As the years have gone by, I have become more and more involved in Kant's metaphysics of nature, thanks to the work of Michael Friedman, Eric Watkins, Karl Ameriks, Dan Warren, Rachel Zuckert, Hannah Ginsborg, and Michelle Grier, among others. I think I have come to revise my views on Kant's empirical realism too. After all, he did believe in what he called the 'natures' of things. His lectures on metaphysics abound with discussions about *grounds* and essential properties and reveal a philosopher who was no stranger to a language reminiscent of what we might call today 'dispositional essentialism' under the veil of the phenomenal world.

Kant barely features at all in this book, although the autobiographical journey that has led me to write it is Kantian through and through. Much as my interest in Kant's philosophy of nature took on a life of its own between 2008 and 2014, the underlying question of realism and perspectivism continued to attract my attention. Among other philosophers of science who had directly engaged with Kantianism, four have played a key role in my journey: Margaret Morrison, Hasok Chang, Philip Kitcher, and Steven French.

I met Margie Morrison in 2007 at a conference I organized at University College London on *Kant and Philosophy of Science Today*. Much as she herself shared a great interest in Kant, with her unrivalled sense of humour, she used to make jokes about both Kant and perspectivism and call the latter 'the view from everywhere'. This book owes a lot to her sharp but always light-hearted criticism. Her work on scientific models and her book *Reconstructing Reality* with its critique of perspectivism has provided much food for thought over the past few years. Sadly, Margie passed away just when I was finishing the edits of this book after a long illness. She is sorely missed. I cannot in all honesty claim that I have answered all her questions or appeased her criticism. But in attempting to do so, I managed to get this book project started. I had to start from somewhere, and Margie's criticism-with-a-smile (which I discuss in Chapter 3) seemed the right place to do so.

With Hasok Chang, I share two decades of conversations that started when I was a student at LSE and continued when we were colleagues in the Department of Science and Technology Studies at UCL. With his pragmatist-tempered interest in Kantianism, C. I. Lewis, the history of science, and scientific practice, Hasok has been a fellow traveller in our HPS field from the time of the UCL conference *Kant and Philosophy of Science Today* to his more recent work on *Realism for Realistic People* (2022). My book shares his 'integrated HPS' approach to the topic of realism and perspectivism by indulging (probably more often than philosophers of science typically do) in historical examples and details of scientific practices.

Philip Kitcher's work on Kant and natural kinds, real realism, and science in democratic societies has had a huge influence on me in more ways that I can possibly recount. I am grateful to him for the enthusiastic support for this book project and for the many conversations over realism, truth, progress, and pragmatism before and during the COVID-19 lockdown. Like Peter Lipton with respect to Giere's book, he too I suspect might not be persuaded by the slight Kantian undertone of my perspectival realism. But his defence of realism in science and of the Galilean strategy behind the *reliability* of scientific knowledge is not only one of the main starting points for my journey, but also a defining feature of the kind of realism I see as compatible with perspectivism.

I cannot stress how much I have learned from Steven French's historically informed approach to the philosophy of quantum mechanics, and the neo-Kantian Cassirer-inspired outlook of his ontic structural realism. I learned from him that answers to the epistemological question need not be outsourced to metaphysics. Or, better, such answers require a different way of thinking about realist metaphysics. He has adopted a structuralist approach. I have followed a perspectivalist one—from data to phenomena, and from phenomena to kinds. But fellow travellers we have remained through thick and thin, walking in the garden of realism in science.

Yet it is not just to my Kantian, neo-Kantian (and quasi-Kantian) fellow travellers that I owe a huge debt of gratitude. In a way, if Peter Lipton's 2007 book review was the starting point, the launch event of this project was the PSA 2014 symposium on Perspectivism in Chicago, with Mazviita Chirimuuta, Ron Giere, Sandra Mitchell, and Paul Teller as co-symposiasts. This was the first and sadly also the last opportunity I had to meet Ron Giere in person. He was enthusiastic about the idea of a symposium on perspectivism and we celebrated afterwards in one of his favourite 'trattorias' in Chicago. The work on colour perception by Mazviita Chirimuuta and on semantic implications of perspectivism by Paul Teller has been inspirational and has opened up a host of philosophical questions about the role of secondary qualities and how exactly to rethink the realist question.

Sandra Mitchell's work on integrative pluralism and laws of nature in biology has accompanied me since 2008, when I first met her at a conference in Heidelberg. She too has become over the years a wonderful fellow traveller in the land of perspectivism. I am hugely grateful to her for the very many conversations on topics as wide-ranging as the nature of phenomena, scientific models, laws, and pluralism, but also protein folding, bees, and flocks of starlings. I have learned from her to better appreciate how perspectivism does not just matter for answering the epistemological question about scientific knowledge. It equally matters for better understanding everyday scientific methodology, be it in biology or in physics or in other scientific fields.

Innumerable friends and colleagues have helped me along the way. In December 2019, I organized an away-day in Edinburgh to discuss some preliminary material from this book with Nancy Cartwright, Hasok Chang, Ana-Maria Creţu, Omar El-Mawas, Franklin Jacoby, Alistair Isaac, Casey McCoy, Mark Sprevak, and Jo Wolff. Many thanks to all of them for all the feedback and pointers that helped me write Chapter 1 and rewrite Chapter 6.

During the long months of COVID-19 lockdown in 2020, I organized a weekly reading group on Zoom to try out preliminary chapter drafts with a small group of colleagues including Julia Bursten, Ana-Maria Creţu, Adrian Currie, Joe Dewhurst, Franklin Jacoby, Catherine Kendig, and Sabina Leonelli. They were an extraordinary group in offering countless suggestions for improvements and identifying problematic passages. I cannot even begin to enumerate the number of changes I made to the manuscript after each reading group. I think the final outcome is a much better-quality book than it would have been if I had not had detailed comments and written feedback by this incredible group of friends and colleagues.

They say it takes a village to write a book. In my case, it felt more like it takes an entire city and all its surroundings, given the number of colleagues across various fields and areas whose feedback has proved important for this volume. Over the years, graduate students at the University of Edinburgh (and visiting postdoctoral fellows too) have been important interlocutors on a number of related topics: I thank Anna Ortín Nadal, Nick Rebol, Giada Fratantonio, Andrea Polonioli, Laura Jimenez, Sander Klaasse, Jan Potters, Nora Boyd, Siska de Baerdemaeker, Laura Bujalance, and Sophie Ritson. I held a number of additional reading groups in the middle of the 2020 pandemic and (in alphabetical order and I hope I am not forgetting anyone) I am hugely grateful to Marialuisa Aliotta, Helen Beebee, Franz Berto, Luigi Del Debbio, Catherine Elgin, Steven French, Roman Frigg, Peter Hawke, Sandra Mitchell, Alex Murphy, John Peacock, Stathis Psillos, Tom Schoonen, and Paul Teller for reading various extracts of the book and offering eye-opening comments.

Special thanks to Franz Berto and Timothy Williamson for helpful pointers and discussions on the material in Chapter 5 on physical conceivability and epistemic conditionals. I thank Helen Beebee, Andrew Clausen, and Wilson Poon for helpful conversations on essentialism, monetary policies, and physical transitions of state, all of which informed my discussion in Chapter 9. Craig J. Kennedy provided helpful references with kelp-making and the Scottish glass industry of the nineteenth century in Chapter 10; and Nathan Brown with chemoinformatics in Chapter 7. Conversations with Rob Rupert offered very helpful pointers for the psychology literature surrounding children's acquisition of natural kind terms that features in Chapter 8. On ethnobiology and local knowledge, I benefited from discussions with and comments by Catherine Kendig and Alison Wylie, respectively: I learned a lot from their work on these topics.

Ken Rice very kindly gave me written comments on two different versions of the material in Chapter 4.b. on climate models, for which I also greatly benefited from comments by Roman Frigg, Benedikt Knüsel, and Wendy Parker. For

Chapter 4.a, I had the privilege of receiving some really helpful references on the history of nuclear physics by Isobel Falconer and Roger Steuwer. (The latter very kindly also posted a book of his for me that I could not find anywhere in the library.) Raymond Mackintosh went over and above any possible expectation by kindly offering detailed written comments on several rewrites of the material in Chapter 4.a. I cannot stress how much I have learned about nuclear models over the past months thanks to my email exchanges with him. I also had the honour of receiving feedback from Uta Frith and John Morton, whose modelling work on dyslexia I discuss in Chapter 4.c. I am immensely grateful to all these colleagues for their generous time and help. Needless to say, and as always, any error or mistake in those three case studies in Chapters 4.a, 4.b, and 4.c and anywhere else in the book remains entirely my own responsibility.

A number of colleagues and friends across different fields over the years engaged with me in conference discussions and offered comments and pointers on innumerable related topics. I thank in particular Theodore Arabatzis, Alan Barr, Homi Bhabha, Alexander Bird, Alisa Bokulich, Abbe Brown, Matthew Brown, Jon Butterworth, Elena Castellani, Tiziano Camporesi, Panagiotis Charitos, Mazviita Chirimuuta, Heather Douglas, Marcelo Gleiser, Rebekah Higgitt, Kareem Khalifa, Michael Krämer, Marcel Jaspars, Ofer Lahav, Walter E. Lawrence, Victoria Martin, Aidan McGlynn, Tom McLeish, Angela Potochnik, Duncan Pritchard, Carlo Rovelli, Juha Saatsi, Simon Schaffer, Justin E. H. Smith, Andrew Schroeder, Mauricio Suárez, Karim Thebault, Nick Treanor, Peter Vickers, Francesca Vidotto and Jo Wolff.

But this book would not have been quite the same without the equally crucial editorial help I received all along from Jon Turney. He patiently read it all and with his eagle eyes trimmed it down and made it more readable. He helped me (and the readers) see the woods for the trees wherever the discussion was getting too cluttered with details and I myself was getting lost down one of the too many rabbit holes. Justin Dyer polished the edited chapters and offered countless suggestions for improvements in the English style as well as correcting my (incorrigible) stylistic infelicities. Peter Ohlin at Oxford University Press has been an enthusiastic supporter of this book project from beginning to end—very many thanks to him too. The staff at the National Gallery (London), Museo del Prado (Madrid), and the Brooklyn Museum (NYC) were very kind in assisting with the copyright of the images reproduced in these pages.

Some of the material here draws on articles I have published elsewhere. In Chapter 3, Section 3.2 and 3.6 are from Massimi (2018b) 'Perspectival modelling', *Philosophy of Science* 85, 335–359. Reproduced with permission from University of Chicago Press, copyright (2018) by the Philosophy of Science Association. And Section 3.3. draws on Massimi (2018a) 'Perspectivism', in J. Saatsi (ed.) *The Routledge Handbook to Scientific Realism*, London: Routledge. Reproduced with permission from Routledge. Chapter 5, Sections 5.3, 5.4, and 5.5, builds on

Massimi (2019a) 'Two kinds of exploratory models', *Philosophy of Science* 86, 869–881. Reproduced with permission from University of Chicago Press, copyright (2018) by the Philosophy of Science Association. Section 5.6 expands on Massimi (2018f) 'A *perspectivalist* better best system account of lawhood', in L. Patton and W. Ott (eds) *Laws of Nature*, Oxford: Oxford University Press. And Section 5.7 draws on Massimi (2018e) 'Four kinds of perspectival truth', *Philosophy and Phenomenological Research* 96, 342–359. In Chapter 6, Sections 6.2 and 6.6. are reproduced from Massimi (2011a) 'From data to phenomena: A Kantian stance', *Synthese* 182, pp. 101–116. Copyright Springer Science + Business Media B.V. 2009. In Chapter 10, Sections 10.2, 10.3, and 10.4 are reproduced (in expanded form for Section 10.2) from Massimi (2019c) 'Realism, perspectivism and disagreement in science', *Synthese* (Open Access), https://doi.org/10.1007/s11229-019-02500-6. The quotes at the start of Chapters 2, 3, 5, 9, 10, and 11 are from *Invisible Cities* by Italo Calvino, published by Secker. Copyright © Giulio Einaudi editore, s.p.a. 1972. English translation copyright © Harcourt Brace Jovanovich, Inc. 1974. Reprinted by permission of The Random House Group Limited. For the US and Canada territories, *Invisible Cities* by Italo Calvino, translated by William Weaver. Copyright © 1972 by Giulio Einaudi editore, s.p.a. Torino, English translation © 1983, 1984 by HarperCollins Publishers LLC. Reprinted by permission of Mariner Books, an imprint of HarperCollins Publishers LLC. All rights reserved.

This book could only have been conceived and executed thanks to a generous and protracted period of research leave made possible by the European Research Council (ERC) under the European Union's Horizon 2020 research and innovation programme (grant agreement European Consolidator Grant H2020-ERC-2014-CoG 647272 'Perspectival Realism: Science, Knowledge, and Truth from a Human Vantage Point'). The same grant pays for the Open Access fees that allow you to read this book entirely for free. I will always be hugely grateful to the ERC for this unique research opportunity over the past five years, and to my ERC team—my PhD students Franklin Jacoby and Lorenzo Spagnesi, my postdocs Casey McCoy and Ana-Maria Crețu, and my project coordinators Deborah Stitt and James Collin—for the intellectual conversations, conferences, and events that made the project behind the book possible at all.

Last but not least, I owe the greatest debt of all to my family. The scattered references to Calvino's *Invisible Cities* at the beginning of various chapters are in loving memory of my father, who adored Calvino and introduced me to him when I was a teenager. But they are also dedicated to my mother, who has been all along my unfailing supporter despite remaining in the village in the Sabine hills where I grew up and in COVID lockdown—by herself—for now almost two years. Writing this book has taken away far too much of my time, time that I could have spent visiting her more often.

My interest in situated and local knowledge is in part autobiographical. I grew up in a rural part of Italy. Fifty years ago, for the generation of my parents, access to higher education was still a privilege for only a few people in that part of the world. And for the generation of my grandparents in the interwar period, education opportunities stopped at the end of primary school in the best case scenario. And yet I have never encountered more reliable knowledge production about a range of natural phenomena than growing up in that local community— including my grandfather Eligio's unfailing knowledge of olive trees flowering and pollination vital for the local rural economy and production of olive oil. It is from them—and from all local epistemic communities past and present (wherever they are)—that I derived inspiration for the discussions on local knowledge in Chapters 8–10 and the non-elitist scientific cosmopolitanism in Chapter 11.

Over the past sixteen years, my husband Mark Sprevak—being a philosopher himself—has been the most dedicated and attentive intellectual interlocutor I could have possibly had. I dedicate this book to him and to our son, Edward, who brought love, joy, and mental sanity amid the madness of finishing a monograph in the middle of a pandemic. I have treasured our countless family walks along the forking paths of nearby Craiglockhart Hill, bordered by foxgloves and wild garlic, and the uncertainty about which path to take, the steadfast climbing of the hill—on rainy muddy days and sunny ones too.

2

The perspectival nature of scientific representation

> The ancients built Valdrada on the shore of a lake, with houses all verandas one above the other, and high streets whose railed parapets look out over the water. Thus the traveler, arriving, sees two cities: one erect above the lake, and the other reflected, upside down....
>
> The twin cities are not equal, because nothing that exists or happens in Valdrada is symmetrical: every face and gesture is answered, from the mirror, by a face and gesture inverted, point by point. The two Valdradas live for each other, their eyes interlocked; but there is no love between them.
>
> Italo Calvino (1972/1997) Invisible Cities,
Cities & Eyes, pp. 45–46[1]

2.1. Introduction

What makes a representation perspectival? Any representation is always from a specific vantage point. But *whose* vantage point? And *in what sense* does the vantage point matter for the representation itself? This chapter starts off by identifying two kinds of perspectival representations in art (Section 2.2). They complement each other very well, although they ultimately exemplify two distinct notions of perspectival representation. (I indicate them with the subscripts perspectival₁ and perspectival₂.)

The analogy with art is functional to drawing a similar distinction between two kinds of perspectival representations in science (Section 2.3). I present a problem affecting perspectival₁ representation in science (Section 2.4), and

---

[1] Copyright © Giulio Einaudi editore, s.p.a 1972. English translation copyright © Harcourt Brace Jovanovich, Inc. 1974. Reprinted by permission of The Random House Group Limited. For the US and Canada territories, *Invisible Cities* by Italo Calvino, translated by William Weaver. Copyright © 1972 by Giulio Einaudi editore, s.p.a. Torino, English translation © 1983, 1984 by HarperCollins Publishers LLC. Reprinted by permission of Mariner Books, an imprint of HarperCollins Publishers LLC. All rights reserved.

Perspectival Realism. Michela Massimi, Oxford University Press. © Oxford University Press 2022.
DOI: 10.1093/oso/9780197555620.003.0002

begin to address it by proposing a novel way of thinking about the perspectival$_2$ nature of scientific representation. The view I advocate has some significant implications for familiar ideas about how science relates to nature, and scientific models to worldly states of affairs. It forces a reorientation of traditional views about scientific modelling, realism, and pluralism.

In art, a representation can be said to be 'perspectival$_1$' because it is a *representation drawn from a specific vantage point*. But it can also be said to be 'perspectival$_2$' because it is a *representation drawn towards one (or more than one) vanishing point(s)*. Colloquially, one does not usually distinguish between these two notions. And clearly an artistic representation is 'perspectival$_1$' as long as it is 'perspectival$_2$'. It is the presence of one or more vanishing points that allows the representation to be drawn as if it were from a particular vantage point. Perspectival representations are often said to represent their object or target system *as it appears from here* or *as it is seen from their point of view there*. The metaphor of the spectator agent whose eye subtends the angle from which the scene is represented has done much to shape intuitive notions of perspectival representation.

But drawing in perspective is a technical skill often executed with the aid of machines such as perspectographs. Perspectival effects depend on vanishing points where lines converge. The easiest examples are one-point perspectival drawings. The addition of a second or a third vanishing point creates more dramatic effects, especially if the vanishing points lie somewhere outside the canvas.

Consider David Hockney's *Woldgate Woods* series, which uses three-point perspective to draw the homely Yorkshire woods of his childhood. The representation is *drawn from the vantage point* of someone who is about to venture into the woods, with three visible paths to choose from. The effect is created by three vanishing points. One is located at the centre of the canvas on the lower horizontal line, giving a sense of depth to the woods and creating the illusion of a central path that extends indefinitely across them. The addition of two more focal points, one located just outside the left margin of the canvas and one at the right margin, creates the wonderful effect of a meadow opening up, giving overall the dynamic effect of a wood approaching and receding from us.

Perspectival representations like this one represent their object *as seen from a particular point of view* only to the extent that the representation is drawn *with respect to one or more vanishing points*. The painter has deftly arranged the lines of composition to create perspectival effects that a flat two-dimensional canvas or panel would lack. By analogy, philosophical discussions of perspectival representations have to start by acknowledging the role of the human agent not just as a spectator, but first and foremost as an *architect* in arranging lines of composition and creating in painstaking detail the perspectival effects of a 'window on reality'.

Articulating the kind of realism in science that emerges out of the perspectival nature of scientific representation is the overarching project of this book. How we skilfully build scientific models, draw inferences about what there is on their basis, and go about reshaping these inferences as more perspectives (more vanishing points, if you like) are added to the original representation is ultimately to tell the story of how we came to model the natural world and of how the thus modelled world reveals itself to us.

Before embarking on this journey, something ought to be said about the two aforementioned ways of locating the perspectival nature of the representation. Despite their Janus-faced nature, emphasis placed on one rather than the other (at least in philosophical discussion) has far-reaching consequences for the way in which perspectivity is understood to offer a 'window on reality'. How one understands the realist commitment that accompanies the modelled world *as it appears to us* depends very much on where one locates (and the emphasis placed on) what is distinctively perspectival in the representation. Two further examples from art provide an apt illustration of the distinction between 'perspectival$_1$' and 'perspectival$_2$'.

2.2. Two varieties of perspectival representations in art

Consider now (Figure 2.1) *Las Meninas*—painted by Diego Velázquez in 1656—and rightly celebrated by Michel Foucault as 'the representation, as it were, of Classical representation' (Foucault 1966/2001, p. 17). At the centre of the scene, illuminated by light from a window, is the five-year-old Infanta Margarita surrounded by four courtiers and maids of honour. Behind this first group stands the painter himself, Velázquez, on the left of the Infanta, gazing at us in the act of painting.

In the background through a carved wooden door a corridor appears, illuminated by another window, against which a male figure in black is standing, posing and gazing at us. Beside the carved wooden door, a flat mirror sits between canvases, near enough centre-stage to act as a focal point for the represented scene (in spite of not being the geometrical vanishing point of the painting).[2] And in the mirror is the feebly reflected image of the King of Spain, Philip IV, and his wife Mariana (Figure 2.2).

---

[2] Technically, the focal point or vanishing point of the painting lies at the elbow of the male figure that appears through the corridor, because the mirror is slightly to the left of where the perspectival lines geometrically converge (Snyder and Cohen 1980 make this point clearly in their critical reply to Searle). Thus, a qualification is needed here. I am not concerned so much about where the geometrical vanishing point of *Las Meninas* is located, but more about the role that the mirror plays in creating a second-order representation of the first-order representation, so to speak.

Las Meninas is a paradigmatic example of perspectival representation. It represents the Infanta Margarita and her entourage as seen *from a specific point of view*, namely the point of view that the Infanta, the male figure in the corridor, and the painter are all gazing at: that of the spectator. But Velázquez's skilful arrangement of figures creates a paradox of perspective that has at once bemused and inspired various commentators.

For the point of view of the spectator is not an empty placeholder for the viewer to occupy. It is in fact presumably occupied by the King and Queen of Spain, who are the very subjects of the portrait that Velázquez is in the act of painting and whose image is reflected in the mirror.

So the scene is represented *from the vantage point* not of the painter but of the spectators themselves (Philip IV and Mariana), who are the very subjects of the represented scene. The mirror in *Las Meninas* creates a meta-representation of the perspectival representation of the Infanta Margarita in her dressing room. The first-order perspectival representation is not just drawn *from a specific vantage point*. It is also *about that specific vantage point*. The Infanta Margarita and her entourage are a distraction. For the real and intended content is the royal couple whose portrait is being painted by Velázquez himself on the canvas.

Las Meninas exemplifies what I am going to call *perspectival₁ representation* in having the following features:

- The representation is situated; it is *from a specific vantage point* (that of the spectator).
- The represented scene is drawn as it appears *from there* (rather than from Velázquez's point of view; or the point of view of one of the maids of honour).
- The perspectival representation is *about* the very vantage point *from which* the representation takes place.

What is distinctive about perspectival₁ representation is not so much that the representation is situated, but instead that the situatedness of the representation affects the representational content, thanks to Velázquez's skilful expedient of the mirror reflecting the feeble image of the royal couple. The content of the representation is affected by the very vantage point from which the representation takes place. The flat mirror is not an abstract vanishing point towards which the geometrical lines of perspective converge, creating an impression of depth, scale, and angle in the reflected scene. On the contrary, it returns a dark background where nothing else aside from Philip and Mariana is visible, themselves barely sketched in broad undefined brushstrokes like the splash of red in the background denoting a curtain. The meta-representation of the mirror is intentionally sketchy. It abstracts from all the details of the scene, and uses daubs of colour to return a feeble idealized image of the couple, as frail and inefficient as was the leadership of Philip IV in a country that had been plagued by the Thirty Years' War.

But a representation can be perspectival without necessarily being self-referential. Consider another famous example: the *Arnolfini Portrait*. Despite prima facie similarities with *Las Meninas*,[3] especially in the use of the mirror to create a meta-representation of the represented scene, there are significant differences that in my view point to the need for distinguishing an alternative notion of perspectival representation.

Painted by the Flemish artist Jan van Eyck in 1434, more than two centuries earlier than *Las Meninas*, the *Arnolfini Portrait* (Figure 2.3) is an extraordinary accomplishment of perspectival representation whose complex symbology and iconography has intrigued and befuddled commentators (Hall 1994; Harbison 1991; Panofsky 1934; Seidel 1993, among others). It is a celebratory painting for the wedding (or betrothal, as some have argued) of a rich Italian merchant, probably Giovanni Arnolfini, although the accompanying question mark in the painting title at the National Gallery in London indicates that the attribution is not certain.

He and his wife stand holding hands in a room lit by a window on the left. A red baldachin bed sits on the right. A few objects catch one's attention in the intimate space of the scene: a dog and some sandals in the foreground; a chandelier above; and, once again, right at the centre of the image, a mirror (Figure 2.4)—convex this time—mounted on the wall beneath the painter's signature ('Jan van Eyck was here 1434').

The convex mirror reflects the couple from the back holding hands, with a flood of light coming from the window (magnified and slightly distorted to follow the curvilinear lines of the convex mirror). A luscious fruit tree—as a counterpoint to the ripe fruits sitting on the bench of the actual scene—can be glimpsed through the window as reflected in the mirror. The chandelier and the wooden beams of the ceiling are all clearly visible too, which gives a sense of space and proportion to the otherwise rather intimate portrait.

---

[3] Searle (1980, p. 484) noted the analogy between *Las Meninas* and the *Arnolfini Portrait*: 'An interesting variation on the mirror self-portrait genre is the van Eyck of Arnolfini and his bride. We see van Eyck in the mirror but he is not in the act of painting, he is witnessing the wedding. Here we are, so to speak, halfway to *Las Meninas*, because in order to resolve the paradox we have to suppose the picture was painted as the artist remembers the scene. . . . Velázquez must surely have seen the van Eyck since it was in the Spanish Royal collection at the time; and though I know of no independent evidence, it is quite possible that it was one of the inspirations for *Las Meninas*'. (For a critical discussion of Searle's interpretation, see Snyder and Cohen 1980.) In what follows, I part ways from Searle in emphasizing not so much the role of the artist in the mirror, but the use of the mirror in both perspectival paintings. I believe the *Arnolfini Portrait* offers a philosophically rather distinct display of what a perspectival representation might achieve.

This time the meta-representation of the mirror creates the illusion of a 'window on reality', to use the expression of art historian Erwin Panofsky,[4] of a space that extends well beyond the represented scene in the canvas, a space seemingly from nowhere, since the painter has disappeared from it. It is as if the scene were seen through bifocals, one for near vision (the spectator's point of view) and one for distant vision (the mirror's point of view offering a reverse second-order image of the scene).

By contrast with the dark, undifferentiated, sketchy image returned by the flat mirror in *Las Meninas*, here the convex mirror is deployed centre-stage to produce magnificent perspectival effects. Only in the mirror can the spectator peruse the full extent of the room, the wooden beams defining the ceiling, the red baldachin bed on the right, and then, at the far back, a door that opens into a corridor. Then right on the door, there is the sketched silhouette of a man with a striking blue robe (maybe van Eyck himself). And behind the blue silhouette, the mirror reveals a further space where a person in red is visible, and behind the person, another point of light, a window, as the last focal point of the scene as seen from the mirror. A space distorted by the convex surface, and yet painstakingly real in the minute details splendidly made possible by van Eyck's use of perspective.

It is this space created by the mirror as a vanishing point that makes the *Arnolfini Portrait* an example of a distinct kind of perspectival representation. I am going to label it *perspectival$_2$ representation* because of the following features:

- The representation has a clear direction; it is *towards a vanishing point* (that of the mirror in this case as a focal point where a meta-representation of the first-order representation takes place).
- The vantage point of the spectator is neither the vantage point of Giovanni Arnolfini and his wife (by contrast with Philip IV and Mariana in *Las Meninas*), nor that of the painter. It is as if the point of view from which the scene is painted had become an empty placeholder for me, you, and everyone else to occupy.[5]

---

[4] The greatest achievement of Renaissance perspectival art, according to Panofsky, was to transform 'the entire picture . . . into a "window". . . . [W]e are meant to believe we are looking through this window into a space [where] . . . all perpendiculars . . . meet at the so-called central vanishing point, which is determined by the perpendicular drawn from the eye to the picture plane. . . . The surface is no longer the wall or the panel bearing the forms of individual things and figures, but rather is once again that transparent plane through which we are meant to believe that we are looking into a space, even if that space is still bounded on all sides. . . . The picture has become a mere "slice" of reality, to the extent and in the sense that *imagined* space now reaches out in all directions beyond *represented* space' (Panofsky 1991, pp. 55 and 60–61).

[5] It is indicative that the blue silhouette visible from the mirror—assuming it is indeed the painter van Eyck—is not represented in the act of painting the scene on a canvas but rather as standing in the doorway at the back of the room, as if he were a witness of the scene instead of a painter representing

– The perspectival representation is *not (self-referentially) about* the vantage point *from which* the representation takes place. The point of view from which the scene is painted does not affect the *content* of the second-order perspectival representation. For there is an entire world in the mirror, not a copy image of the Arnolfini themselves.

This whole book is an endeavour to understand how perspectival representations in science, despite being always *from a specific vantage point*, can nonetheless give us a 'window on reality' that—fleeting as the mirror image in the *Arnolfini Portrait*—is our most solid and well-grounded commitment to what there is. To do that, I must first couch the distinction here only sketched between *perspectival₁* and *perspectival₂ representations* in a more familiar philosophical idiom.

2.3. Locating the perspectival nature of scientific representation

In the notion of perspective, as so often, we have a cluster concept, with multiple criterial hallmarks. There is no defining common set of characteristics, only family resemblances among the instances. Whether or not something is aptly called perspectival depends on whether some appropriate subset of these hallmarks are present, but what amounts to 'appropriate' we cannot delimit precisely either.

—Bas van Fraassen (2008) *Scientific Representation: Paradoxes of Perspective*, p. 59

In his landmark book *Scientific Representation: Paradoxes of Perspective*, Bas van Fraassen includes among the criterial hallmarks of perspectival representation '*occlusion, marginal distortion, texture-fading (grain), angle*, and, with special importance, *explicit non-commitment* and the "horizon of alternatives". These are all characteristics that relate to the content of the representation' (van Fraassen 2008, p. 59, emphases in the original).

Occlusion concerns the shape and size of the object being represented *from a particular vantage point*. For example, in *Las Meninas*, the large canvas in the foreground occludes a significant portion of the room, and the canvas is slightly tilted to the left as seen from the particular point of view of the spectators (in this case the royal couple). Marginal distortion is the outcome of representing the object from a particular vantage point. Perspectographs used among

the scene. It is as if the painter's point of view had disappeared from the perspectival representation, leaving only a signature on top of the mirror as the last vestige of his presence.

Renaissance painters produced perspectival representations as seen through a pinhole in a plain frame, with the perspectival lines subtended to the angle of aperture, resulting in a marginal distortion of the image: for example, in *Las Meninas*, the disproportionally large gown of the maid of honour to the right of the Infanta Margarita. Texture-fading is evident in the use of colours in perspectival representations to coarse-grain details as the eye moves from the foreground to the background of the picture: from the exquisite details of the Infanta Margarita's rose-shaped red laces to the barely distinguishable black breeches worn by the male figure silhouetted in the far end corridor. The angle from which the scene is being depicted determines and affects these other effects.

Then there are two of the most important hallmarks in the cluster concept of perspectival representation. Van Fraassen refers to Dominic Lopes's notion of *explicit non-commitment* (van Fraassen 2008, pp. 38–39): a perspectival representation represents its subject as having some property (or properties) that precludes being committal in some other respect.[6] For example, Velázquez represents the Infanta Margarita, right in the centre of the picture, as curiously gazing at the spectator while being dressed and her parents as looking at her with a sort of anodyne expression. He might have equally represented her as a toddler playing with the dog and the parents smiling at her. The perspectival nature of the representation—understood along the lines of *perspectival$_1$*—affects the *content* of the representation: the painting represents the Infanta Margarita and her parents *as* a composed and timid five-year-old girl watched over by her parents; and it is explicitly non-committal as to whether, for example, Margarita was or was not a playful toddler cheered by her parents.

The situated, partially occluded, and explicitly non-committal nature of perspectival representation points to the need for a broader viewpoint, one which—barring any view from nowhere—can encompass different points of view, without reducing, merging, or overlapping them. Van Fraassen calls this broader viewpoint the 'horizon of alternatives':

> When we think of a picture as being drawn from one point of view (the location of the eye and direction of vision), we are attending to its alternatives: thinking of it as set in a 'horizon' of other perspectives on the same objects. . . . Thus,

---

[6] Catherine Elgin too stresses the *explicitly non-committal* nature of perspectival representation in her book *True Enough*, where she observes that 'A representation is *explicitly noncommittal* with respect to a given property if its representing the having of one feature precludes its taking a stand on the having or lacking of another. A representation of a man wearing a hat is explicitly noncommittal with respect to whether he is bald, because representing him as hatted makes it impossible for the picture to commit itself on the question of his baldness. . . . [B]ecause of these limitations, there cannot be a single, comprehensive perspectival representation that represents everything from a single point of view. The "God's eye view" cannot be a point of view' (Elgin 2017, pp. 154–155, emphasis in original).

as to the hallmarks of perspective: the characteristics listed [occlusion, grain, non-commitment] will not suffice if applied 'piecemeal'. The content of the picture must be related to a 'horizon' of alternatives that we can think of as coming from 'different points of view', if these characteristics are to count as marks of perspectivity—and the explicit or implicit reference to such a horizon of alternatives is what is most important in the concept of perspectivity. (van Fraassen 2008, p. 39)

The accompanying footnote refers to the work of Nelson Goodman and Catherine Z. Elgin on pictorial symbol systems. And Elgin herself has returned to this topic in her book *True Enough* (2017). Instead of 'horizon of alternatives' she refers to a 'logical space' as a 'multidimensional array of possibilities' where to

locate an item in a logical space is to determine which of the possibilities de-fined by that space it realizes. . . . Representations in a logical space, like representations in a physical space, can be perspectival. They can show how occupants of that space appear from a certain vantage point. . . . I take the term 'perspectival' to refer to any representation that represents how things appear from a particular point of view. (Elgin 2017, p. 155)

Thus, a representation is deemed to be perspectival because it is *from a particular point of view*; and, most importantly, the particular point of view is deemed to affect the *content* of the representation. Occlusions, marginal distortion, angle, grain, explicit non-commitment, and 'horizon of alternatives' are hallmarks relating to the *content* of a perspectival representation, by picking out one among other available options.

This is well expressed by the familiar notion of *representing-as* (see Elgin 2017, pp. 252–257; Frigg 2010b; Hughes 1997; van Fraassen 2008), which has often been contrasted with the notion of *representation of*. In brief, *x* is a representation *of y* insofar as *x* denotes *y*: *Las Meninas* is a representation *of* the Spanish royal family because it denotes it (by fiat or by stipulation). But *Las Meninas* is an *x*-representation because it represents the Spanish royal family *as* a composed and timid family.

Representation-as captures, in my view, some salient features of an inferential way of thinking about the perspectival nature of the representation. Not only is the representation *from* a particular point of view that—as such—affects the *content* via the occlusion caused by objects in the foreground, the angle subtended to the spectator's eye, the marginal distortion of the resulting representation, the coarse-graining of the sketchy and idealized image, and so on. In addition to all these, to classify a representation (in art or in science) as perspectival is often understood as implying that its *content* is *represented-as y* (rather than as *z*, or as k, or as j) *from* that specific vantage point. And y, z, k, or j are all attributes or properties belonging to the 'horizon of alternatives' that the representation (qua *representation-as*) ascribes to the target. Crucially, these attributes or properties may well be incompatible with one another, and this incompatibility is usually accepted as unproblematic in artistic representations, where an element of interpretation is inevitably present.[7]

When it comes to perspectival$_1$ representations in science, one possible way of thinking about them is to say that many models for the same target system represent selected features of the target system *as y*, or *as j*, or *as k*, where y, j, and k are different properties belonging to a 'horizon of alternatives' and often enough incompatible or inconsistent with one another. For instance, to use Margaret Morrison's (2011) example, one might say that the shell model of the nucleus *represents* the target system *as* a set of concentric 'shells', so to speak (orbitals more precisely), when viewed *from the point of view* of isotopic phenomena. But from the point of view of quantum chromodynamics, the quark model represents the nucleus *as* a bunch of valence quarks exchanging gluons. And from the point of view of stellar nucleosynthesis, the cluster model represents the nucleus *as* a tetrahedron where nucleons are clustered in equal and even numbers. And from the point of view of nuclear fission, the liquid drop model represents the nucleus *as* a drop of incompressible nuclear fluid.

One may think of the 'horizon of alternatives' along these lines in art or science. But this way of thinking about perspectival representations raises formidable challenges in dealing with a plurality of scientific models for the same target system. Unsurprisingly, critics of perspectivism have cast doubts on the ability of perspectivism to deliver on realism, if the perspectival nature of the scientific representation is understood first and foremost along the lines of this particular way of thinking about what it means to be perspectival$_1$.

But, I contend, it does not have to be. For we may think of models, or better modelling more broadly, along the less familiar—yet no less relevant—lines of perspectival$_2$ representation. Think of perspectival representation not so much in terms of how the content is represented *from a specific vantage point as* having some property y rather than z or k. Think of it instead as opening up a 'window on reality' that extends well beyond the boundaries of the representation itself and where the depth, angle, and scale of the representation leave enough room to *make inferences* about the space and what's in it.

---

[7] Van Fraassen clarifies this point in *The Empirical Stance* (2002, pp. 148–149): 'The painting represents the subjects as arrogant or as complacent, and the fact is that their comportment, as displayed to the painter, allowed both interpretations. So at the very least we must think of the representations of nature that science gives us as representations of nature as thus or so. The "as" indicates an interpretive element, and our example shows that the facts leave sufficient leeway to allow for representations as A or as B, where A and B are incompatible (in that nothing could at the same time be both A and B)'.

PERSPECTIVAL$_2$ REPRESENTATION

(a) The representation is not just situated. It also has a directionality. It is *towards* one or more vanishing points that create the effect of a 'window on reality' extending beyond the boundaries of the representation itself.

(b) The representational *content* is not itself perspectival in that it is not affected by the vantage point *from* which the representation takes place: it is not an instance of *representing-as* understood as ascribing alternative and incompatible attributes or properties. Van Fraassen's hallmarks still apply, but the 'horizon of alternatives' does not capture rival or incompatible property ascriptions to the same target system. It instead allows for a plurality of lines of inquiries and inferences about the target system.

The full implications of perspectival$_2$ representation will become clear in the next chapters, where I unpack the positive role I see a plurality of scientific models to fulfil in particular contexts of inquiry. But let me first attend to a more general problem that seems to arise from the perspectival nature of representations in science.

2.4. A problem with perspectival representations in science

Intriguing as it might be, the analogy between art and science has its limits. In what sense, if any, can a scientific model of, say, the nucleus, a protein, or something else be labelled as perspectival? Van Fraassen himself (2008, p. 86) warns us: '[G]eneral scientific theories, in their "official" formulation, are not perspectival descriptions and their models—if we consider the entire range of models for a given theory—are generally not perspectival representations'.

The remark deserves a qualification, which van Fraassen articulates in his earlier book *The Empirical Stance* (2002, pp. 148–151). There, referring to Newtonian celestial mechanics, he envisages a critic—a 'conservative philosopher'—who might reply that 'science, like art, interprets the phenomena, and not in a uniquely compelled way. Science itself, however, does not admit of alternative rival interpretations' (p. 149). Although the phenomena did not compel Newton's interpretation (to the extent that one by Einstein was eventually adopted), science (by contrast with art) does not easily lend itself to be interpreted in a plurality of ways.

Against the 'conservative' philosopher, van Fraassen rejoins that the 'history of science put the lie to this story, and in successively more radical ways', for 'in science too, we find interpretation at two different levels. The theory represents the phenomena as thus or so, and that representation itself is subject to more than one tenable but significantly different interpretation' (2002, p. 151).

A detailed analysis of this point will have to wait until Chapter 3, where I turn to the problem of inconsistent models. But, in the meantime, a bigger and more general problem looms.

That scientific representations afforded by models are perspectival is not a problem. *How* scientific theories qua families of models produce perspectival representations is not problematic either. From models of measurement to data models; from phenomenological models to theoretical models; from diagrams, histograms, charts, all the way up to more abstract models: our images of nature are inevitably always situated, from a specific vantage point that affects the content of the representation itself (as when we measure temperature in Kelvin rather than in Celsius, or length in metres rather than feet, or the passage of time with atomic clocks rather than sundials). Scientists know how to reliably use instruments, measurement techniques, and models to deliver perspectival representations, just as Renaissance artists knew how to reliably use a perspectograph to draw in perspective.

The problem with perspectival representations concerns, instead, how to explain *our experience* of a three-dimensional perspectival space that extends well beyond the boundaries of the canvas and its flat two-dimensional drawings. How is it that from perspectival representations in science, one *can experience* a world teeming with electrons, Higgs bosons, proteins, eukaryotic cells, DNA strands, and so on and so forth? In other words, why is it that *in spite of* the perspectival$_1$ nature of the scientific representation one can nonetheless legitimately claim to experience a 'window on reality'?

A critic might say that what one experiences is in fact just an illusion—as illusory as the space created by van Eyck's mirror. I reply that perspectival as they might be, these scientific representations are in fact representations of a world (or a window thereon) that is independent of the particular models, experimental techniques, and so on, necessary to deliver them. A world that is as real as is the fruit tree one can reliably infer from what is partially visible from the window in the *Arnolfini Portrait*, no matter if we do not have a way of coming to know it outside the boundaries of van Eyck's representation. Is not the natural world teeming indeed with electrons, Higgs bosons, proteins, DNA strands, eukaryotic cells, hellebores, bees, and chemical elements, regardless of how perspectival$_1$ the scientific representations of them might all be?

The perspectival nature of the scientific representation is not in itself a hurdle to realism in science.[8] Philip Kitcher (2001, p. 51), for example, has convincingly

---

[8] For a non-exhaustive list of examples that somehow go in a kindred direction, see Psillos's (2012) causal descriptivism; Chakravartty's (2011) ontological pluralism; Teller (2019) on reference-continuity; Cartwright (1999), Chang (2012a), and Dupré (1993), for a variety of realist commitments that are compatible with varieties of pluralism; and French (2014) and Ladyman and Ross (2007) for sustained articulations of structuralist responses to a similar problem.

warned against conflating the possibility of constructing representations with that of constructing the world. Our selected vantage point—if understood, for instance, as the selective language with its taxonomic categories that we use to describe 'the world *that we care about*' (p. 46, emphasis in original)—is just one among many possible ones:

> Different ways of dividing nature into objects will yield different representations of reality. Users of different schemes of representation may find it difficult to co-ordinate their languages. Properly understood, however, the truths they enunciate are completely consistent. (p. 47)

Resisting the grip of Plato's metaphor ('carving nature at its joints') puts centre-stage 'human capacities and human purposes' (p. 48). Yet recognizing that ep-istemic communities use alternative classifications[9] and engender different perspectival representations is not denying that 'Like maps, scientific theories and hypotheses must be true or accurate (or, at least, approximately true or roughly accurate) to be good' (p. 61). If anything, it is only rejecting the notion of a 'completed science, a Theory of Everything, or an ideal atlas' as a myth and proposing a rival vision whereby

> what counts as significant science must be understood in the context of a par-ticular group with particular practical interests and with a particular history. It further suggests that just as maps can play a causal role in reshaping the ter-rain that later cartographers will depict, so too the world to which scientists of one epoch respond may be partially produced by the scientific endeavours of the past—not in any strange metaphysical sense but in the most mundane ways. (p. 61)

Navigating between the realist stance crisply articulated by Kitcher and the inevi-tably perspectival nature of scientific representation established in van Fraassen's wake leads me to *perspectival realism*.

A natural worry immediately arises: namely that our representational vantage point risks somehow moulding the content of our representations and that ei-ther perspectivism is impermeable to questions about realism or that realism can

---

[9] 'Those in the grip of Plato's metaphor suppose not only that there is a privileged way to divide na-ture into objects but that there are natural ways to assort those objects into kinds. They recognise that, at different times and in different places, people have used alternative classifications. . . . Faced with this variety, ambitious realists argue that the sciences teach us how to abandon faulty taxonomies and to recognise the "real similarities" in nature. Here too I see a dependence on human capacities and human purposes' (Kitcher 2001, p. 48).

only be jettisoned by such an encounter. The arguments behind this worry are familiar from, for example, Thomas Kuhn's 'living in a new world' thesis.[10]

At the heart of this influential tradition lies, in my view, the conflation between the notions of perspectival$_1$ and perspectival$_2$ representation; or, better, the assumption that perspectivity univocally means perspectival$_1$ representations. Under the Kuhnian tradition, whether one can assert that there is Galilean free fall rather than Aristotelian motion towards a natural place; Supernova Ia rather than fixed stars; or kinetic energy rather than caloric has a lot to do with historical contingency. Realism gains no traction on this ground. Although perspectivism should not be conflated or equated with Kuhn's view, the aforementioned worry about perspectival realism shares some of the spirit of this Kuhnian debate.

Realism in science has typically been associated with the expectations that scientific models (and scientific theories as families of models) 'accurately-albeit-partially' represent relevant aspects of the target system. Some of the concerns about a perspectivalist kind of pluralism in relation to realism originate, in my view, from rather stringent readings of this expectation, which I am going to call:

> *The representationalist assumption.* Scientific models (partially) represent relevant aspects of a given target system S. A scientific model, or, better, the claims of knowledge delivered by a scientific model, are true (or approximately true) when the model provides a partial yet accurate representation of the target system.

The representationalist assumption is in tension with another assumption that is often at play in discussions of scientific modelling, what I am going to call:

> *The perspectivalist assumption.* Scientific models offer perspectival representations of relevant aspects of a given target system S. This is (often implicitly) understood along the lines of *perspectival$_1$ representation* rather than *perspectival$_2$ representation*. Namely, scientific model M_1 *perspectivally$_1$ represents S as z*; but model M_2 *perspectivally$_1$ represents S as y* (where z and y are properties belonging to a 'horizon of alternatives').

---

[10] 'But when seen through the paradigm of which these conceptions were a part, the falling stone, like the pendulum, exhibited its governing laws almost on inspection. . . . [Galileo] had developed his theorem on this subject together with many of its consequences before he experimented with an inclined plane. That theorem was another one of the network of new regularities accessible to genius in the world determined jointly by nature and by the paradigms upon which Galileo and his contemporaries had been raised. *Living in that world*, Galileo could still, when he chose, explain why Aristotle had seen what he did. Nevertheless, the immediate content of Galileo's experience with falling stones was not what Aristotle's had been' (Kuhn 1962/1996, p. 125, emphasis added).

The perspectivalist assumption—if understood primarily along the lines of this particular reading of perspectival$_1$ representation (qua representing-as)—suggests that rival and often incompatible property ascriptions are at play for the same target system S across several models.

The tension between these two assumptions underlies the concern about perspectival realism. Upholding realism about science by being able to *come to know* that the world *is* thus or so would seem to be precluded within the bounds of a plurality of perspectival$_1$ representations ascribing incompatible or even inconsistent properties to the same target system. This chimes with a variety of antirealist views and arguments that I will not review because my task is constructive: namely to show how the perspectival nature of scientific representation does *not* necessarily preclude realism.

But before proceeding, something ought to be said about a classical move to deflate this concern. One way of thinking about property ascription inherent in representing-as is in purely pragmatic terms. For example, scientific representations may be said to represent the target system *from a specific vantage point* as that vantage point relates to an agent who is *using* the representation *for a specific goal*. Is there not an easy way forward for perspectival realism once one takes into account the pragmatic use of scientific representation for specific goals? In other words, can the perspectival nature of scientific representation be made realist-friendly on pragmatic grounds?

2.5. Indexicality: the pragmatics of perspectival representations

Indexicality refers to the context-sensitive *use* that epistemic agents make of perspectival representations. As van Fraassen (2008, pp. 76 and 80) rightly observes, '[T]he activity of representation is successful . . . only if the recipients are able to receive that information through their "viewing" of the representation. . . . The recipient must be in some pertinent sense able to relate him- or herself, his or her situation, to the representation'.

Measurements, maps, models, all offer perspectival representations for specific *uses*, goals, purposes of agents. Van Fraassen (2008, p. 176) distinguishes between '*Measuring is Locating* and *Measuring is Perspectival*', whereby 'what is perspectival is not the action of measuring but the content of the measurement outcome, and locating is an action, not a content'. Representing requires that the agents locate themselves with respect to the horizon of alternatives so as to see the representation *from a specific vantage point* and to *use* it for specific purposes.

Consider, for example, measurements as one example of perspectival$_1$ representation. Room temperature can be measured on either the Fahrenheit scale or the Celsius scale. Yet it would be hasty to conclude that these two measurement scales locate the same object in two different logical spaces. For there is a 'single logical space in which the two scales are, in effect, two coordinate systems' (van Fraassen 2008, p. 174) and a linear transformation allows agents to seamlessly move from one to the other. Although one might say that the Fahrenheit thermometers and the Celsius thermometers offer 'different perspectives on the same magnitudes', this would only be a *façon de parler*, rather than a genuine difference, since the underlying theory is the same.[11]

Along similar lines, Ron Giere discusses what I call the perspectival$_1$ representations associated with measuring the intensity and distribution of gamma rays coming from the centre of the Milky Way '*as indicated by COMPTEL or OSSE*' (Giere 2006a, p. 48, emphasis in original). Here again the perspectival content afforded by each of these different measurement techniques and instruments is functional to specific uses and purposes by the epistemic agents.[12] This view is in accord with Giere's overarching agent-based conception of scientific representation (Giere 2010), whereby representation is understood as a four-step process: agents (1) intend (2) to use model M (3) to represent a part of the world W (4) for some purpose P. Like van Fraassen and Giere, Catherine Elgin too (2017, pp. 158–159) stresses the pragmatics of perspectival representations:

> Measurement is always indexical and perspectival. Hence for science to do its epistemic job, it must involve indexical, perspectival representations. . . . Those representations are objective in that they contain information that is invariant across representations of the same object. They are testable in that multiple representations of the same objects from the same perspective yield equivalent information, and in that information can be accessed from multiple perspectives.

[11] Cf. van Fraassen (2008, p. 173, emphasis in original): 'If a thermometer is used by one person to locate the air in his room on the Fahrenheit scale, and by another to locate it in the space of possible mean kinetic energies of its molecules, the two are locating the same thing, by means of the same instrument, in two different logical spaces. That we should not at once call a change of perspective, though it certainly marks a change in ways of thinking about the air. What if a theory then equates temperature and mean kinetic energy? In that case we should say that *relative to the theory* it is appropriate to call this a change of perspective. This qualification is important, though it may be left tacit in a context where the theory has been entirely accepted'.

[12] Giere (2006a, p. 48, emphasis in original): 'Humans and various other electromagnetic detectors respond differently to different electromagnetic spectra. Moreover, humans and various other electromagnetic detectors face the same spectrum of electromagnetic radiation and yet have different responses to it. In all cases, the response of any particular detector, including a human, is a function of *both* the character of the particular electromagnetic spectrum encountered and the character of the detector. Each detector views the electromagnetic world from its own *perspective*. Every observation is perspectival in this sense'. And in the associated endnote Giere specifies: 'Instruments interact directly with objects in the world. They typically *produce* representations of the objects in question. But the representation is a *product* of the interaction between instrument and object, not part of the process by which the instrument detects aspects of the objects' (p. 126, n. 7, emphasis in original).

A common theme runs through indexicality as a pragmatic feature of perspectival$_1$ representation: namely that the ability of the agents to locate themselves in the logical space of alternatives and to pragmatically use a multitude of (presumably) perspectival$_1$ representations for different purposes is not only necessary for testing and measuring in science, but it is also metaphysically and epistemically innocent.

It is metaphysically innocent because it does not affect reality itself: perspectival$_1$ representations—if understood in pragmatic terms—do not engender a multitude of perspectival facts, somehow reminiscent of Kuhnian 'living in a new world'.

Moreover, indexicality seems to be epistemically innocent. Perspectival$_1$ representations presumably share the same underlying theory; or yield information that can be accessed from multiple perspectives;[13] or are subject to linear transformations that allow agents to seamlessly move from one perspective to another.

I think that the point about metaphysical innocence is well taken (and I will expand on it in Chapter 3). However, I am less sanguine about the point on epistemic innocence in situations where:

- the partiality of the representations at issue concerns incompatible or even inconsistent property ascriptions to the target system;
- the representation is not just intended to *exemplify* or be *similar* in some relevant respects to selected aspects of the target system (while being false in others); but instead, it is intended to ascribe to that target system property *y* rather than property *z* (where those properties are incompatible or inconsistent with one another);
- there is no seamless transformation across perspectives or single unifying theory behind the plurality of perspectival$_1$ representations.

---

[13] I take this to be Elgin's own view about why shell models and liquid drop models for the nucleus do not constitute a logical or conceptual difficulty. As I read Elgin, she links representation with what she calls 'exemplification': '[W]hen *x* represents *y* as *z*, *x* is a *z*-representation that *as such* denotes *y* . . . in being a *z*-representation, *x* exemplifies certain properties and imputes those properties or related ones to *y*' (Elgin 2017, p. 260, emphasis in original). Thus, perspectival$_1$ representations (e.g. the shell model and the liquid drop model) might 'exemplify' different sets of properties each of which might be displayed by the represented target system in the relevant context for specific purposes: 'Sometimes, although the target does not quite instantiate the features exemplified by the model, it is not off by much. Where their divergence is negligible, the models, although not strictly true of the phenomena they denote, are true enough of them. . . . Where a model is true enough, we do not go wrong if we think of the phenomena as displaying the features that the model exemplifies. Whether a representation is true enough is a contextual question. A representation that is true enough for some purposes or in some respects is not true enough for or in others. This is no surprise' (p. 261). Or, better, it is no surprise as long as the similarity (or exemplification, to use Elgin's terminology) is not interpreted as a genuine 'horizon of alternatives' against which different perspectival$_1$ representations (and related perspectival contents) are pitched.

These are the situations that I think cannot be helped by appeal to pragmatic considerations. Indexicality seems to me to leave untouched the *epistemic problem* of how one reliably comes to know the world from a plurality of scientific perspectives.[14] The answer to this question cannot be delegated to pragmatic considerations about how agents relate to or use perspectival$_1$ representations for specific purposes in specific contexts.

However, indexicality conceals two very important insights. First, the *use* that particular communities make of their perspectival$_1$ representations for particular purposes is absolutely key to understanding the *local* and always *situated* nature of scientific knowledge claims. There will be plenty of examples of local and situated knowledge claims in this book, starting from my case studies in Chapters 4a, 4.b, and 4.c. And I shall return to the topic of the situatedness of scientific knowledge at the very end of this book in Chapter 11, where I tease out the connection with multiculturalism in science.

Second, the very action of representing perspectivally is directed to a specific goal. The answer to the problem of how wonderfully diverse human beings like us come to reliably know the world from a plurality of situated scientific perspectives ultimately lies in how one understands the nature of perspectival$_2$ representations delivered by scientific models. I attend to this task in Part I of this book.

In the next chapter, I return to the *representationalist assumption*. I show how the tension it causes when combined with the *perspectivalist assumption* is at work in the problem of inconsistent models as an *epistemic* (not a metaphysical) problem. I analyse three perspectival answers to the problem of inconsistent models, lay out the argument for it, and ultimately urge resistance to some of the premises behind it.

---

[14] Intentionally so, of course, in van Fraassen's antirealist stance and similarly in Elgin's defence of models as 'felicitous falsehoods'.

3

Pluralism and perspectivism

In Eudoxia, which spreads both upwards and down, with winding alleys, steps, dead ends, hovels, a carpet is preserved in which you can observe the city's true form. At first sight nothing seems to resemble Eudoxia less than the design of that carpet, laid out in symmetrical motives. . . .

But you could, similarly, come to the opposite conclusion: that the true map of the universe is the city of Eudoxia, just as it is, a stain that spreads out shapelessly, with crooked streets, houses that crumble one upon the other amid clouds of dust, fires, screams in the darkness.

Italo Calvino (1972/1997) *Invisible Cities,*
Cities & the Sky, pp. 86–87[1]

3.1. Varieties of scientific pluralism

Scientific pluralism has been part of the philosophical discourse on science since at least the 1960s. Pluralism has become an irresistible philosophical stance in the wake of historical and social studies of science that have rightly rediscovered the very many ways of knowing displayed by scientific communities across the history of science and across cultures. Ron Giere described the historical background against which scientific pluralism emerged in the 1960s and 1970s:

During this period, many who grew up after WWII found themselves horrified by the use of B-52s and other high-powered military technology against Vietnamese peasants riding bicycles and armed with little more powerful than an AK-47. Around the same time, it became clear that modern industries and, particularly, agricultural technologies were degrading the environment in

---

[1] Copyright © Giulio Einaudi editore, s.p.a 1972. English translation copyright © Harcourt Brace Jovanovich, Inc. 1974. Reprinted by permission of The Random House Group Limited. For the US and Canada territories, *Invisible Cities* by Italo Calvino, translated by William Weaver. Copyright © 1972 by Giulio Einaudi editore, s.p.a. Torino, English translation © 1983, 1984 by HarperCollins Publishers LLC. Reprinted by permission of Mariner Books, an imprint of HarperCollins Publishers LLC. All rights reserved.

Perspectival Realism. Michela Massimi, Oxford University Press. © Oxford University Press 2022.
DOI: 10.1093/oso/9780197555620.003.0003

many ways. Additionally, some women began to regard new household technology as more enslaving than liberating. Some people whose fundamental attitudes were formed during this period became university professors. And a few of these focused their attention on the sciences, not as scientists themselves, but as critics of science. Not surprisingly, many of these critics found their way into the humanities and social sciences, such as history, philosophy, sociology, literature, or, eventually, cultural studies. (Giere 2006a, pp. 1–2)

There was an intellectual reaction against what many perceived as the enslaving rather than liberating power of science and technology, in a political and social context where minorities and marginalized communities challenged scientific orthodoxy. Unsurprisingly, scientific pluralism went hand in hand with the emergence of standpoint theory in epistemology.[2] Two main themes feature in both. The first is an anti-foundationalist stance regarding scientific knowledge. Anti-foundationalism is often born out of discussions about underdetermination of theory by evidence (see Longino 2002) and sociological considerations about traditional epistemology being developed in an exclusionary framework of Europe and North America (see, e.g., Harding 1995).

The second and related theme is what Alison Wylie (2003) has characterized as the *situated knowledge thesis*: how being situated affects and limits what one can know. Within standpoint theories, this situatedness takes a distinctive sociopolitical dimension.[3] In Helen Longino's words (1994, p. 474), 'a standpoint is a perspective afforded by a social location' (e.g. the standpoint of women, of the working class, of ethnic minorities).

While sharing both the anti-foundationalist stance and the situated knowledge thesis, scientific pluralism has developed its own distinctive outlook in the philosophy of science. Its anti-foundationalist stance has often translated into a programmatic anti-essentialism. Likewise, the situated nature of human knowledge is often understood as ranging more widely than social location in the

[2] It goes well beyond my expertise and the remit of the present book to engage with the voluminous literature on standpoint theories. For some representative examples of the relevant discussions, see Harding (1986, 1991, 1995), Hartsock (1997, 1998), Hekman (1997), Longino (1994, 2002), Smith (1974, 1997), and Wylie (2003). In what follows, I tease out two broad themes from this literature—with special reference to the work of Longino and Wylie—and highlight their relevance to my ongoing discussion of pluralism and perspectivism.
[3] In the words of Harding (1995, pp. 341–342, emphases in original), 'Standpoint theories argue that what we *do* in our social relations both enables and limits (it does not determine) what we can know. Standpoint theories, in contrast to empiricist epistemologies, *begin* from the recognition of social inequality . . . in contrast with the consensus model of liberal political philosophy assumed by empiricists'. And as Haraway (1988, p. 590) puts it, 'The only way of finding a larger vision is to be somewhere in particular. The science question in feminism is about objectivity as positioned rationality. Its images are not the products of escape and transcendence of limits (the view from above) but the joining of partial views and halting voices into a collective subject position that promises a vision of the means of ongoing finite embodiment . . . of views from somewhere'.

literature on scientific pluralism. Taking a cue from Kuhn's influential 'working in a new world' thesis (Kuhn 1962/1996, p. 121; see Massimi 2015 for a discussion), occupying a view *from somewhere* (rather than 'nowhere' or 'everywhere') invites us to ask how the conceptual resources, technological and experimental tools, and theoretical assumptions typical of a given epistemic community affect their ability to come to know the world *this* way or *that* way. Work in integrated history and philosophy of science (HPS) has elaborated this Kuhnian idea, with interesting results for scientific experimentation and practice.[4] Situated knowledge becomes an invitation to set more modest goals for scientific inquiry. Some examples are van Fraassen's (1980) empirical adequacy; Kitcher's (2001) objectivity-with-no-ideal-atlas; Solomon's (2001) view on consensus forming; Chang's (2012a) coherence in scientific practice; or Daston's (1992) discussion of aperspectival objectivity.

Scientific pluralism is usually pitched against scientific monism, the view that science aims to establish a single, complete, and comprehensive account of the world. But what the position precisely amounts to (and whether or not it might be compatible with realism in science) has invited a variety of responses.

For example, Longino (2001, 2006, 2013) has persuasively argued that to successfully explain aspects of human behaviour—such as aggression—it is necessary to abandon the presumption that there is one single correct approach and acknowledge the advantages of adopting a form of theoretical pluralism. Miriam Solomon (2015, p. 206) refers to it as 'methodological pluralism' to distinguish it from other 'ontological' varieties of pluralism.

In Carla Fehr's (2001, 2006) explanatory pluralism, more than one explanation is required to account for a given phenomenon (in her case study, the evolution of sex), and these multiple explanatory accounts cannot be integrated without loss of content.[5] Alisa Bokulich (2017, 2018a) has defended a similar type of pluralism, and argued that, in relevant contexts, a plurality of scientific models offer explanans that are decoupled from different possible kinds of causal mechanisms and are not even intended as idealized representations of any causal mechanism (Bokulich 2018a, p. 149).

Sandra Mitchell (2003, 2009) has shown how in complex systems the emphasis is on 'how multiple explanatory factors operating at different ontological levels enter into explanation in the biological and psychological sciences' (Mitchell 2009, p. 114). John Dupré's 'promiscuous realism' (1981, 1993) has underlined how the plurality of taxonomic classifications available for the same entities is a function of specific interests of different epistemic communities.

---

[4] See, for example, the work of Ankeny and Leonelli (2016), Arabatzis (2006), Chang (2012a), Feest and Steinle (2012), Schickore and Steinle (2006), and Steinle (2002).
[5] Other examples of explanatory pluralism are Jackson and Pettit (1992) and Sterelny (1996) in evolutionary biology.

Anjan Chakravartty (2007, 2011, 2017) has advocated what he calls 'sociable properties' involving ontological pluralism about how to package or repackage collections of properties as constituting different kinds of things. His variety of ontological pluralism is pitched against two specific readings of perspectivism (Chakravartty 2011), understood either as a claim about our scientific knowledge being irreducibly perspectival (P1: non-perspectival facts are beyond our epistemic grasp), or as a stronger claim that perspectival knowledge delivers perspectival ontology (P2: there are no non-perspectival facts to be known). We shall return to this in Section 3.5.

But ontological pluralism need not necessarily be about how properties get packaged or repackaged. Stéphanie Ruphy (2010, 2011, 2016, and Massimi 2017b for a discussion) has articulated what she calls 'foliated pluralism' inspired by Hacking's styles of reasoning (Hacking 2002). It holds that questions about what kinds of entities there are remain relative to different styles of reasoning, and how they 'widen and diversify the classes of propositions that can be true or false about them' (Ruphy 2016, pp. 30–31). And Hasok Chang, in turn, has championed what he calls active normative pluralism 'advocating the cultivation of multiple systems of practice in any given field of science, where a 'system of practice' is a coherent and interacting set of epistemic activities performed with a view to achieve certain aims' (Chang 2012a, p. 260).

What is striking about this non-exhaustive list is the way discussions of epistemic varieties of pluralism often branch out into ontological pluralism: from the methods of science to the nature of scientific entities. In the next section, I zoom in and clarify the link between, on the one hand, pluralism as an epistemic stance prompted by scientific practice and, on the other, some of the (far from innocent) ontological implications that many philosophers have seen at play in the pluralist stance. This has become known as the problem of inconsistent models.

3.2. The 'pluralist stance' and the problem of inconsistent models

Is there a common ground in these many varieties of pluralism?[6] Stephen Kellert, Helen Longino, and C. Kenneth Waters describe the 'pluralist stance'—a minimal, empirically motivated commitment to

[6] In what follows, I draw on Massimi (2018b) 'Perspectival modelling', *Philosophy of Science* 85, 335–359. Reproduced with permission from University of Chicago Press. Copyright (2018) by the Philosophy of Science Association.

kinds of situations produced by the interaction of factors each of which may be representable in a model or theory, but not all of which are representable in the same model or theory. . . . A more complete representation of some phenomena requires multiple accounts, which cannot be integrated with one another without loss of content. (Kellert et al. 2006, p. xiv)

The pluralist stance admits that 'if two models distort some of the same aspects, they might distort these aspects in different ways, giving rise to inconsistencies. This is just one kind of situation in which a plurality of inconsistent approaches might be defended' (Kellert et al. 2006, pp. xiv–xv). It is this inconsistency-friendly feature that some critics find problematic.

The pluralist stance acknowledges that—given the *representational* role of models—whenever there are situations requiring a plurality of models for a given phenomenon, each model might accurately represent some aspects of the phenomenon at the cost of blurring, idealizing, or distorting other aspects. The outcome is that different models may deliver different, sometimes incompatible, or even inconsistent images for the same phenomenon.

Consider, for example, inconsistent families of models for the atomic nucleus, discussed by Margaret Morrison (2011), among others. In high-energy physics, the nucleus is accurately represented by quantum chromodynamics as consisting of hadrons (protons and neutrons), each of which in turn consists of coloured quarks. But in the context of radioactivity and chemistry, the nucleus is best represented by the shell model, which represents the nucleons as occupying 'shells'/orbitals in some 'magic numbers'. If the phenomenon under study is nuclear fission, the liquid drop model best represents the nucleus and its behaviour. Finally, if the phenomenon of interest is the clustering of light nuclei (important in stellar nucleosynthesis), then the cluster model best represents the nucleus, with nucleons clustered in even and equal number (e.g. ^{4}He, ^{8}Be, ^{12}C).[7]

Analogous situations abound. In chemistry, there exist several different submodels for water within continuum physics, classical atomistic physics, and quantum physics, respectively (Izadi et al. 2014). And when it comes to biology, predicting protein folding from amino acid sequences can be obtained through a plurality of experimental practices *in vivo* or *in vitro* that while not forming

---

[7] A ground-clearing remark is in order at this point. A possible reaction here would be to draw a line somehow between theory-driven models, such as quantum chromodynamics (QCD), on the one hand, and phenomenological models, such as the shell model or the liquid drop model, on the other. Or maybe to supplement this distinction with the further point made by Newton da Costa and Steven French (2003) that phenomenological models are somehow heuristic devices more than representational models of the target system in their own right. Yet we should not underestimate the representational value of phenomenological models of the nucleus. Such models do seem to have a representational value of their own, even if it is not downstream from a scientific theory like QCD. I thank Demetris Portides for helpful conversations on this topic (see also Portides 2005 and 2011).

a linear combination can nonetheless be integrated (see Mitchell 2020 and Mitchell and Gronenborn 2017). What is to be said about this plurality of modelling practices broadly understood (to include experimental methods)?

Suppose we take the primary role of models to be representational, *and* we understand such a representational role in terms of the model mirroring, accurately describing, or mapping onto 'states of affairs' about the target system. It comes then as no surprise that the pluralist stance is inconsistency-friendly. And this is a programmatic commitment that several scientific pluralists are happy to make. They regard it as a positive outcome of thinking about models as offering partial, idealized, and abstract representations of the target system (on the idealized nature of models' representations, see Bailer-Jones 2003; Cartwright 1999; Potochnik 2017; Rice 2018; Rohwer and Rice 2013; Weisberg 2007, 2013). The literature that has emphasized the role that abstractions and idealizations play in scientific representations has also paved the way to a model pluralism that accepts the partial, incomplete, and idealized nature of each representation. The inconsistency-friendly pluralist stance is a consequence of this more general trend in philosophy of science.

Yet the pluralist stance is not confined to a purely epistemic claim. The bone of contention has to do instead with the *metaphysical gloss* that such an inconsistency-friendly pluralist stance seems to invite.[8] This problem has attracted attention under the name of *the problem of style* (to use Frigg and Nguyen's terminology, 2020), or *the problem of inconsistent models* (or PIM—see Morrison 2011, 2015, Ch. 5). PIM is ubiquitous in the sciences. It does not just apply to physical models. And it poses a serious problem for realism.

Let us return to what in Chapter 2 I called *the representationalist assumption:* namely that one of the main tasks of any scientific model M is to represent (in part at least) relevant aspects of a given target system S, and that claims of knowledge based on the scientific model are true (or approximately true) when the model provides a partial yet accurate representation of the target system.

Consider now situations where there is more than one model M that fulfils this representational role for the same target system S. Let us call this the *pluralist assumption* as a sufficiently generic and unqualified assumption. This is not to be confused with the *perspectivalist assumption* introduced in Chapter 2, which qualifies already the kind of pluralism involved, and to which I shall return in Section 3.4.

These two assumptions create a tension for realism. For how can a plurality of equally representational models *all* latch onto *real features* of the same target

---

[8] Unsurprisingly, some pluralists advocate a more cautious approach that restricts the pluralist claim to the epistemic domain. For example, some scientific pluralists (from Ruphy with her foliated pluralism to Mitchell with her integrative pluralism) have expressly reacted against inconsistency-friendly and non-integrative varieties of pluralism.

system (no matter how partially and idealized each representation might be)? What really *is* the atomic nucleus? And can one answer this question by appealing to models, given their mutually incompatible representations?

Situations like this typically invite two kinds of answers. The first is to go instrumentalist: models are useful to get calculations done, but their representational content should not be taken literally (see Hacking 1982 for a classic example). The second is to defend realism by introducing a series of caveats. One such caveat is that one would have to demonstrate that models enjoy equal explanatory and predictive success before drawing any conclusion about their being epistemically on a par. This first caveat takes care of situations such as Ptolemaic vs. Copernican models of the solar system, where the former did not enjoy the same predictive success as the latter (e.g. no phases of Venus being predicted, among others).

A second caveat is that each model can only represent veridically some *parts* or *portions* of the target system while idealizing and misrepresenting others. A scientific realist might take the liquid drop model as providing an approximately true story of how the binding energy can be released in nuclear fission, while inevitably idealizing and misrepresenting the atomic nucleus as a drop of incompressible nuclear fluid.

These caveats, however, do not entirely succeed in sheltering realism from the problem of inconsistent models. For they do not address what I take to be the real bite of PIM. That comes if different models (partially and idealized though they might be) represent relevant properties of the target system *and* these properties are understood as both *essential* and *inconsistent* with one another.

In other words, PIM arises when the representation afforded by each model is understood as a *de re* representation of the target system, where a *de re* representation is a representation that ascribes essential properties as opposed to, say, nominal properties to the target system. One can take different attitudes towards the *representational content* of scientific models. For example, models can be regarded as representing *de re* relevant aspects of the target system when they map onto properties that are regarded as not just real but also essential. But models can also be regarded as representing *de dicto* relevant aspects of the target system when they map onto properties that are regarded neither as real nor as defining the essential nature of the target system.[9] Thus, PIM engenders metaphysical inconsistency because it envisages the following scenario:

---

[9] For example, one might take Maxwell's honeycomb model of the ether as offering a *de dicto* representation of electromagnetic induction in the sense that the representation does not map onto real and essential properties (for there is no hexagonal ether and the electric displacement cannot be interpreted as being essentially constituted by rotating idle wheels among hexagonal vortices). Thus, in a way, fictionalism about models is less vulnerable to PIM (unless the representational function of fictional models is itself understood along the lines of essential properties attribution to the target

Model M_1 delivers a partial yet *de re* representation of selected relevant properties a_1, b_1, c_1. These are real-qua-essential properties of the target system in that they capture *essential* features of it.

Model M_2 delivers a partial yet *de re* representation of selected relevant properties a_2, b_2, c_2, which also capture *essential* features of the same target system.

And a_2, b_2, c_2, are inconsistent with a_1, b_1, c_1.

Bearing in mind the discussion of perspectival representations in Chapter 2, this situation sounds familiar. That different models ascribe different properties to the same target system should not be surprising given that the representation afforded by each model can be regarded as perspectival$_1$: that is, *from a specific vantage point*. It is then tempting to suggest that the problem of inconsistent models originates from the *perspectival$_1$ representation* that each model affords of the same target system. Each model can be regarded as offering a *de re* representation of the nucleus that is perspectival$_1$ in that the representational *content* is *represented-as y* (rather than as *z*, or as *k*, or as *j*) *from* the specific vantage point of the model. Properties *y*, *z*, *k*, or *j* are regarded as essential attributes or properties belonging to the 'horizon of alternatives' that the situated representation (qua *representation-as*) ascribes or imputes to the target. And they may be incompatible or inconsistent with one another.

Where to go from here? Perhaps one might reply that PIM is not a real problem because there is one and only one model among these that genuinely provides a veridical (no matter how partial and perspectival$_1$) *de re* representation of the target system. For example, one could insist that only the model that is directly linked to a theory such as quantum chromodynamics (QCD) gives the *best de re representation* of the nucleus and that the other models—being more phenomenological models than theory-driven ones—are not really intended as being epistemically on a par with the QCD model.

But this reply misses an important point. What is the purpose of a plurality of models if only one of them purportedly offers a veridical *de re* representation? Pluralism about models cries out for an explanation. Recall the situated knowledge thesis. The QCD model is useless for explaining nuclear fission, or the stability of radioactive isotopes. Depending on the contexts and problems at hand, other models have to be used. The scientific image is always *from somewhere*, and if that somewhere is anywhere near stable isotopes, say, one would do better to avail oneself of the shell model rather than attempting an analysis in terms of quarks and gluons.

system via analogy with the fictional model). For a recent discussion on this topic, see Frigg and Nguyen (2016).

If, on the other hand, there is not one and only one model that can legitimately claim to provide a partial yet veridical *de re* representation of the target system, then PIM has a genuine bite for realism. In the next two sections, I explain how scientific perspectivism enters this debate as a possible way of answering PIM.

3.3. Two varieties of scientific perspectivism

Scientific perspectivism is a philosophical view with a long-standing history dating back to Leibniz and to the Kantian idea that our scientific knowledge is always from a human vantage point (see Crețu and Massimi 2020). Ron Giere (2006a) has advocated perspectivism as a reaction against both the monist 'view from nowhere' and a broad family of anti-realist views.[10]

Giere's view is still realist. It implies belief in a world that, neither constructed by us nor relativized to scientific perspectives, is nonetheless the object of modelling practices and their representations. And representations are always perspectival. In Giere's own example, the White House is to the right of the Washington Monument if viewed from the steps of the US Capitol Building. But it would be to the left of the Monument from the steps of the Lincoln Memorial (Giere 2006a, p. 81). The vantage point does not affect the reality of either the White House or the Washington Monument. Hence, there is a clear and almost mundane sense in which realism can be upheld while also acknowledging the perspectival (in the sense of perspectival$_1$) nature of representations.

Giere presents his own view by analogy with Thomas Kuhn, whereby 'Claims about the truth of scientific statements or the fit of models to the world are made within paradigms or perspectives' (Giere 2006a, p. 81). Giere's scientific perspectives are akin to Kuhn's disciplinary matrices (although they do not include beliefs, values, and techniques shared by a scientific community).[11]

'Scientific perspectives' (e.g. Giere refers to the Newtonian perspective, or the Maxwellian perspective)[12] are thought of as hierarchies of scientific models, under this view. Starting bottom up, models of the data (or data models) are first. Starting top down, scientific principles (say, Newton's laws of motion) and initial conditions come first. In the middle of this hierarchy, Giere locates what he calls 'representational models'. These are designed to offer a way of fitting scientific

---

[10] In what follows, I draw on Massimi (2018a) 'Perspectivism', in J. Saatsi (ed.), *The Routledge Handbook to Scientific Realism*, London: Routledge. Reproduced with permission from Routledge.

[11] For a detailed discussion of the analogy between Giere's perspectivism and Kuhn's view, see Giere (2013). I have analysed Giere's perspectivalist reading of Kuhn in Massimi (2015). For a novel discussion of the role of perspectivism in scientific instrumentation see Crețu (forthcoming).

[12] 'Newton's laws characterize the classical mechanical perspective; Maxwell's laws characterize the classical electromagnetic perspective; the Schrödinger Equation characterizes a quantum mechanical perspective' (Giere 2006a, p. 14).

principles to data models via specific hypotheses and generalizations. For example, the pendulum model offers a way of fitting Newton's laws of motion plus initial conditions to models of the data (the specific observed motion of the pendulum) via tailored hypotheses (about the length of the rope, the mass of the bob, etc.).

Giere understands 'fitting' via a notion of scientific representation that relies on similarity rather than on any traditional notion of correspondence qua 'exact fit' or mapping along the notion of isomorphism. The representational models are perspectival in idealizing some factors and abstracting from others (e.g. in the model of the harmonic oscillator, mass is idealized as a point-mass, and displacement from equilibrium abstracts from disturbing factors). Finally, the general principles that inform the choice of the representational models are themselves perspectival (switch perspective, switch laws and principles).

Giere's perspectivism takes the Kuhnian view that there is no cross-paradigmatic (or cross-perspectival) notion of truth in play here. What claims of knowledge count as true (or false) is simply a function of how particular data models fit particular theoretical models. And since both kinds of models are perspectival,, any model-based knowledge claim is true or false only within the boundaries of the chosen scientific perspective. Giere's *epistemological argument* for perspectivism can be summed up as follows:

(A) Our scientific knowledge is perspectival (in the sense of *perspectival,*) because scientific knowledge claims are only possible within a (historically) well-defined family of models (e.g. the Newtonian perspective, the Maxwellian perspective), which constrains both the data available (via data models) *and* the interpretation of those data (via theoretical models and principles). No knowledge of nature is possible outside the boundaries of historically well-defined scientific perspectives.

Perspectivism so understood is, then, inherently pluralistic because there are multiple perspectives on the same target system (e.g. the Newtonian perspective and the Einsteinian perspective when it comes to gravitational phenomena). Different scientific perspectives are at play both diachronically (across the history of science) and synchronically (as with rival models for the same target system at any given time).

The synchronic version of perspectivism has also been advocated by Alexander Rueger and Paul Teller. Rueger (2005) has addressed how scientific realism can handle the problem of inconsistent models by making a different perspectival move. For different models, Rueger argues, seem to offer different perspectives on the same target system, whereby *intrinsic* properties of the system turn out to be in fact *relational* properties. Accordingly, models should not be regarded as delivering inconsistent images. Instead, they deliver *partial and perspectival* (qua again perspectival$_1$ in my idiolect) images that can still be unified into a final coherent account, as realism would have it.

Along similar lines, Teller (2011) has argued that since models are idealizations (hence, inevitably imprecise), it is possible to have more than one model for the same target system without having to forgo realism (e.g. hydrodynamical vs statistical mechanical models for describing the properties of water). According to Teller, scientific perspectives should be understood as idealized representational schemes that do not get things exactly right.

Thus, false precise statements (e.g. 'John's height is six feet precisely') can be transformed into true yet imprecise ones (e.g. 'John's height is six feet close enough'). Teller calls such pairs 'semantic alter-egos' and contrasts what he calls the 'conditions of application' with the distinct, traditional notion of truth conditions. These are different in as much as traditional truth conditions are taken to be completely precise while Teller's conditions of application are ineliminably open-ended.[13]

The Rueger–Teller argument for perspectivism is, then, primarily a *methodological argument* from modelling practices (rather than scientific perspectives as Giere understands them) that concludes:

(B) Scientific knowledge is perspectival (in the sense of *perspectival$_1$*) because scientific knowledge claims are only possible within well-defined families of models of any given scientific perspective at any given time (e.g. hydrodynamics and statistical mechanics for water). No knowledge of nature is possible outside the boundaries of well-defined scientific perspectives with their pluralism about models.

The epistemological and methodological arguments share a somewhat Kantian-flavoured insight about conditions of possibility of knowledge. Our human (historical and cultural) vantage point makes possible scientific knowledge claims. However, the two arguments differ in the way they explicate the perspectival nature of scientific knowledge. They diverge in relation to scientific truth as well.

Version (A) argues that the truth of scientific knowledge claims is *relative* to historically defined scientific perspectives. (What was true for Ptolemy proved false for Copernicus.) By contrast, version (B) takes pluralism about models as an indication *either* (Bi) that truth, as traditionally understood in terms of truth conditions, is an idealized and in practice unobtainable goal (Teller); *or* (Bii) that truth can be preserved if one understands property ascription to a target system as *perspectival$_1$-qua-relational* (Rueger). In the next section, I show how

---

[13] I thank Paul Teller for helpful conversations and feedback on this material.

each version of scientific perspectivism can provide a possible answer to PIM. In Section 3.5, I review why some critics have been unimpressed.

3.4. Three perspectival answers to the problem of inconsistent models

Can either Giere's or Rueger–Teller's distinctively perspectival forms of scientific pluralism help with the problem of inconsistent models? Let us take a closer look at how each version might answer PIM.

Giere's version begins with the analogy with colour vision. The difference between human trichromats and animal monochromats is presented as a case of perspectival pluralism: there is no way of claiming that the coloured perspective is 'objectively correct, or in some sense uniquely veridical, while the black-and-white perspective is incorrect or nonveridical' (Giere 2006b, p. 28). Presuming otherwise would imply that colours are somehow intrinsic properties of objects rather than perspectival properties of how light rays reflected from different kinds of surfaces impinge on the retinas of different animals.[14]

A similar line of reasoning applies to scientific observations. Here too, observations are relativized to the perspective of the relevant instrument and there is 'no such thing, for example, as *the* way the Milky Way looks. There is only the way it looks to each instrument' (Giere 2006b, p. 30, emphasis in original). Different perspectives are consistent and complementary because they can all be regarded as observations of a unique world.

But things get more complex in the case of scientific theorizing. Here the tension between, say, quantum mechanics and general relativity has arguably invited scientists to try to search for a solution that could unify the two perspectives (quantum gravity). Giere sees this tension as unproblematic, however, because '[g]ood theoretical science does not *require* finding genuinely universal principles. Well-fitting models, based on a variety of principles, are good enough' (2006b, p. 33, emphasis in original). He concludes:

Employing a plurality of perspectives has a solid *pragmatic* justification. There are different problems to be solved, and neither perspective by itself provides adequate resources for solving all the problems. Of course, a metaphysical realist will ask, 'But what is water, really?' assuming that the answer must be molecules. But perspectivism yields the desired answer without giving in to monism. Nothing in perspectivism dictates that all perspectives are created

---

[14] For a discussion of the literature on colour vision and an analysis of the philosophical debate surrounding the status of colours as objective vs perspectival properties, see Chirimuuta (2015).

equal. Some are better than others in many different respects. In this case, there is a clear asymmetry in favour of a molecular perspective. . . . So we can say that there is, *in principle*, a molecular model for all of the many manifestations of water. In *practice*, there are many manifestations of water that can only be modelled within other perspectives. In this case, while pragmatism dictates a pluralist attitude towards theoretical perspectives, the intuitions, though not the metaphysics, of the metaphysical realist can be accommodated. (Giere 2006b, p. 34, emphases in original)

Thus Giere's perspectivism is pluralist on pragmatic grounds, while also sharing the metaphysical realist's intuition that there is a single world after all. The metaphysical intuition gives rise to a *methodological maxim*: 'Proceed as if there is a single world with a unique structure' (Giere 2006b, p. 36). Therefore, in reply to PIM, Giere seems to maintain that pluralism about perspectives does not imply pluralism about metaphysical facts. Yet truth is always relative to perspectives, a point he reiterates in a later article (Giere 2013), where he interprets Kuhn as a perspectival realist in taking scientific paradigms/lexicons as essential for evaluating the truth or falsehood of any statement.[15]

A similar pragmatist flavour can be found in Teller's version of perspectivism. Returning once more to the example of water, he notes:

Yes, the model does not tell us *precisely* what truth is, more carefully, exactly what kind of thing one has done when one positively evaluates a statement for representational success. Neither does chemistry, when it tells us that water is H_2O, tell us *precisely* what water is: at what pressures and temperatures do collections of H_2O molecules count as water? And how many H_2O molecules need one have to have some water? Neither model tells us precisely what their target is; but they tell us a lot nonetheless. (Teller 2011, p. 470, emphases in original)

Inconsistent models for water (i.e. continuum hydrodynamics vs statistical mechanics) are 'semantic alter-egos'. Accuracy is traded for imprecise representations that are nonetheless still true of an independently operating world. Like Giere's best-fitting models, Teller maintains that it is ultimately the practical representational success of our models in mapping the world (no matter how imprecisely and partially) that secures the realist commitment to truth in his non-traditional sense.

---

[15] 'But to claim that evaluation of the truth of a statement presupposes a lexicon is already to embrace a form of perspectival realism. A lexicon defines a "perspective" within which to formulate truth claims. Claims to truth are relative to that perspective' (Giere 2013, p. 55).

Rueger's perspectivism addresses PIM's metaphysical challenge another way, head on, for he treats inconsistent models explicitly as 'perspectival models':

> Even though they look as if they ascribed different intrinsic properties to the system (in different regions, ultimately), in truth, these models describe the system relationally; from *this* perspective, the system looks as if it had intrinsic property x, from *that* perspective it looks like it has property y. This interpretation of the formalism should be pleasing to the realist because the fact that a system seems to have incompatible properties when viewed from different points of view is obviously no more of an embarrassment for the realist than that one and the same table can look like a square and a trapezoid. (Rueger 2005, p. 580, emphases in original)

Rueger observes (correctly in my view) that it will not do to try to deflate PIM by appealing to the idealized and partial nature of the representation afforded by each model. What is still missing is an account of why models do not conflict; in what ways they are complementary rather than contradictory.[16] Rueger's relationalist strategy is meant to provide such an account. For example, in studying the viscosity of a fluid, models make use of the Navier–Stokes equations. Sometimes, the fluid is treated as an ideal (or non-viscous) substance, when the flow is at a distance from the solid boundary. At other times, viscosity is assigned to the fluid whenever dealing with regions close to the boundary walls. These two models, however, do not ascribe inconsistent intrinsic properties to the same target system (where I take 'intrinsic' to be interchangeable with what I called above 'essential' properties, and I spell out what I mean by essential properties in Section 3.6). They assign relational properties to the system— Rueger argues—and the *prima facie* metaphysical inconsistency disappears as soon as one considers the two perspectives from which the phenomena manifest themselves.[17]

---

[16] 'Understanding the models as idealisations or as vague representations is not enough by itself— not all vague models can plausibly be compatible, or "fit together" just in virtue of being vague rather than exact representations. With the notion of perspectival models, I provide, for a class of models, an interpretation that goes further in explaining the constraints on models that need be satisfied in order for the models to "fit together" into a more complete picture of a system' (Rueger 2005, p. 593).

[17] 'The important point for the realist here is that the two different spatial scales or perspectives are different assumptions with respect to a *relational* property of the system rather than with respect to one of its *intrinsic* properties, such as viscosity. Relational properties can vary without change in the system itself. "Incompatible" assumptions about relational properties are harmless for the realist because they do not imply incompatibility with respect to intrinsic properties ... what look like incompatible intrinsic properties in our models are only *ascriptions* of intrinsic properties from different perspectives' (Rueger 2005, p. 586, emphases in original).

3.5. Is perspectivism redundant, inconsistent, and unstable?

Can Rueger's relational reading of property ascriptions dissolve PIM? Or can Giere's or Teller's perspectival answers deflate PIM either by going pragmatist or by appealing to imprecise-yet-successful representations of an independently operating world? Critics of scientific perspectivism have been unimpressed. Some have argued that scientific perspectivism is a platitude and hence redundant. Or, worse, it perpetrates precisely the kind of metaphysical inconsistency that it was supposed to dissolve. In the best case, the view has been regarded as unstable in falling back onto a form of dispositional ontology (see Chakravartty 2017, Ch. 6).

Margaret Morrison (2011, 2015, Ch. 5) has, for example, argued that there is no genuine middle ground for perspectivism. Perspectivism is already embedded in our current modelling practices and, as such, is redundant. Or, worse, it is unhelpful in situations where there might be several incompatible (or inconsistent) models. Let us elaborate on these criticisms. Redundancy first. Latching onto Rueger's way of couching perspectivism in relational terms, Morrison (2015, pp. 159–160) points out that perspectivism seems to be the view that

> from the perspective of theory T, model M represents system S in a particular way. While this sounds like an appealing way to address the problem of inconsistent models, some nagging worries remain; in particular, how we should answer the general question 'Is model M an accurate representation of system S?' But epistemically it is not clear that anything significant follows from this. . . . For example, there is only one nucleus, but if we say that from perspective x it looks like y, and from perspective a it looks like b, we are no further ahead in finding out its real nature.

Morrison asks an important question: What does it mean for model M to represent system S from a particular perspective? What is involved exactly in the perspectival nature of representation afforded by a plurality of models? Morrison agrees with Rueger that PIM is not a problem about the inevitable idealizations involved in modelling. Idealized models, she says, are 'unproblematic since the alleged gap between the model and reality can be closed by the addition of parameters and the use of approximation techniques when applying the model to concrete systems. If it is achieved in a non-ad-hoc way, then it is typically taken as evidence for a realistic interpretation of the model' (Morrison 2015, p. 157).

PIM arises instead when there are radically different models of the system being proposed 'each of which describes it in ways that contradict the assumptions of the original model. . . . In this context we usually have no way to determine which of the many contradictory models is *the more faithful representation* especially if each is able to generate accurate predictions for certain features of the phenomena' (Morrison 2015, p. 158, emphasis added). Faced with this question, perspectivism seems to balk. The question 'Is model *M* an accurate representation of system *S*?' has 'no meaning unless a perspective has been specified' (p. 159). Moreover, 'adopting a particular perspective does not help in these contexts. . . . If we take perspectivism seriously, then we are forced to say that the nucleus has no nature in itself and we can only answer questions about it once a particular perspective is specified' (p. 160).

Let us take stock. This way of understanding perspectivism is common currency in the literature and finds its expression in what I called in Chapter 2:

> *The perspectivalist assumption.* Scientific models offer perspectival representations of relevant aspects of a given target system *S*. This is (often implicitly) understood along the lines of *perspectival$_1$ representation* rather than *perspectival$_2$ representation*. Namely, scientific model M_1 *perspectivally$_1$ represents S as* z; but model M_2 *perspectivally$_1$ represents S as* y (where z and y are properties belonging to a 'horizon of alternatives').

The underlying idea is that scientific models offer *de re* representations of relevant aspects of the target system that are perspectival$_1$ by ascribing rival and often inconsistent properties to the same target system *S*. This is problematic if one wants to live up to the promise of realism in science. Perspectivism would reduce to 'a nontrivial version of instrumentalism, particularly since we have no obvious way of eliminating the less profitable or useful perspectives' (Morrison 2015, p. 160). A realist—Morrison presses on—would like to know the nature of nucleons: 'are they probability waves, point particles, or space-occupying objects?' (p. 179). Ultimately, perspectivism 'either adds nothing that is not already implicit in the modelling context or it fails to provide a solution to the problems of inconsistency' (p. 195).

This twofold charge of redundancy and inconsistency is echoed by Anjan Chakravartty (2010, p. 407, and 2017, Ch. 6), who presents perspectivism as committed to either of these claims:

> P1. We have knowledge of perspectival facts only, because non-perspectival facts are beyond our epistemic grasp.

> P2. We have knowledge of perspectival facts only, because there are no non-perspectival facts to be known.

Chakravartty argues that this way of reading what is at stake in perspectivism is justified by arguments typically invoked in the literature: what he calls the argument from the partial and conditioned nature of experimental detection.

However, he continues, 'the fact that detectors generally yield information relating to limited aspects of the target system does not by itself imply any philosophically interesting sort of perspectivism' (Chakravartty 2010, p. 407). Hence, perspectivism is indeed redundant. Neither are prospects any rosier for the ability of perspectivism to solve PIM. When

> extended into the domain of ontology, it is hard to see how it [viz., perspectivism] can be made coherent at all. . . . The perspectivist who puts Kant on wheels in the manner described here . . . proposes something significantly more radical: it would seem that on this view, scientists interact with a fundamentally conflicted reality. It is not merely that there is an empirically accessible world, about which one has only perspectival beliefs. It is furthermore that this world comprises a tortured assembly of Frankenstein facts or states of affairs. Water would, on this view, be both a continuous medium *and* a collection of discrete particles. Light would be both a classical wave entity *and* an excitation state of a field. . . . [E]mpirical reality itself consists of a hodgepodge of contradictory states of affairs, created (in part) by the human act of theory use and model construction. (Chakravartty 2010, p. 411, emphases in original; these objections are also in Chakravartty 2017, Ch. 6)

Chakravartty defends what he calls a form of non-perspectival knowledge, which sometimes takes the form of a knowledge of 'categorical facts' (Chakravartty 2010, p. 410), namely non-perspectival facts about properties of objects that manifest themselves via dispositions. For example, what seems a *prima facie* perspectival fact about salt being soluble or insoluble in water (depending on whether or not water is already saturated) proves to be a dispositional fact based on properties of salt: that is, the inter- and intra-molecular forces underpinning salt's solubility. Hence, Chakravartty's conclusion that scientific perspectivism is on closer inspection also unstable because perspectival facts turn out to be different manifestations of dispositions.

Whether or not salt dissolves in water is a matter of dispositional manifestations. Metaphysicians of science may disagree as to whether or not dispositions are grounded on categorical properties. As Chakravartty stresses (p. 412), 'One may or may not have an underlying theory of categorical facts with which to underwrite the dispositions of water, but quite independently of whether one does, one may have genuinely non-perspectival knowledge of their manifestations, nonetheless.'

Dispositional realism is a metaphysical realist view about relevant properties of entities being identified by dispositions without any risk of falling back onto an ontology of perspectival facts. What makes a property the property that it is—in other words what constitutes its 'essence'—is the set of dispositions associated with it. This is the 'dispositional identity thesis' that Chakravartty (2007) defends. Some versions of the view take dispositions as manifestations of more fundamental categorical properties; others treat dispositions as themselves fundamental without any further grounding. In either case, dispositional realism shows that there is no need to postulate perspectival facts because the ontology of nature is an ontology of dispositions. This seems bad news for perspectivism and for the prospects of a kind of perspectival pluralism that might be able to address PIM and ultimately deliver on realism.

In the final section, I scrutinize some of the explicit and hidden premises behind PIM and I raise some questions about what I take to be the unduly demanding realist gloss these premises seem to carry. I sum up the discussion so far by drawing a series of preliminary conclusions about how we should *not* understand perspectivism. Ultimately, I agree with Anjan Chakravartty that there are no perspectival facts and that perspectivism should not be understood as licensing any 'tortured assembly of Frankenstein facts'. I part my way from him in thinking nonetheless that scientific knowledge is fundamentally perspectival; and that once one takes perspectival knowledge on board, realist ontology needs be reconsidered along new lines. Dispositional realism, in its various flavours, would not do. In the next chapters, I further motivate, clarify, and substantiate a positive view of perspectival modelling that can make sense of the peculiar kind of pluralism present in some modelling practices. But for now, let us go back to the drawing board one more time.

3.6. Resisting some of the premises behind PIM

There is still hope for saving perspectivism, and realism, from PIM in spite of these objections. It turns on the fact that what I called above the *representationalist assumption*—that scientific models (partially) represent relevant aspects of a given target system S—hides, on closer inspection, two implicit (and more controversial) premises, which have to be in place for PIM to work as an argument. These premises are also implicitly at work in the controversy surrounding perspectivism as I summarized it in the previous section. I call these two hidden premises *representing-as-mapping* and *truth-by-truthmakers*:[18]

> *Representing-as-mapping.* The true model is the one that offers an accurate, partial, *de re* representation of relevant essential features of the target system.

[18] In what follows, I draw on Massimi (2018b) 'Perspectival modelling', *Philosophy of Science* 85, 335–359. Reproduced with permission from University of Chicago Press. Copyright (2018) by the Philosophy of Science Association.

Offering an accurate, partial, *de re* representation means to establish a one-to-one mapping between relevant (partial) features of the model and relevant (partial)—actual or fictional—states of affairs about the target system.

Truth-by-truthmakers. States of affairs ascribe essential properties to particulars, and, as such, they act as ontological grounds that make the knowledge claims afforded by the model (approximately) true.

Representing-as-mapping is implicit in the emphasis placed sometimes in this literature on how models represent selected aspects of the target system and how realism seems to commit us to look for *the* representational model: that is, the model that offers the best (i.e. most accurate or most veridical) *de re* representation of relevant features of the target system. Similarly, *truth-by-truthmakers* is implicit in the often tacit assumption that a realist commitment should involve ascribing intrinsic essential properties to the target system. Such intrinsic essential properties can either be irreducibly dispositional or can be thought of as categorical properties upon which dispositions are grounded. For example, the solubility of salt in water is a disposition to behave in a certain way that might be regarded as either grounded or not on categorical properties about salt—its molecular structure and chemical properties.

The qualification 'essential' for properties (be they categorical or dispositional) is an important one for my reconstructed argument for PIM. The kind of realism that is at odds with scientific perspectivism is a certain kind of metaphysical realism that encompasses a number of views ranging from the aforementioned dispositional realism (dispositions are real whether or not they are grounded in categorical properties) to dispositional essentialism, which further qualifies the reality of the dispositions as follows: 'Essentially dispositional properties are ones that have the same dispositional character in all possible worlds; that character is the property's *real* rather than merely nominal essence' in contrast with categorial properties that may 'change their dispositional characters (and their causal and nomic behaviour more generally) across different worlds' (Bird 2007, p. 44; see also Ellis 2001, among others).

But obviously there are other varieties of realism that would not endorse dispositional realism and dispositional essentialism and, as such, they are less vulnerable to PIM and easier to reconcile with perspectival pluralism. Ultimately, the kind of realism that I shall unpack and articulate in the second part of this book endorses the *anti-foundationalist anti-essentialist* attitude I mentioned at the start of this chapter and, as such, it is at a distance from dispositional realism and dispositional essentialism. But let me go back and clarify key aspects of these two aforementioned premises.

First, *representing-as-mapping* is key to certain accounts of scientific representations that have emphasized mapping-onto-a-target-system (e.g. Giere's agent-based account of representation—Giere 2010). But it is less compatible with alternative accounts of scientific representation that have deflated the importance of representing-qua-mirroring or mapping-onto-a-target-system, such as Suárez's (2004, 2015a) inferentialist view of representation. More generally, an inferentialist view of how models offer perspectival₂ representations of the target system bypasses altogether *representing-as-mapping*. This is the route I shall pursue from Chapter 4 onward.

Second, 'states of affairs' has become a term of art in a huge literature (for an excellent introduction, see Textor 2020). I use the term 'states of affairs' (loosely) in David Armstrong's sense (1993): that is, I take states of affairs to be the truthmakers (or ontological grounds) that make the knowledge claims afforded by a model true (if only partially or approximately). However, I offer an important caveat. For Armstrong, states of affairs must be actual—there cannot be a non-existent state of affairs because universals must be instantiated. However, in discussing models and what makes model-based claims of knowledge true, I make room for non-actual states of affairs as well. Think of fictional models of the ether, for example, where the states of affairs in question are often taken as fictional or imaginary by the modellers. Thus, states of affairs should be understood loosely to include those that are the product of recombining some particulars and properties in fictional, non-actual ways (see Armstrong 1989, pp. 45–49).

For example, one might think that the fictional state of affairs *electrons are idle wheels in an elastic ether* is a recombination of particulars (*electrons*) and essential properties (*rotating frictionless*). Although it is not an actual state of affairs, it can nonetheless act as the truthmaker of the knowledge claim expressed by the sentence 'electric displacement is generated by the magnetic field within Maxwell's ether model of electromagnetic induction'.

We are now in a position to see how the two tacit premises (*representing-as-mapping* and *truth-by-truthmakers*) are at play in the seemingly innocuous *representationalist assumption*, which enters into an argument for PIM. Let us call this argument for PIM the *Have-Your-Cake-And-Eat-It* argument (or HYCAEI).

(HYCAEI)

1. Realism about science is the view that scientific theories (qua families of models) are approximately true, in the partial and qualified sense explained earlier. (*Realism*)
2. A scientific model is true when the model provides a partial yet accurate representation of the target system. (*Representationalist assumption*)
   2a. The true model is the one that offers an accurate, partial, *de re* representation of relevant essential features of the target system. Offering an accurate, partial, *de re* representation means to establish a one-to-one mapping between relevant (partial) features of the model and relevant

(partial)—actual or fictional—states of affairs about the target system. (*Representing-as-mapping*)

2b. States of affairs ascribe essential properties to particulars, and, as such, they act as ontological grounds that make the knowledge claims afforded by the model (approximately) true. (*Truth-by-truthmakers*)

3. Scientific perspectivism is the view that from the perspective of theory T, model M_1 represents system S in a particular way (say z); but from the perspective of theory A, model M_2 represents system S in a different way (say b).

4. Scientific perspectivism implies that different models provide different accurate, partial, *de re* representations for the same target system S. (Via 2a)

5. Different accurate, partial, *de re* representations entail different states of affairs—actual or fictional—as the respective truthmakers of knowledge claims afforded by different models. (Via 2b)

6. But different states of affairs—actual or fictional—ascribe different essential properties for the same particulars.

7. It follows that there is metaphysical inconsistency in supposing that one and the same target system is *de re* accurately represented (even partially) by different perspectival models. (PIM via premises 3 and 4)

8. Hence, realism (1) is incompatible with scientific perspectivism (3).

I take HYCAEI to be the main argument for PIM (although nowhere clearly stated in the literature). This is the argument that underpins PIM: for example, the atomic nucleus cannot *essentially* be a set of orbitals where nucleons sit according to magic numbers *and also* essentially be a bunch of strongly interacting quarks. These two *metaphysical images* are inconsistent with one another, PIM defenders would argue

A defender of PIM might reply that maybe *truth-by-truthmakers* is not needed for PIM to stand. Could a more modest theory of truth do the job for PIM? But what would such an alternative PIM-friendly theory of truth look like? Deflationism about truth would not help. A deflationist about truth would not see rival models as giving rise to any metaphysical inconsistency, because the whole point about deflationism is that truth comes with no metaphysical baggage. Tarskian theories of truth would not help because they operate a purely formal apparatus that does not discriminate between realism and anti-realism as such. And a correspondence theory of truth along more modest metaphysical lines (such as Austin 1961) would regard the correspondence between propositions and facts as purely conventional (rather than having metaphysical import of the type required for PIM). Thus, *truth-by-truthmakers* is required for PIM (and, if my argument is correct, *truth-by-truthmakers* is indeed surreptitiously assumed in HYCAEI as the argument for PIM).

Another possible line of reply would be to play down my emphasis on *essential properties* at play in *truth-by-truthmakers*. Why insist on essential properties? Could not real properties, maybe natural properties of Lewisian memory, serve just as well? In reply, I do not think so. As I see PIM qua a problem of *metaphysical inconsistency*, it requires that the properties ascribed to particulars ought to be essential in the 'deep essentialist' sense described by L.A. Paul whereby such properties determine the nature of the particular and 'give sense to the idea that an object has a unique and distinctive character, and make it the case that an object has to be a certain way in order for it to *be* at all' (Paul 2006, p. 333, emphasis in original).

An important consequence of this essential property ascription is that if an object possesses such properties, it possesses them in an absolute and non-contextual way. One cannot possibly maintain that the nucleus has essential property *y and* that property *y* applies only in particular contexts of inquiry or modelling practices. If it is an essential property of the nucleus that it consists of 'shells'/orbitals, the holding of such an essential property cannot be at the whim of the context of inquiry and modellers' choice of using the shell model rather than the liquid drop model for particular phenomena but not others. The charge of metaphysical inconsistency posed by PIM banks on this stringent essentialist reading of properties, as HYCAEI reconstructs it. A range of metaphysical realist views would be open to such essentialist reading of property ascription. But not all of them are.

For example, Humean-inspired accounts of properties as David Lewis's natural properties offer no traction for PIM. For it is perfectly compatible with taking natural properties ascriptions as dependent on the specific context of inquiry. Real natural properties are not the properties in virtue of which an object is what it is in an absolute way. It is perfectly possible to think of an electron as having two natural properties—position and momentum—while also accepting the Copenhagen interpretation of quantum mechanics that there are different contexts of inquiry in which each of these properties might not be sufficiently well defined to be ascribed to the electron *simpliciter*. Relatedly, it is perfectly possible to accept that there are claims of knowledge about the momentum of the electron or its position whose truth or falsity is well defined but only in particular experimental contexts and not in others.

This is good news for the Lewisian account of natural properties and varieties of realism in science that do not endorse deep essentialism, and as such are unscathed by PIM. It is also good news for perspectival realism for it shows the way forward in cashing out a variety of realism that can be made compatible with perspectival pluralism about models. Ultimately, the realism I will defend does not hinge on Lewisian natural properties, and it will instead take phenomena— rather than properties—as its ontological unit (see Chapter 6, Section 6.7.2, for a phenomena-first approach to ontology). I shall return in more detail on this issue and why I am not a 'deep essentialist' in Chapter 9, Section 9.2.

But for the time being all I need to point out is simply the unduly stringent essentialist reading at play in premise 2b *truth-by-truthmakers* behind HYCAEI and how PIM does not take off unless such a premise is endorsed. HYCAEI functions as an argument against perspectival realism only under the unwarranted assumption that the realism that one is after here has to be of a 'deep essentialist' nature. But that is neither the only nor the most desirable variety of realism that perspectival pluralism might avail itself of. Indeed, the whole point of this book is not to retrieve that kind of realism but to show that an alternative variety of realism—one more congenial to perspectival pluralism—is in fact possible.

What about the other premise, 2a? A defender of PIM might similarly claim that *representing-as-mapping* is not needed for PIM. Even if no representing-as-mapping applies, PIM might still arise because, after all, PIM is a problem about models making contradictory claims about the target system.

But under which conditions would such metaphysically contradictory claims of knowledge arise? Would they arise if the models were not interpreted as 'representing accurately' in the aforementioned sense? I return to the distinction made above about representing *de re* vs *de dicto*. A fictionalist about models would argue that models represent *de dicto* relevant aspects of the target system because, say, the representation afforded by Maxwell's honeycomb model of the ether does not latch onto any actual state of affairs. Similarly, rival models of the atomic nucleus represent *de dicto* relevant aspects because they invite us to play a make-believe game about the target system.

Thus, in a way, fictionalism about models (with its less stringent notion of representing) is less vulnerable to PIM. The problem, however, is that fictionalism about models does not help with realism either (for what would a fictionalist say about it?). And that does not help if the overarching goal is precisely to demonstrate that perspectival pluralism about models might after all be compatible with some suitable form of realism.

Where to go from here? How is scientific knowledge produced through a plurality of models? How does realism fall out of these modelling practices if not because models partially represent-by-mapping onto relevant states of affairs about the target system? In Chapter 4 and subsequent Chapters 4.a–4.c, I offer three case studies that show perspectival varieties of model pluralism in action. What's distinctive in all three cases is the non-mapping nature of the perspectival representations afforded by each model. Through these case studies, I pave the way to an alternative way of thinking about perspectival$_1$ representations neither qua representing-as y or z or k, nor as mapping onto different partial states of affairs but as *exploring possibilities* instead.

Perspectival modelling is first and foremost an exploratory variety of scientific modelling. Perspectival models represent a given target system—phenomenon of interest—to the extent that they allow different epistemic communities to make relevant and appropriate inferences about what is possible concerning the phenomenon. Without denying the representational role of models, I articulate a different view of them as *inferential blueprints* (Chapter 5) that—by contrast with isomorphism, homomorphism, and similarity accounts—does not hinge upon *representing-as-mapping* in some form or another.[19]

The heavy lifting of the representational role is not done by how well or accurately the model gives some kind of mapping onto the target system, but by intersecting scientific perspectives —and their modelling practices broadly construed. Such intersecting improves and refines the overall reliability of the relevant and appropriate inferences about the phenomena of interest. This is how I see perspectival models as ultimately offering perspectival$_2$ representations. In analogy with the *Arnolfini Portrait*, the perspectivity of representation opens up a 'window on reality'. As I argue in Chapter 5, perspectival models open up windows on reality by acting as inferential blueprints for advancing claims about what is possible.

If the line of reasoning so far is correct, we can glimpse what else is wrong with PIM. The problem is not that different models provide different partial and incomplete *perspectival$_1$ representations* for the same target system. The real bite of PIM lies in the assumption that *different perspectival models ascribe different essential properties to the same target system*. But the charge of metaphysical inconsistency (and also that of instability) is ultimately based on an unduly stringent realist reading of the *representationalist assumption*, captured by the two hidden premises 2a and 2b. I have here only made the point that one does not have to accept either of them. *Representing-as-mapping* and *truth-by-truthmakers* are not forced upon us. And since HYCAEI relies on them, resisting them is a way of resisting PIM. Or, better, it is a way of showing that PIM has traction only as long as 2a and 2b apply.

But a different way forward is possible. My task ahead is to articulate a different way of thinking about what good a plurality of perspectival models is. What sets perspectival pluralism aside from other varieties of model pluralism? In the next chapters, I explore this by plunging into three modelling practices that I see as genuine examples of perspectival pluralism: models of the atomic nucleus (*encore*), climate modelling, and developmental contingency models for dyslexia.

---

[19] For an interesting discussion of some issues arising from the homomorphic account, see Pero and Suárez (2016).

4
Perspectival modelling as modelling possibilities

4.1. Where to go from here?

The goal of this chapter and the next four ones is to carve out a positive role for perspectivism. I have dealt already with the charge of metaphysical inconsistency that has been levelled against perspectivism by laying out what I take to be the main argument for it (the Have-Your-Cake-And-Eat-It argument, or HYCAEI). I highlighted its additional and surreptitious premises and offered reasons to resist them. Perspectival realism should not be fazed by HYCAEI. For the culprit for the problem of inconsistent models is not the perspectival nature of the representation offered by models but instead an unduly demanding realist gloss on the *representationalist assumption*.

In Chapter 2, I hinted at a different way of thinking about the perspectival nature of the representation which I labelled *perspectival₂*. I stressed how I see these two notions (*perspectival₁* and *perspectival₂*) as complementary. A scientific representation is perspectival in being both situated (*from a vantage point* or *perspectival₁*) and also in being directed towards one or more vanishing points (*perspectival₂*). Let me expand on this idea here. I'd like to think of perspectival models as offering perspectival₂ representations of the relevant target system, which—like the mirror in the *Arnolfini Portrait*—open up 'windows on reality'. The realism I see as compatible with perspectivism is downstream from this exercise of perspectival modelling as opening up 'windows on reality'. In this and the next four chapters (4.a–4.c and chapter 5), I unpack the artistic analogy with the *Arnolfini Portrait*.

First, though, there is still a question looming large here about what perspectivism may contribute to the long-standing debate about pluralism in science. Is perspectivism just another name for scientific pluralism? What (if anything) is distinctively perspectival about model pluralism? My answer develops in five steps:

(1) *Why is perspectivism not just another name for model pluralism?*

Perspectival Realism. Michela Massimi, Oxford University Press. © Oxford University Press 2022.
DOI: 10.1093/oso/9780197555620.003.0004

Answer: because perspectivism captures a subset of model pluralism where models are best characterized as exploratory.

(2) *What makes perspectival modelling 'exploratory'?*
Answer: perspectival modelling enables a particular kind of inferential reasoning that proves fruitful when one wants to explore what is *possible* (instead of mapping-onto-what-is-actual).

(3) *Who cares about what is possible? Is not science after what is actual?*
Answer: of course, science is about finding out what is actual. But you should care about what is possible because, in the absence of a God's-eye access to reality, knowing what is possible is an important (dare I say, it is the only) guide to find out what is actual. Like Marco Polo on his journeys into uncharted territories, we are not equipped with an ideal scientific atlas of the Realm-That-Is-Actual. We have to find it out for ourselves walking along inferential paths that resemble Jorge Luis Borges's (1941/ 2000) 'garden of forking paths'. Perspectival modelling guides epistemic communities over time across such a garden where at every twist and turn new paths can be explored and old ones left behind.

(4) *What is to be said about this inferential garden of forking paths and perspectival modelling?*
Answer: perspectival models, as I argue in detail in Chapter 5, guide communities across time along inferential paths by acting as 'inferential blueprints'. From an epistemic point of view, perspectival-models-qua-inferential-blueprints deliver modal knowledge claims by inviting us to physically conceive particular scenarios. From a semantic point of view, perspectival-models-qua-inferential-blueprints support a particular kind of epistemic conditionals, namely indicative conditionals with a suppositional antecedent.

(5) *Where is the realist element in this story? (aka 'How can walking along the inferential garden of forking paths warrant realism?')*
Answer: perspectival-models-qua-inferential-blueprints help a plurality of situated epistemic communities to navigate the inferential space of what is possible. Along the way, these communities come to reliably identify *modally robust phenomena* (more on this notion in Chapter 6). Often such an identification proceeds through data-to-phenomena inferences that are entirely perspectival. Each epistemic community may avail itself of experimental and technological resources for harvesting the data and of justificatory principles that are genuinely diverse and belong to different

scientific perspectives (qua historically and culturally situated scientific practices as defined in Chapter 1). Thus, perspectival modelling (as I use the term here) is not narrowly confined to scientific models. It refers instead more broadly to the situated modelling practices of particular epistemic communities, including the way they use particular data to make inferences about phenomena of interest. As such, perspectival modelling is a defining feature of what a 'scientific perspective' is for. It captures how situated epistemic communities across a number of intersecting scientific perspectives come to know the world as being a certain way. Spoiler alert: the Realm-That-Is-Actual is nothing but the realm of modally robust phenomena displaying lawlike dependencies and inferred via perspectival modelling across a number of scientific perspectives (more on this in Chapters 5 and 6).

These questions will guide my journey into perspectival modelling in what follows. And I will have a lot more to say in Chapter 5 about the answers I have simply sketched here to Q4 and Q5. They will also pave the way to Part II of the book (Chapter 6 onwards), where the realist promise behind the title of my book is waiting to be delivered. But, for now, first things first: (1) *Why is perspectivism not just another name for model pluralism?*

4.2. Non-perspectival varieties of model pluralism

What good is a plurality of models in any given area of scientific inquiry, anyway? The pluralist stance has long acknowledged that different models might accurately represent some aspects of the target system at the cost of distorting others. Inherent in this stance is a commitment to different models delivering incompatible, or even inconsistent, images for the same target system, a commitment that scientific pluralists by and large are happy to make. *That* there is pluralism about models is a fact about scientific inquiry. *What good it is for* remains debatable. Even more debatable is *what perspectivism has got to do with it*.

Many realists would like to think of model pluralism as a 'means to an end': the end is realism about science; the means are as diverse as scientific practices typically are. There are two widespread attitudes to be found about model pluralism (the first especially among scientists, the second mostly among philosophers of science). Neither of them in my view captures what really goes on with *perspectival varieties of* model pluralism:

(a) *Model pluralism as a transient stage in scientific inquiry.* 'We want to try out as many models, theories, and explanations as possible before settling on the right one'. On this view, model pluralism becomes like someone going to buy an evening dress and trying on every one in the shop before choosing the right one. Similarly in science, it might be tempting to think that one ought to be pluralist for the sake of making progress and choosing the *best* model. Or one might proceed in a semi-Popperian mode and argue as follows: let a number of conjectures (i.e. model hypotheses) come forward (the more, the merrier); and let us eventually refute them one by one until the one that survives severe testing is identified as the *corroborated* one (i.e. *best = corroborated*). Others may want to equate the *best* model with the most *empirically adequate* one (to borrow van Fraassen's terminology); or with the model that has the higher *puzzle-solving* power (after Kuhn), and so forth. The point is that no matter how one defines 'the best model', the underlying intuition is that model pluralism is a means to an end, under this view.

Some may think this is a common situation in science. Consider Galileo's experiments with inclined planes. Through the medieval works of Abu'l-Barakāt, Oresme, and Buridan, among others, free fall was revealed as a kind of accelerated motion, rather than motion towards a natural place as Aristotle had maintained. Galileo went through several attempts at explaining the phenomenon in the Pisan treatise *De motu antiquiora* ca. 1590 (see Massimi 2010 for the historical details) before he hit on the best model that relates the distance traversed to the square of the time (as he eventually demonstrated it in *Two New Sciences* in 1638). Some scientific realists may argue that Galileo's model pluralism was a typical example of 'means to an end': try out models until you find the *best* one (in this case, the one that matches experimental data). It is a default methodological stance that virtually any scientific realist would endorse.

However, this is not an example of a *perspectival* variety of model pluralism. Perspectivism is best seen as capturing a subset of model pluralism, where models are best characterized as *exploratory*, enabling a particular kind of inferential reasoning that explores what is *possible*. Galileo's studies on free fall were aimed at finding out the *actual* nature of free fall by matching a number of hypotheses with the observed data. Of course, once the best model has been found, one might as well use it to give how-possibly explanations (Bokulich 2014; Verreault-Julien 2019) of how free fall works, how it might get replicated with bodies made of different materials and with different densities, how errors might be introduced, and so on. But providing how-possibly explanations is—in this example—downstream from finding the best model in the first instance. Common wisdom that actuality is a guide to possibility applies here. The kind of exploratory exercise that matters to perspectival modelling goes in the opposite direction: exploring what is possible to find out what is actual. And exploring what is possible turns out to be a lot more complex and nuanced than just trying as many models as one can think of until one hits on the best one.

(b) *Model pluralism as the acknowledgement of the existence of different communities with different epistemic aims.* Another common stance on model pluralism goes as follows: pluralism is a fact about scientific inquiry. It is the expression of the existence of different epistemic communities, each with their own epistemic aims and needs. It is hard to disagree. Accordingly, one may think that scientific pluralism is a gentrified expression for the mundane fact that there are a variety of views and voices in any given area of science.

But perspectivism is not just another name for there being 'many points of view'. Nor is it a generic proxy for 'there are as many models as there are epistemic communities with different research interests'. Perspectivism captures a well-defined subset of a larger family of model pluralism where the searches are exploratory in distinctive ways.

Thus, on my view, one can be a model pluralist but not necessarily a perspectivist. But one cannot be a perspectivist and not be a model pluralist of some sort. The charge of redundancy misplaces perspectivism in the wider landscape of scientific pluralism. There are significant areas of scientific inquiry where pluralism is a defining feature while perspectivism is absent. But using the term 'perspectivism' as if it were interchangeable with 'model pluralism' does a disservice to both.

Since I see perspectival models as *exploratory*, let me briefly do some more philosophical landscaping. First, I do not use 'exploratory' in the way in which it is sometimes used colloquially to denote something transient (e.g. exploratory talks). My usage bears family resemblances with the more recent literature on 'exploratory experimentation' (see Burian 1997; Elliott 2007; Peschard 2012; Steinle 2005/2016; Fisher, Gelfert and Steinle 2021, and related articles in this edited journal special issue, among others) as a way of studying phenomena, and their rules and laws, even in the absence of a fully fledged theoretical framework.

The exploratory nature of scientific models in general has only recently begun to attract attention. Axel Gelfert (2016, pp. 83–97) has presented exploratory models as fulfilling four distinct (not exhaustive) functions:

- they may function as a *starting point* for future inquiry (as with car-following models of traffic flow);
- they may feature in *proof-of-principle* demonstrations like the Lotka–Volterra model of predator–prey dynamics;

- they may generate a *potential explanation* of observed (types of) phenomena, as with Maxwell's honeycomb model of the ether;
- they may lead to assessments of the *suitability of the target*.

What I call perspectival models add to this list the specific task of modelling what is possible (rather than mapping-onto-what-is-actual). Conflating the exploratory role of perspectival modelling with its representational role is therefore like staring at the finger while pointing at the Moon. Representation is a means to an end, not the end itself. Despite being complementary, one should not confuse perspectival$_1$ representation (representation *from a vantage point*) with perspectival$_2$ representation (representation *towards one or more vanishing points*, as in the *Arnolfini Portrait*). To understand how a plurality of perspectival models can open up a 'window on reality', we should concentrate on how they fulfil their exploratory role, rather than on how the representational content might be affected by the vantage point from which the representation takes place.

Let me briefly articulate how the exploratory role of perspectival modelling relates more directly to its ability to model possibilities. I see this process not so much as situating a particular case within an already 'given' space of modal facts (the possibilities) but as *figuring out what the space of possibilities looks like* in the first instance.

4.3. Perspectival modelling and its exploratory role

Once in a seminar, philosopher of physics Tim Maudlin asked me why I was placing all this emphasis on modelling what is possible: 'Is not science after what is actual rather than what is possible?' Fair question. Science is after what is actual. Climate scientists want to find out what the global mean surface temperature is going to be in 50–100 years. Nuclear physicists around the 1930s–1950s were keen to find out what was responsible for observed isotopic abundances. And developmental psychologists are interested in finding out the specific pathways of language development in children.

There is an old dictum that 'actuality is a guide to possibility': if something is actual, it must also be possible, otherwise it could not be actual. But when it comes to scientific modelling, one needs first to find out what is actual. Actuality does not come served on a silver platter. Scientific modelling is the necessary scaffolding for getting at what is actual. My route to realism necessarily has to go through such scaffolding, or, to use a better metaphor, through the garden of forking paths (to echo Borges once more) that perspectival modelling qua modelling possibilities opens up. There is no shortcut to knowing what is actual. It is by moving along the inferential forking paths opened up by perspectival modelling broadly understood that epistemic communities gain over time 'windows on reality'.

Scientific modelling aimed at exploring the space of possibilities has begun to attract attention in the philosophy of science literature. Models are said to probe what *might be* the case in given situations. For example, models can sometimes lead to understanding how certain processes might happen in nature. At other times a model can lead scientists to canvass possible economic situations (see Grüne-Yanoff 2009 and Grüne-Yanoff and Marchionni 2018); or gain knowledge via concrete artefacts (Knuuttila and Boon 2011), or help scientists figure out important details about what kind of material, shape, and structure might be more resistant for a bridge (Weisberg 2007). Sometimes models are built with the hope of finding new entities in nature (Hartmann 1999), or to provide how-possibly explanations for social phenomena like segregation in urban areas (think of Schelling's model discussed in Verreault-Julien 2019). In yet other cases, the emphasis is on how engineered models in synthetic biology might be built (see Kendig 2016b; Knuuttila 2017; Knuuttila and Loettgers 2013, 2017, 2021); or how laboratory experiments might help in the understanding of 'ecological possibility' (Currie 2020; Kendig 2016a); or how to explain the way in which phenotypic traits in a population may optimize fitness (Rice 2015, 2019, 2021). The list is pretty much open-ended.

The modal aspect of scientific modelling has been presented in various ways. Sometimes it is articulated in terms of multiple model idealizations (see Weisberg 2007). Other times it is explained by appealing to counterfactual conditionals (Rice 2019); or how-possibly explanations (Bokulich 2014); or modal understanding (Elgin 2017; Grimm 2012; Le Bihan 2017; Potochnik 2017); or in more foundational domains, in terms of Lewisian possible worlds (Wilson 2020).

I share the spirit of all this existing literature in stressing the role of modality (and possibilities, in particular) for grasping what scientific models are really for. My specific task here is to carve out three *distinctively perspectival* varieties of model pluralism where 'exploring the space of possibilities' acquires a specific meaning. For the possibilities in question in perspectival modelling *do not* concern either

- mere variations in initial conditions that might affect the modelling outcome (e.g. change nucleotide sequence as an input and the DNA modelling outcome is changed as a result);

or

- how jiggling one parameter might (causally) affect a connected one (e.g. change the kind of material used for modelling the bridge and the resistance to strain gets changed);

or

- the counterfactual reasoning about whether had C been the case, E would have been the case (e.g. had certain constraint C been the case, the trait distribution in a population would have been E).

The possibilities of perspectival modelling have to do instead with *modelling modally robust phenomena* that *could occur* in more than one way and *could* be elicited via a number of *perspectival data-to-phenomena inferences*. It is this modal robustness over time and across domains that makes perspectival modelling distinctive as an exercise in modelling possibilities. I will have more to say on this ontological aspect of the view in Chapter 6. But here and in Chapter 5 my attention is on laying out perspectival varieties of model pluralism through which such modally robust phenomena are explored.

Therefore, I see perspectival modelling as enabling a variety of situated epistemic communities over time to collaborate—either within the same perspective or across a number of scientific perspectives—and make *relevant and appropriate inferences* to explore what's possible (not knowing yet what is actual) about their object of study.[1] The view I shall defend is broadly inferentialist. But it departs in specific ways from traditional views where models have been seen as supporting inferences, or what is sometimes called 'surrogative reasoning'.[2]

By and large, I do not see perspectival models as autonomous entities mediating between the theory and the experimental data along the lines of Morgan and Morrison's (1999) 'models as mediators'. I am operating with an inflated inferentialist view, whereby perspectival modelling (broadly understood) allows

---

[1] The '(not knowing yet what is actual)' caveat is important. For it is not always or necessarily the case that the outcome of this exploratory exercise is to stumble into a modally robust phenomenon. In some cases, scientists might be looking for those phenomena but not necessarily find them. This is the case with hypothetical modelling when the target system is hypothetical (neither known to be actual nor known to be fictional), as with supersymmetric models in high-energy physics, which I discussed in Massimi (2018b, 2019a). But not all perspectival modelling is hypothetical. In the three case studies I discuss in what follows, nuclear stability, global warming, and difficulties with reading are all examples of modally robust phenomena that could be evinced from a plurality of perspectival data-to-phenomena inferences. The purpose of the perspectival modelling is to explore the very many ways in which each of these phenomena might robustly occur (depending on assumptions about nuclear structure, greenhouse gas concentrations, or the contingent pathway breakdown in language development, respectively).

[2] For a discussion, see Frigg and Nguyen (2020). On the inferentialist view of scientific representations for models, see Contessa (2007), Hughes (1997), and Suárez (2004, 2015a).

epistemic communities over time to make inferences from a number of datasets to what I call *modally robust phenomena*.

Consider the three examples to whose details I turn to in the following Chapters 4.a, 4.b, and 4.c. To start with, consider the atomic nucleus. Scientific knowledge of it is inevitably perspectival, subject to the specific technological, experimental, and theoretical resources that were available to different communities at different times. How did physicists gain knowledge of the nucleus, its nature and structure? In Chapter 4.a, I argue that such knowledge accrued via perspectival modelling that around the 1930s–1950s enabled a variety of epistemic communities to make inferences from data about the Earth's crust and meteorites, among others, to relevant modally robust phenomena such as nuclear stability.

The flurry of nuclear models (especially shell, liquid drop, and odd-particle models around the 1930s–1950s) that accompanied such practices were all perspectival in being exploratory. They allowed atomic physicists, physical chemists, spectroscopists, et al. to collaborate, make inferences, and deliver knowledge about what is possible about the nucleus, the isotopic stability of some nuclides, the nature of nuclear rotational spectra, and so on. They offered perspectival$_2$ representations of the nucleus in that they jointly allowed scientists over time to gain a 'window on reality'.

In this case, the 'window' on the nature of atomic nuclei required modelling what is possible about a number of modally robust phenomena (e.g. nuclear stability, slow neutron capture, nuclear fission, nuclear rotational spectra) identified over time through data-to-phenomena inferences that were perspectival every inch of the way. By 'being perspectival', I mean that the specific epistemic communities involved in each of these data-to-phenomena inferences proved remarkably diverse, ranging from petrology to cosmochemistry, from spectroscopy to atomic physics.

My second case study concerns climate modelling. This is also an example of exploratory modelling because the task is to explore both that global warming has historically occurred compared to the pre-industrial era *and* how it might accelerate in the future unless action is taken to cut greenhouse gas (GHG) emissions. Climate scientists need to model the multifactorial phenomenon called 'global warming'. This is another example of what I call a modally robust phenomenon, which can be reliably studied through a plurality of perspectival data-to-phenomena inferences from dendroclimatology, geothermal physics, and palaeoclimatology, as I reconstruct in Chapter 4.b.

Knowing the past global temperature of the planet and how it has changed over the past 100 years is only the first step in this process. To make future climate projections that can inform climate policy, a distinctive type of perspectival model pluralism is adopted by the Intergovernmental Panel on Climate Change

(IPCC) in the so-called Coupled Model Intercomparison Projects (CMIP). A plurality of models is required to make robust climate projections under a range of conceivable GHG concentration scenarios. Here again possibility is our guide to actuality. To find out what the climate will be like in 100 years, climate scientists have to model a plurality of factors concerning land surface temperature, changes in glaciers, and ocean heat content, among others, under the suppositions of various GHG scenarios.

And to give a third example, consider language development in children. To learn how to read and write, children develop a range of subskills and go through different learning stages as they move from the pre-school years to primary and secondary school. A child who displays difficulties with reading at the age of, say, 8 does not necessarily remain an underperforming reader at the age of 13 if early educational interventions are put in place. The phenomenon of what I shall call 'difficulties with reading' is another example of a modally robust phenomenon in that it might occur for a number of different reasons, and it is of interest to diverse epistemic communities, including cognitive psychologists, educationalists, and neurobiologists, all trying to study and explain dyslexia.

To this end, developmental psychologists Uta Frith and John Morton have articulated what they call developmental contingency modelling (DCM) for dyslexia. DCM is another example of perspectival modelling in that it models *possible causal pathways* for language development setbacks across different domains (i.e. neurobiological, cognitive, behavioural, and environmental). A child who displays difficulties with reading at the age of 8 might be a child who could develop successful compensatory strategies by the age of 13. Being able to perspectivally model the relevant phenomena is key for effective school interventions in early years and vital to giving each child the best educational opportunities.

To sum up, there are specific contexts in which modelling possibilities matters and the model pluralism in those contexts is perspectival in that it is designed to be *exploratory*. Or better, the perspectival nature of the representation is best understood along the lines of *perspectival$_2$*. Cognitive neuroscientists do not build developmental contingency models for dyslexia because they want to represent dyslexia, or the mechanisms thereof *from a specific vantage point*. They build those models because they want to explore how breakdown points might contingently occur in the long journey of developing language skills during the early years. The function of those models is to provide a framework that covers all possible contingencies and may facilitate diagnoses and appropriate educational interventions at different stages and across different natural languages.

The nuclear physicists who came up with models of the nucleus were not primarily interested in 'representing' the nucleus *from a particular point of view*. They aimed to explore instead how a range of phenomena might be related to one another: for example, how patterns of isotopic abundances might be related to the phenomenon of slow neutron capture.

And the climate scientists who build ensemble models and run CMIP for refining and improving the robustness of global warming projections over the next 50–100 years do not do so because they want to 'represent' climate change *from the point of view of different GHG scenarios*. They want to tackle instead climate change by offering to policymakers and politicians actionable points that are informed by evidence, state-of-the art modelling techniques, and a plurality of perspectival data-to-phenomena inferences (from tree rings, corals and boreholes, among others).

These are examples of situations where modelling what is possible is a necessary step to find out what is actual. If the goal is to find out how global warming will evolve over the next 50 years; or how the nucleus will behave in radioactive chains; or how reading skills acquisition will be hampered in particular cases, modelling possible long-term GHG concentration scenarios, possible nuclear reactions, or possible language development paths, respectively, is the way forward.

In handling situations that involve dynamic change over time and across domains, the modelling in question is bound to be perspectival, that is, pluralistic and exploratory. This is a distinctive kind of modelling possibilities, very different from other examples in the existing literature on scientific modelling. How to better characterize this exploratory exercise in each case and how it does provide us with a 'window on reality' is something I explore in more detail in the three following case studies of Chapters 4.a, 4.b, and 4.c.

THREE CASE STUDIES

4.a

A tale from the atomic nucleus,
ca. 1930s–1950s

4.a.1. Scientific perspectives on atomic abundances, ca. 1900–1924

Historical work on early nuclear models has drawn attention to the unexpected role played by geochemistry, earth sciences, and cosmochemistry (see Johnson 2004; Kragh 2000). I draw on this work, retrace some of these surprising connections, and look into new ones too. These multi-disciplinary contributions should not be consigned to the infancy of nuclear physics. I see them as pivotal to its successful historical development in the 1930s–1950s. They furnish a perfect example of a plurality of scientific perspectives in dialogue, and of perspectival modelling in action.

What chemical elements compose the earth's crust and atmosphere? Why are some but not others abundant? How did chemical elements form in the universe? Which *new* chemical elements *could be* found in nature or created in a lab? The story of the atomic nucleus begins with these questions. And answers to them around the 1900s–1930s did not necessarily come from the burgeoning quantum theory of Max Planck and Bohr's model of the atom, but instead from the daily work of petrologists, mineralogists, meteorologists, and geochemists.

4.a.1.1. Volcanoes, igneous rocks, and the relative abundances of elements in the earth's crust

The Italian region from Naples to north of Rome (near Lake Bolsena), with its volcanic origins and active Phlegraean Fields, has always been of great interest for volcanologists, geologists, mineralogists, and petrologists. In 1894, a young American petrologist, Henry Stephens Washington[1]—just graduated from the University of Leipzig—embarked on a journey in the area to study the mineral composition of volcanic rocks. Samples were carefully labelled according to their geographical origin and classified by chemical composition. Igneous rocks were

[1] On Washington, see Belkin and Gidwiz (2020), on which I draw here.

Perspectival Realism. Michela Massimi, Oxford University Press. © Oxford University Press 2022.
DOI: 10.1093/oso/9780197555620.003.0005

divided into class, order, range, and subrange, following what became known as the 'Cross, Iddings, Pirsson and Washington (CIPW) norm classification' (see Cross et al. 1902) marking the beginning of normative mineralogy. The idea behind the CIPW norm was to offer chemical analysis of the elements and their estimated percentages in rocks that had formed via complex geochemical processes.

The outcome of long years of petrological fieldwork in Central Italy was a volume that Washington published in 1906, *The Roman Comagnatic Region*. The work established a common origin for the volcanoes of the area and defined the region as 'potassic' owing to the high percentage of potash. Percentages for silica, aluminia, lime, and others were carefully calculated. Via petrological studies of rocks, the relative abundances of chemical elements in the earth's crust entered the public domain. But it was not just the study of the earth's crust and volcanism that provided data about elements and their abundances.

4.a.1.2. Noble gases, meteorology, and the relative abundances of elements in the atmosphere

Data about the percentages of chemical elements in the atmosphere increased at the start of the twentieth century. In 1895, William Ramsay had discovered helium in a radioactive mineral called cleveite (Ramsay 1895). Samples of cleveite placed in an exhausted glass flask with boiling acid yielded a gas whose spectrum had a bright yellow line. The gas was eventually named 'helium' after Lockyer's first identification of it in the spectrum of the sun's ('helios' in Greek) chromosphere a few years earlier. This marked the beginning of a long series of experiments extracting gases from radioactive substances.

Rutherford, Barnes, and Soddy ran experiments to extract inert gases like helium and argon from radium (see Ramsay 1904) and thorium. By 1908, Ramsay had published estimates of percentages of krypton and xenon in the atmosphere (Ramsay 1908). Over the following two decades, percentages of chemical elements present in the earth's atmosphere became of increasing interest for meteorologists all over the world to better understand the troposphere and the stratosphere. Data about percentages of oxygen, hydrogen, helium, neon, krypton, and argon came in from Bavaria, Paris, and Moscow using balloon flights able to go as high as 8–9 km in the atmosphere. Refinements of these percentages continued into the late 1930s to eliminate errors due to contamination from the balloon gas. By 1937, the Austrian-British chemist Friedrich Paneth announced that the percentage composition of the air was independent of height throughout the whole troposphere and in the first kilometres of the stratosphere (Paneth 1937).

4.a.1.3. Meteorites, isotopes, and the mass spectrograph

These data about percentages of chemical elements in the earth's crust and atmosphere raised new questions. Was any such pattern specific to the terrestrial distribution? Or was it identifiable also in outer space and in meteorites? Between 1915 and 1921, the American chemist William Draper Harkins concluded that elements with even atomic number Z were predominant. He concluded that 89% of atoms on the surface of the earth and 98% in meteorites had an even atomic number (Harkins 1921), with even-Z atoms having on average twice as many isotopic varieties as odd-Z atoms.

But these were figures from massive aggregates of atoms and molecules. What about a more precise estimate of the percentages of *individual* atomic species with their isotopic varieties? How many isotopes are there for *each* chemical element? Answering this question requires going beyond geochemistry and studying the atomic mass of individual atoms. A new instrument was necessary: the mass spectrograph.

Working on the same principles as cathode rays (ionization at low pressure in a strong field), the mass spectrograph earned Francis Aston the Nobel Prize for chemistry in 1922 (Aston 1922/1966). By ionizing a sample of a chemical element and using a strong magnetic field to deflect the ions, Aston was able to measure individual isotopic varieties in samples of non-radioactive elements. This reopened the search for new patterns.

4.a.2. From silicate crust and meteorites to the phenomenon of 'nuclear stability'

In a 1924 paper, Francis Aston referred to the work of Henry S. Washington on composition of igneous rocks and Ramsay on the atmosphere, and plotted the relative abundances of atomic species for the first 39 elements of the periodic table (Aston 1924). The table showed on the x-axis the so-called mass number (or atomic weight) of each atom (which Aston identified with protons since the neutron had not yet been discovered) and on the y-axis the logarithm to base 10 of the total number of gram-atoms on earth.

The task was nothing less than trying to identify 'the relative stability of nuclei during the evolution of the atoms', assuming 'a lithosphere of mass 5.98×10^{27} gm having the average composition of the igneous rocks, a hydrosphere of mass 1.45×10^{24} gm of water and an atmosphere of mass 5.29×10^{21} gm of ordinary air' (Aston 1924, p. 394). The graph did not reveal any regular pattern but a stark abundance of elements of even-atomic number Z, and peaks around oxygen (O) with $Z = 8$, silicon (Si) with $Z = 14$, calcium (Ca) with $Z = 20$, and iron (Fe) with $Z = 26$.

That tables like Aston's could provide evidence for the stability of the inner nucleus—even in the absence of an explanation—became key to an entire programme of *cosmochemistry* that flourished between 1926 and 1937 thanks to the work of Victor Moritz Goldschmidt, among others.[2] Like Washington before him, Goldschmidt was not a nuclear physicist. He was the Director of the Mineralogical Institute of the University of Oslo and Head of the Mineralogical Institute in Göttingen from 1929 before returning to Norway and eventually having to flee to Sweden during World War II. Mineralogy was a thriving field in Norway. As Chairman of the Norway Government Commission for Raw Materials (Rosbaud 1961), Goldschmidt had an important task, for establishing the relative abundances of elements such as nickel ($Z = 28$) in rocks had far-reaching economical-industrial consequences for the production of alloys and for minting coins.

But there was a more far-reaching interest as well. What chemical elements are most concentrated in the silicate crust of the earth? And what can the relative abundances of the same elements inside meteorites tell us about the origins of chemical elements in the primordial universe? To answer these questions, X-ray crystallography offered a powerful new instrument, for it allowed the gathering of data about the crystallization of molten rocks and silicate melts and which atoms and ions might have escaped the process. Goldschmidt saw the basic problem of geochemistry as that of determining 'the quantitative chemical composition of the earth and to find the laws which underlie the frequency and distribution of the various elements in nature' (Rosbaud 1961, p. 361). When Goldschmidt plotted the data about relative abundances, now against neutron number N rather than the proton number Z, a regular pattern began to emerge around $N = 2, 8, 20, 50, 82, 126$ (see Figure 4.a.1).

'Much of what we know today about the origin of the elements has been derived from chemical analysis of meteorites performed by Goldschmidt and his students' (as summarized in Figure 4.a.1), writes physical chemist Hans E. Suess (1988, p. 385). While the considerations behind Goldschmidt's analysis—like Washington's and Ramsay's before him and Suess's after him—were mostly geochemical and cosmochemical, the numbers so identified had already attracted attention elsewhere, among different epistemic communities.

From the atomic theory point of view, analogies between the structure of nucleons and that of electrons had been explored since the discovery of the Pauli exclusion principle in 1924. In 1930–1932, James Bartlett (1932) had suggested

---

[2] On the role of Goldschmidt's cosmochemistry for the history of nuclear physics, see Kragh (2000).

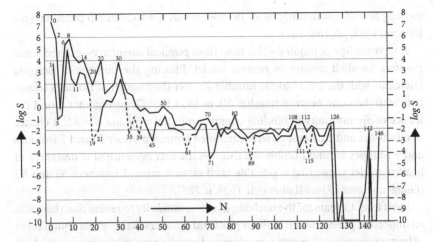

Figure 4.a.1 Relative abundances of chemical elements in the solar system in log *S* plotted over their respective neutron number *N*. From V.M. Goldschmidt, 'Geochemische Verteilungsgesetze der Elemente, IX. Mengenverhältnisse der Elemente und der Atomarten', *Skrifter av det Norske Videnskaps-Akademi i Oslo, Mat.-Nat. Klasse* (1937), Fig. 5. Reprinted from *Applied Geochemistry* 3, Hans E. Suess, 'V.M. Goldschmidt and the Origin of the Elements', Fig. 2, p. 386, Copyright (1988), with permission from Elsevier. https://doi.org/10.1016/0883-2927(88)90119-9

that nucleons might be arranged a bit like electrons in *s*, *p*, and *d* shells (or orbitals) with closed shells of 2, 8, and 18 nucleons. Some evidence for these speculations came from light nuclei with mass number up to 36, but the data were lacking for heavier nuclei and the nature of the nuclear force binding protons and neutrons remained to be explored.

In 1933–1934, Walter Elsasser published two papers (Elsasser 1933, 1934; for a discussion, see Mladjenovic 1998; pp. 287–305). Using Aston's data for light nuclei and invoking Pauli's principle, Elsasser assumed that protons and neutrons moved independently as particles in a field. Nucleons would occupy independent closed shells, whose energy levels were defined by a set of quantum numbers including azimuthal quantum number *l* and spin quantum number *s*. Pauli's principle dictated that there could not be more than $2(2l + 1)$ nucleons for each shell and hence 2, 6, 10 . . . nucleons; but Bartlett's data for light nuclei suggested a higher number of nucleons (2, 8, 18 . . .) per closed shell.[3] Moreover, the nature of the potential in which the allegedly independent nucleons were

---

[3] For a discussion of Elsasser's and Guggenheimer's early shell models, see Johnson (1992, 2004), on which I draw here.

moving proved a stumbling block because it was not a Coulomb potential (see Johnson 1992, pp. 166–167).

A separate line of inquiry—this time from physical chemistry—offered some pointers for shell closure in heavier nuclei. Plotting the number of isotopes (elements with the same atomic number Z) over the number of isotones (elements with the same neutron number N), in 1934 Karl Guggenheimer found evidence of discontinuity in binding energy around $N = 50$ and $N = 82$. A cluster of 6 isotones and 11 isotopes were identified around number 50 and 7 isotones and 9 isotopes around number 82: this was the first evidence of abundances in heavier nuclei suggesting a possible shell closure around numbers 50 and 82 (Guggenheimer 1934; Mladjenovic 1998, p. 289).

But it took 13 years for the nuclear physics community to realize that there was an important phenomenon in those plots and numbers: namely that abundances of heavier chemical elements were clustered around particular neutron numbers N and proton numbers Z such as 20, 28, 50, 82, and that those numbers were special (*ausgezeichneten Zahlen*, as Hans Suess called them) or 'magic numbers', as they became known. It was Hans Suess, with the help of Otto Haxel, who in 1947 realized the importance of these special numbers as key to the atomic structure (Suess 1947a, 1947b).

The story goes that Otto Haxel had discussed the matter with nuclear theorist Hans Jensen in Hannover and

> Both Haxel and Suess tried to convince Jensen that their 'special numbers' were some sort of key to nuclear structure, but he seemed reluctant to pursue the issue. He saw absolutely no theoretical way to account for the regularities. Then, in August 1948, Maria Goeppert Mayer's paper appeared in *The Physical Review*, setting out extensive evidence for the same numbers that Haxel and Suess were so excited about, and referring to those numbers as 'magic numbers'. (Johnson 2004, pp. 303–304)

Hans Jensen (1965) recalls how he came across the work of Maria Goeppert Mayer while visiting Bohr in Copenhagen and how it was Bohr's interest that encouraged him to pursue the matter further. But what was so special about those 'magic numbers' that Maria Goeppert Mayer saw in 1948? The answer is revealing both about the way scientific perspectives intersect, and about the importance of a plurality of perspectival models.

To recap the early part of the story, data from meteorites, ores, silicate melts, and so forth, provided evidence for the relative abundances of some chemical elements. Data plots like Bartlett's, Aston's, Goldschmidt's, or Suess's gradually revealed the abundance of some nuclides in the earth's crust and in the universe.

Such plots provided evidence for the phenomenon of 'nuclear stability'. The inference from data about cosmic abundances to the phenomenon of nuclear stability was *perspectival* in distinctive ways. It required a number of *experimental, theoretical, and technological resources* spread out across a number of scientific communities. For example, while Aston's mass spectrograph was central to the work of atomic theorists like Elsasser, X-ray crystallography was pivotal to Goldschmidt's mineralogical research building on Washington's earlier petrological work.

This plurality of scientific perspectives allowed *reliable* scientific knowledge claims about the percentages of particular elements in rocks, alloys in meteorites, and gases in the atmosphere, among others. The *methodological-epistemic principles* that justified the reliability of the claims were themselves perspectival, ranging from principles in normative petrology to those of crystal structure and nuclear physics.

This plurality of 'intersecting scientific perspectives' made it possible to establish that there was a modally robust phenomenon about the *stability* of *some nuclides* with *special or magic numbers*. This phenomenon could be teased out from Goldschmidt's data about cosmic abundances as well as from Suess's later data. Chemical considerations about the periodicity of chemical elements (like inert gases) and analogy with Pauli's principle for electron shells were important too. And slow neutron-capture data, as we shall see later, played a key role in the rest of this story.

The fact that it took a decade to realize that these data were evidence for nuclear stability is revealing of the role and importance of perspectival modelling. Data D *by themselves* are not enough to draw conclusions about any specific phenomenon P. Being able to *reliably* infer that P from D required a plurality of *scientific models*. What was needed in particular was the shell model of the nucleus introduced independently by Jensen and Goeppert Mayer in 1948. They succeeded where Elsasser's and Guggenheimer's shell models had failed in explaining the interactions among nucleons inside the atomic nucleus and shedding light on the 'magic numbers' associated with the abundant nuclides.

The main stumbling block for the emergence of the shell model was the popularity of another model of the nucleus: the liquid drop model. Bohr's 'papal blessing' was required for Hans Jensen to take up the modelling challenge and look more closely into Suess and Haxel's data. In Maria Goeppert Mayer's case, her experimental training allowed her to see something in the shell model that others had not.

But another (now forgotten) model, the so-called odd-particle model, explored by Theodore Schmidt and Katherine Way in the late 1930s, was also important in bridging the gap between the fashionable liquid drop model and the shell model. In the next section, I take a look at these three models for the nucleus and the reasons why they are good examples of what I call *perspectival models*.

4.a.3. Perspectival models of the nucleus around the 1930s–1950s: the liquid drop model, the odd-particle model, and the shell model

4.a.3.1. From Gamow's liquid drop model to Bohr's compound model of the nucleus

In the early 1910s, the alpha-particle experiments led by Ernest Rutherford at the University of Manchester provided evidence that almost the entire mass of the atom was compressed within a tiny core, the nucleus, contradicting J.J. Thomson's earlier 'plum pudding model'. Yet these experiments did not give any conclusive answer about the nature of the nucleus, its constituents, or the force binding them. Indeed, in the late 1920s, Erwin Schrödinger was still cautioning that: 'Just because you see alpha particles coming out of the nucleus, you should not necessarily conclude that they exist inside it in the same form!' (quoted from Jensen 1965, p. 1420).

Roger Stuewer (1994) has reconstructed the development of the liquid drop model, starting with George Gamow's model first presented at the Royal Society in 1929. Well before the emergence of Planck's quantum hypothesis, a liquid drop model was originally applied to the study of electrically charged liquid droplets (Rayleigh 1879). A half-century later, Gamow proposed it could also explain the nuclear binding energy.

Taking a cue from Rutherford's alpha-particle scattering experiments and speculations that those particles must have been inside the nucleus with protons and electrons, Gamow came to conceive of the nucleus as a 'small drop of water in which the particles are held together by surface tension' (see Rutherford et al. 1929, p. 386). The main evidence for this model was once again found in the data from Aston's mass spectrograph.

As Aston himself explained at the same 1929 Royal Society meeting, by measuring accurately the atomic mass number for various elements and plotting against it the 'percentage excess or defect from a whole number on the oxygen scale'—known as the 'packing fraction' (Aston in Rutherford et al. 1929, p. 383)—a curve was found (known as the mass defect curve for it highlighted the discrepancy between the total mass of an atom and the sum of the masses of its alleged constituents). Gamow was able to derive a formula for the total energy of nuclei based on number of alpha particles they presumably contained. But his efforts came to a halt because of poor agreement with Aston's mass defect curve.

The 1932 discovery of the neutron by Chadwick as detailed in Heisenberg's (1934) paper for the Solvay Congress laid the foundations for a model of the nucleus as consisting of neutrons and protons. One of the main theoretical problems was to understand the nature of the nuclear binding force as a function of number of nucleons, or atomic mass A. In 1935, Heisenberg's student Carl Friedrich von Weizsäcker gave a semi-empirical formula for the nuclear binding energy of the liquid drop model, which included a series of terms (volume or total number of particles, surface energy proportional to the surface area of the nucleus, Coulomb repulsive energy acting among protons, among others) showing that for nuclei with Z less than or equal to 20, the greatest stability for any given atomic mass occurred when number of neutrons N equals number of protons Z (see Portides 2011 for a philosophical discussion). For each Z, there was a range of N for which the nucleus was stable. Nuclei outside that range were subject to beta-decay. Understanding nuclear stability versus beta-decay required making a series of assumptions about the shell structure of the nucleus. But after the setback experienced by Elsasser's and Guggenheimer's earlier shell models, that took more than a decade.

While Heisenberg and von Weizsäcker were laying foundations for the liquid drop model, Enrico Fermi and colleagues in Rome were irradiating rhodium with a neutron source (Amaldi et al. 1935). They expected that neutrons would undergo scattering (with associated nucleus excitation). But the experiments revealed instead a new phenomenon: neutron capture. The neutrons would attach to the nucleus rather than being scattered.

There was more. If the apparatus was immersed in water, neutrons would interact with the hydrogen of water and be slowed down, and slow neutrons were more easily captured by nuclei. Fermi examined the absorption of slow neutrons in different elements and identified 'anomalously large absorption coefficients for the slow neutrons' (Amaldi et al. 1935, p. 525). Collision cross-sections of slow neutrons were much larger than those for fast neutrons and exceeded the expected scattering cross-sections. This raised interesting questions about the nature of the force acting in neutron capture. In 1936, Bohr published a paper in *Nature* where he referred to Fermi's results to conclude that

the phenomena of neutron capture thus force us to assume that a collision between a high-speed neutron and a heavy nucleus will in the first place result in the formation of a compound system of remarkable stability. The possible later breaking up of this intermediate system by the ejection of a material particle, or its passing with emission of radiation to a final stable state, must in fact be considered as separate competing processes which have no immediate connection with the first stage of the encounter. (Bohr 1936, p. 344)

This was the beginning of Bohr's 'compound nucleus model'—an evolution of Gamow's 1929 liquid drop model—which treated the incoming neutron hitting the nucleus as if it were absorbed by the nucleus, exciting states of it. As James Rainwater pointed out many decades later in his Nobel Prize speech, Bohr's compound model

> is not necessarily incompatible with a shell model, since the shell model refers mainly to the lowest states of a set of fermions in the nuclear 'container'. However, when combined with the discouragingly poor fit with experiment of detailed shell model predictions . . . the situation 1948 was one of great discouragement concerning a shell model approach. (Rainwater 1975, p. 262)

Bohr's attack on the shell model as 'unsuited to account for the typical properties of nuclei for which . . . energy exchanges between the individual nuclear particles is a decisive factor' (Bohr 1936, p. 345) played a central role in shifting attention away from the shell model for about a decade. Bohr argued that in the case of the nucleus the 'procedure of approximation, resting on a combination of one-body problems, . . . loses any validity' when dealing with 'essential collective aspects of the interplay between constituent particles' (p. 345). Bohr referred once more to Aston's precise measurements of isotopic varieties to conclude that (contrary to the shell model) the excitations of heavy nuclei should be attributed to the 'quantised collective type of motion of all the nuclear particles' (p. 346) rather than the excitation of individual nucleons.

In the meantime, Gregory Breit and Eugene Wigner (1936) were also working on slow neutron capture and were able to derive formulas for neutron-capture and neutron-scattering cross-sections, which agreed with Fermi's results (see Johnson 1992, p. 168). And between 1936 and 1937, Hans Bethe co-authored three substantial review papers that became known as the 'Bethe Bible' and contributed to the popularity of the liquid drop model (see Johnson 1992, p. 169). In his first paper co-authored with R. F. Bacher, Bethe pointed out that while neutron and proton shells provided the 'basis for a prediction of certain periodicity in nuclear structure for which there is considerable experimental evidence', the 'assumption can certainly not claim more than moderate success as regards the calculation of nuclear binding energies' (Bethe and Bacher 1936, p. 171). Referring to the work of Bartlett, Elsasser, and Guggenheimer, they warned against 'taking the neutron and proton shells too literally . . . with the effect of discrediting the whole concept . . . among physicists' (p. 176).

In his second paper, Bethe compared the shell model with the liquid drop model. Both offered statistical treatments of the nucleus. The former, Bethe said, started with the assumption of free individual particles and treated the nucleus as if it was a mixture of two Fermi gases of protons and neutrons. The latter did not regard nucleons as individual particles and treated the interaction among nucleons as larger than the individual kinetic energies as if the nucleons behaved like particles in a drop of liquid.[4] He concluded that the latter model seemed to 'come nearer the truth' (Bethe 1937, p. 80) as the estimated nuclear energy levels compared 'very favorably with the average spacing of neutron levels estimated from experimental data' from slow neutron experiments (p. 90).

A turning point in the history of the liquid drop model came in December 1938, when Otto Hahn and Fritz Strassman found that slow neutrons interacting with uranium led to barium. To explain this, Lise Meitner and her nephew Otto Frisch resorted to the liquid drop model. They reasoned that the nucleus might have become deformed after absorbing a neutron, with the surface area increasing and the surface tension opposing this deformation and trying to keep the nucleus spherical. However, Frisch and Meitner realized that under the repulsive force among protons the deformation would eventually split the nucleus in two.

Roger Stuewer (1994) argues that the joint work of Frisch and Meitner represented the coming together of two different strands in the history of the liquid drop model. Meitner, familiar with Heisenberg and von Weizsäcker's work, approached the problem of the nuclear mass defect. Frisch, familiar with Bohr's work, focused instead on the dynamic features of the model and how it fared vis-à-vis nuclear excitations. Bohr heard of the Frisch–Meitner interpretation of the phenomenon while visiting Princeton. He began to work there with John Wheeler to develop the liquid drop model into a full-blown theory of nuclear fission—the resulting Bohr–Wheeler paper (1939) laid the foundations for nuclear fission with uranium-235 and plutonium-239.

But, surprisingly, the same neutron-capture phenomenon that had been an incentive for the liquid drop model proved also a key factor in the revival of the shell model. In 1948, Gamow sent a letter to *Physical Review* jointly written with R.A. Alpher and H. Bethe proposing that all chemical elements could have formed via neutron capture from an overheated primordial 'neutral nuclear fluid', which eventually produced protons and electrons via beta decay in an expanding early universe (Alpher et al. 1948). The relative abundances of individual atoms were ascribed to neutron-capture cross-sections rather than mass defect, and once again Victor Goldschmidt's geochemical data about abundances offered a benchmark.

Nine years later, the same geochemical data—improved by more recent measurements by Suess and Urey—were key for the interpretation of how

---

[4] Bethe attributed authorship of the model to Bohr and Kalckar, effectively identifying Bohr's compound model with the liquid drop model and not crediting Gamow (Rutherford et al. 1929) for the genesis of the model.

chemical elements might have formed inside stars in the seminal paper by E.M. Burbidge, G.R. Burbidge, W. A. Fowler, and F. Hoyle (1957). The phenomena of neutron capture which had been the original trigger for Bohr's 'compound model' also prompted a revival of interest in the shell model around 1939–1949 in the continuing attempt to understand nuclear stability and isotopic abundances.

4.a.3.2. The odd-particle model: a bridge between the liquid drop model and the 1949 shell model

In 1963, Maria Goeppert Mayer shared half of the Nobel Prize in Physics with Hans Jensen (the remaining half went to Eugene Wigner). The prize was given for her discovery concerning the nuclear shell structure. A year later, she wrote a review article in *Science* where she described two approaches to nuclear physics:

> There are essentially two ways in which physicists at present seek to obtain a consistent picture of the atomic nucleus. The first, the basic approach, is to study the elementary particles, their properties and mutual interaction. Thus one hopes to obtain a knowledge of the nuclear forces. If the forces are known, one should in principle be able to calculate deductively the properties of individual complex nuclei. Only after this has been accomplished can one say that one completely understands nuclear structures. . . . But our knowledge of the nuclear forces is still far from complete.
>
> The other approach is that of the experimentalist and consists in obtaining by direct experimentation as many data as possible for individual nuclei. One hopes in this way to find regularities and correlations which give a clue to the structure of the nucleus. . . . The shell model, although proposed by theoreticians, really corresponds to the experimentalist's approach. It was born from a thorough study of the experimental data, plotting them in different ways and looking for interconnections. This was done on both sides of the Atlantic Ocean and on both sides one found that the data show a remarkable pattern. (Goeppert Mayer 1964, p. 999)

The breakthrough came in 1947 when Goeppert Mayer found an explanation for the surprising stability of certain heavier nuclei. Only a few possible combinations of neutrons and protons exist in nature as stable nuclei that do not decay by beta decay. The most stable nuclei tend to have an even number of protons and neutrons, as Suess had already noted. 'Eighty-two and fifty are "magic" numbers. That nuclei of this type are unusually abundant indicates that the excess stability must have played a part in the process of the creation of the element', declared

Goeppert Mayer (1964, p. 999). Magic numbers were found elsewhere: 2, 8, 20, 28, 50, 82, and 126 were all magic numbers.

Goeppert Mayer published two articles in *Physical Review*. Her first, on 1 August 1948, established the stability of nuclei with 20, 50, 82, and 126 neutrons or protons (Goeppert Mayer 1948). Referring to the earlier work of Elsasser, the paper covered experimental evidence for the existence of nuclear stability, including familiar items such as:

1. isotopic abundances relative to even atomic number Z with e.g. Calcium (Z = 20) having five isotopes and lead (Z = 82) having four stable isotopes, the heaviest of which ^{208}Pb with 126 neutrons is stable;
2. a high number of isotones (to use Guggenheimer's terminology) for neutron numbers 50 and 82;
3. Goldschmidt's plot of abundances over the neutron number N.

But the list also included unexpected new pieces of evidence that up to this point had fallen within the province of the liquid drop model, such as:

4. the unusually low neutron absorption cross-sections for nuclei with 50, 82, or 126 neutrons emerging from the experiments of Griffith on rare earth such as yttrium (N = 50) and Mescheryakov on lanthanum and barium ^{138}Ba (the latter with Z = 56 and N = 82);
5. asymmetric fission of ^{235}U into fragments of 82, 50, and 11 neutrons adduced as evidence for the closed shells at 82 and 50.

A footnote indicated that: 'The author is indebted to Dr Katherine Way, who pointed out the connection of the closed shells with neutron absorption cross sections' (Goeppert Mayer 1948, p. 238). This connection was surprising and a welcome addition to the experimental data. More importantly, this was the kind of evidence needed to convince physicists that the shell models had some legs, despite Bohr's influential opposition and Bethe's verdict in favour of the liquid drop model.

Katharine Way was a former PhD student of John Wheeler and a later member of the Manhattan Project, who had herself worked on the liquid drop model. In a 1939 paper (Way 1939), she pointed out that the liquid drop model was in 'very poor' agreement with experimental evidence about nuclear magnetic moments found by spectroscopist Theodore Schmidt in 1937 (Schmidt 1937; see also Schüler and Schmidt 1935).[5] Moreover, Way pointed out, using data for heavy nuclei from Wheeler and Teller, that even the largest nuclear spin I would be

---

[5] For a discussion see Johnson (1992, p. 165).

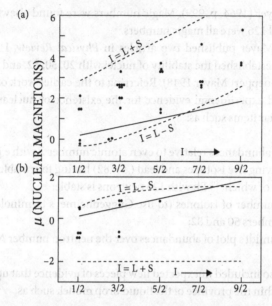

Figure 4.a.2 Experimental data (dots) on nuclear magnetic moments vis-à-vis the magnetic moments calculated on the basis of Schmidt's odd-particle model (dotted lines) and those based on the liquid drop model (full line) for a spinning drop of $Z/A = 50 / 119 = 0.42$. (a) is the odd-proton scenario and (b) the odd-neutron scenario. I is the nuclear spin quantum number in units of $\hbar$. Reprinted Fig. 1 with permission from K. Way (1939) 'The liquid-drop model and nuclear moments', *Physical Review* 55, 964. Copyright (1939) by the American Physical Society. https://doi.org/10.1103/PhysRev.55.963.

too small to justify the identification with a uniformly charged spinning drop. A better agreement with the measured data, Way concluded, was given by the so-called odd-particle model (known also as single-particle model; see Figure 4.a.2)

In this alternative model, the magnetic moments for nuclei with odd-Z and odd-N were non-zero and were attributed entirely to the single extra odd nucleon 'moving outside a central momentless core' (Way 1939, p. 964). Now long forgotten, the odd-particle model was not a full-blown model as such and it cannot be regarded as a shell model either,[6] but it provided nonetheless an important

---

[6] The same consideration applies to a precursor of Way's model, as Johnson remarks: 'Schmidt's single-particle model has been commonly associated with shell models, but this is misleading for two reasons. First, it was not really a nuclear model in the conventional sense of the term. . . . Second, this does not qualify as a shell model. In their first paper Schmidt and Schüler do explicitly assume the existence of some sort of shell. . . . However, they pay no further attention to those shells, focussing instead on the nucleons that lie outside closed shells. The real significance of this scheme lies in the explicit treatment of nucleons as discrete particles with individual values of spin and magnetic moment' (Johnson 1992, p. 165).

bridge between the early shell models pre-1947 and the fashionable liquid drop model of the late 1930s. It showed the epistemic limit of the liquid drop model. And it offered reasons as to why a quasi-shell model could provide better agreement with experimental data on nuclear magnetic moments and neutron absorption cross-sections. The model allowed scientists to make relevant and appropriate inferences about features (e.g. the nuclear magnetic moments and neutron absorption cross-sections) of the phenomena under study (e.g. stable nuclides). Such inferences were in turn pivotal to establish clear shifts in the stability line (and hence binding energy) of nuclei, as Maria Goeppert-Mayer concluded in her 1948 paper:

> Between $Z = 50$ and $N = 82$, however, the experimental values of Z seem to be below the theoretical curve. The disagreement can be explained by a definite shift of the stability line at 82 neutrons. This shift of the stability line can be explained by a change in binding energy of about 2 MeV. . . . Whereas these calculations are undoubtedly very uncertain, they may serve as an estimate of the order of magnitude of the discontinuities in the binding energies. Since the average neutron binding energy in this region of the periodic table is about 6 MeV, the discontinuities represent only a variation of the order of 30 percent. This situation is very different from that encountered at the closed shells of electrons in atoms where the ionization energy varies by several hundred percent. Nevertheless, the effect of closed shells in the nuclei seems very pronounced. (Goeppert-Mayer 1948, p. 239)

Despite the limits of the analogy with electronic shells, clear indications of shifts in the stability of nuclei had been found by 1948. That 50, 82, and 126 were special or 'magic' numbers for nuclear stability was now established. What was still missing was an explanation for what made those numbers 'magic'.

4.a.3.3. The Nobel Prize-winning shell model of 1949 and the 'unified model' of Rainwater, Bohr, and Mottelson

In her second *Physical Review* article, Goeppert Mayer stated that the magic numbers occur 'at the place of the spin-orbit splitting of levels of higher angular momentum' (Goeppert Mayer 1949, p. 1969). The idea that spin-orbit coupling might explain the stability of heavy nuclei was suggested to her by Fermi. But what was needed was a *model* that could explain the magic numbers along the lines of how the spin-orbit coupling had been helpful to explain the closure of electronic groups. This was the shell model for which Goeppert Mayer and Jensen shared half of the Nobel Prize for Physics in 1963.

The model treated each proton or neutron as a fairly independent particle occupying orbitals, rather like those of electrons in atoms, whose orbital angular momentum (indicated by the quantum number l) is quantized so that for each l, there was a discrete number of states of different orientation in space given by the magnetic quantum number m_l. The only problem was that following the atomic prescription for the magic numbers led to particularly stable nuclei for heavier nuclei that had the wrong numbers of protons and neutrons, as Elsasser and Guggenheimer had already found. Goeppert Mayer and Jensen saw that, assuming an additional degree of freedom with the spin (quantum number m_s which can be $m_s = \frac{1}{2}$ for spin up and $m_s = -\frac{1}{2}$ for spin down), Pauli's principle could be applied to the structure of the nucleus and dictate the maximum number of nucleons that could be sitting in any shell. Goeppert Mayer postulated a particularly strong spin-orbit interaction that led to a reordering of the energies of the proton and neutron orbitals.

As the protons or neutrons increased to fill orbitals to capacity, energy gaps appeared (see Figure 4.a.3). This led to stability so that orbitals below the energy gaps were full while orbitals above the energy gaps were empty. The numbers of protons or neutrons are the magic numbers. Nuclei with magic numbers of both protons and neutron, like ^{208}Pb, are said to be doubly magic. Such nuclei are not only markedly more stable than those with more or fewer nucleons, but they are also always spherical, meaning that they never show evidence of rotational properties.[7]

As we now know, not every conceivable combination of protons and neutrons can exist in nature (e.g. ^{40}C or ^{100}H). The limits to the number of protons or neutrons for any given mass number A correspond to the driplines in the Segrè chart that maps atomic nuclei on the basis of their proton number Z and their neutron number N. Iron, for example (with $Z = 26$ and an average atomic mass A of 55.8 due to the various isotopes ^{54}Fe, ^{56}Fe, ^{57}Fe, ^{58}Fe, among others), is one of the most stable elements in nature—and one of the most abundant elements in the earth's core and stars.

The heavier and larger the nuclei, the more sensitive they become to the electrostatic repulsions among protons. The radioactive element barium Ba ($Z = 56$) and its isotope ^{137}Ba mark (with $A = 137$) the bottom of what physicists call the

---

[7] The orbitals defining a shell should not be thought of as localized at a single radius, of course. Given a shell with quantum numbers l, m_l, and m_s, no more than $2 (2l + 1)$ nucleons of the same kind can occupy it. Thus, the lowest energy shell, designated $1s$ where the s indicates $l = 0$, can have just two neutrons and two protons (^4He). The next shell has $1p$, $l = 1$ with six states, which, in addition to the $1s$ states, add up to eight nucleons of each kind. All this was in agreement with Wigner's work on the light nuclei, which, however, could not similarly be extended to heavy nuclei. In fact, using a three-dimensional oscillator shell to describe the available states above $Z = 20$, Goeppert Mayer noted how a different series of numbers were generated (40, 70, 112) which are not magic unless the spin-orbit coupling is introduced. I am very grateful to Raymond Mackintosh for helpful discussions and comments on the shell model.

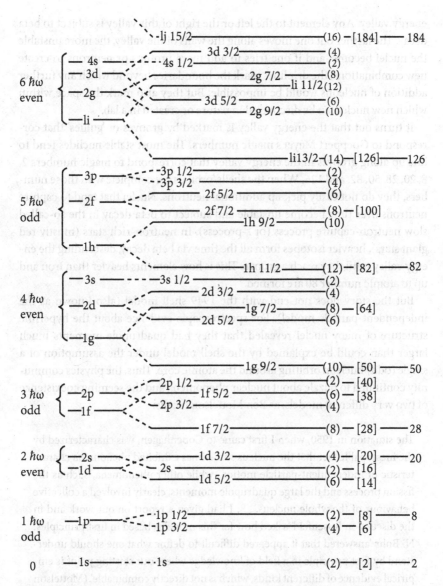

Figure 4.a.3 Magic numbers for heavier nuclei and their relation to quantum numbers and spin-orbit coupling in Maria Goeppert Mayer shell model. See e.g., how the spin-orbit coupling makes it the case that the 1f 7/2 states lie at lower energies than the 1f 5/2 states, creating a gap that corresponds to magic number 28. Reproduced with permission. Fig. 7 from Maria Goeppert Mayer, 'The Shell Model', *Nobel Lecture*, 12 December 1963. © The Nobel Foundation.

energy valley. Any element to the left or the right of this valley is subject to beta decay. The farther out one moves along the walls of the valley, the more unstable the nuclei become, and if one tries to add more protons or neutrons to create new combinations, the driplines mark the boundaries beyond which any further addition of nucleons would be impossible. But they also mark the space within which new nuclei can be discovered in nature or created in a lab.

It turns out that the energy valley is marked by grooves or 'gullies' that correspond to Goeppert Mayer's magic numbers. The most stable nuclides tend to line up along grooves in the energy valley that correspond to magic numbers 2, 8, 20, 28, 50, 82, and 126. When the shells/orbitals are complete with those numbers, they do not easily pick up additional neutrons. Nuclei that tend to capture neutrons over time become unstable and subject to beta decay in the so-called slow neutron-capture process (or s-process). In neutron-rich stars (mostly red giant stars), heavier isotopes form all the time via beta decay, descending the energy valley until they reach a groove. This is how elements heavier than iron and up to atomic number 80 are formed.[8]

But the story does not end with the 1949 shell model (also known as the independent-particle model). For spectroscopic evidence about the hyperfine structure of many nuclei revealed that they had quadrupole moments much larger than could be explained by the shell model under the assumption of a single (odd) nucleon orbiting around the atomic core. Thus, the physics community continued to puzzle about nuclear phenomena and the seeming coexistence of two very different models, as Ben Mottelson recalls:

> The situation in 1950, when I first came to Copenhagen, was characterized by the inescapable fact that the nucleus sometimes exhibited phenomena characteristic of independent-particle motion, while other phenomena, such as the fission process and the large quadrupole moments, clearly involved a collective behaviour of the whole nucleus. . . . I had given a report on our work and in the discussion Rosenfeld 'asked how far this model is based in first principles'. N. Bohr 'answered that it appeared difficult to define what one should understand by first principle in a field of knowledge where our starting point is empirical evidence of different kinds, which is not directly combinable'. (Mottelson 1975, pp. 236–237)

---

[8] But to form heavier elements like uranium ($Z = 92$) for example, more violent and more neutron-rich events such as Type II supernova explosions and neutron star mergers are required. When huge amounts of neutrons are released in these explosions, nuclides are subject to the so-called rapid neutron-capture process (or r-process). In this process, very short-lived, highly unstable nuclei form by rapidly capturing neutrons until they too tumble down towards the energy valley. Many thanks to Marialuisa Aliotta for helpful discussions on this topic.

A solution to the problem of large quadrupole moments was glimpsed in 1949 by John Wheeler, who 'realized that in big nuclei, a single nucleon, constrained by liquid-drop tension, could travel around the rest of the nucleus in a large orbit, deforming the nucleus substantially' (Thorne 2019, p. 9). Wheeler sent the paper to Bohr, and while waiting for comments, the same idea was discovered independently by James Rainwater at Columbia University. Rainwater understood that the large nuclear quadrupole moments could be explained if the nucleus (and hence nuclear charge) could be deformed and take the shape of a spheroidal liquid drop under the action of the outer nucleons orbiting the atomic core.

Rainwater shared the 1975 Nobel Prize for Physics with Aage Bohr (Niels Bohr's son) and Ben Mottelson for the 'discovery of the connection between collective motion and particle motion in atomic nuclei and the development of the theory of the structure of the atomic nucleus based on this connection'.[9] Rainwater's contribution consisted in working out the exact physical details of the spheroidal distortion of the atomic nucleus with a prolate potential (Rainwater 1950). Bohr and Mottelson (1953, 1969, 1975), in turn, assumed a non-spherical potential in which particles moved and were able to show how nuclear rotational spectra were the outcome of the coupling between the outer particles' motion and the motion of the deformed nucleus, offering in this way a 'unified model' that combined features of the liquid drop model and key insights of the shell model (for a recent review of these developments, see Caurier et al. 2005; see also Mackintosh 1977).

4.a.4. Concluding reflections on perspectival modelling

All of the nuclear models discussed here are examples of perspectival models in being *exploratory*. They allowed nuclear physicists to gain knowledge of the nuclear structure at a time (in the early 1930s) when neutrons had just been discovered; speculations still abounded that the nucleus might consist of alpha particles; the quantum chromodynamic nature of the strong interaction binding nucleons was still unknown. The exploratory nature of the 1930s–1950s nuclear models is rooted in their *historical evolution* in response to new data becoming available over time (e.g. from neutron-capture cross-sections to large quadrupole moments) and new phenomena (e.g. nuclear fission, nuclear rotational spectra) being inferred from these data over time.

These nuclear models are, then, *not* perspectival$_1$ in representing the nucleus from different points of view. They are not perspectival$_1$ representations in

---

[9] https://www.nobelprize.org/prizes/physics/1975/rainwater/facts/. I am very grateful to Isobel Falconer for helpful comments on this topic.

offering incompatible or inconsistent images of the nucleus with conflicting essential properties ascriptions. They offer instead perspectival$_2$ representations in opening up a 'window' on the reality of nuclear structure despite the partial, limited, and inevitably piecemeal epistemic access to it. They delivered knowledge of *what is possible* about the nucleus, its internal structure, isotopic stability, nuclear spectra, and the range of possible combinations of protons and neutrons (either to be found in nature or to be created in a lab).

Could alpha-particle natural radioactive chains end with thallium ($Z = 81$) rather than lead ($Z = 82$)? No, because lead has proton number $Z = 82$ (magic number), marking a groove in the energy valley (and making lead a stable and abundant element on earth). Could there be in nature (or be artificially produced) a nucleus like, for example, ^{100}H? No, because it would fall out of the dripline of the energy valley. Could there be new very short-lived nuclei with a very large neutron excess along the neutron dripline? Yes, there could be such nuclei, and large investments have gone into searching for them.[10]

To be a perspectival realist about the atomic nucleus is, then, to engage with an open-ended series of modally robust phenomena (e.g. nuclear stability, neutron capture, nuclear rotational spectra) at the experimental level and with the many exploratory models that over time have allowed physicists to gain knowledge about *what is possible* concerning each of these phenomena.

Perspectival models of the atomic nucleus, ca. 1930–1950, were therefore exploratory in enabling a variety of epistemic communities to make *relevant and appropriate inferences* about the nucleus. Elsasser's early shell model allowed inferences from data for light nuclei via Aston's mass spectrograph to the possible number of nucleons per shell. Schmidt's odd-particle model enabled inferences from data about nuclear magnetic moments to the possible zero-moment atomic core, which in turn informed Way's research about the unusually low neutron absorption cross-sections for nuclei with 50, 82, or 126 neutrons. Goeppert Mayer's shell model in turn allowed inferences from low neutron absorption cross-sections (traditionally within the remit of the liquid drop model) and Goldschmidt's plot of isotopic abundances to the possible existence of shifts in the stability lines (or grooves in the energy valley) corresponding to the magic numbers. Perspectival models of the nucleus around the 1930s–1950s allowed exploration of what is possible about nuclear structure by acting as *inferential blueprints*—a notion I elaborate in Chapter 5.

Second, these models show the *collaborative and social nature* of scientific knowledge production, the seamless flow through which model-based

---

[10] See, for example, the recently announced measurements of the so-called Gamow–Teller strength distribution of isotopes ^{116}Sb and ^{122}Sb at the Research Centre for Nuclear Physics (RCNP), Osaka University (Douma et al. 2020).

knowledge claims are historically put forward, modified, corrected, and re-enacted. To what extent was the Nobel Prize-winning 1949 shell model an evo-lution of (instead of an abrupt shift from) the 1934 shell models? How to classify Schmidt's odd-particle model in this lineage? (Technically, as Johnson (1992) remarks, it was not a full-blown nuclear model, yet it assumed that the atomic core consisted of shells.) What about the relation between Gamow's 1929 liquid drop model and Bohr's 1936 compound model? How to locate the Rainwater–Bohr–Mottelson 'unified model'—with its combination of liquid drop model and shell model—in this model genealogy?

One thing is clear. Models at play in perspectival pluralism—as in this histor-ical case study—are not the static entities representing-qua-mapping one-on-one relevant aspects of the target system, as a somewhat impoverished picture of them (often found in philosophy of science) has suggested. These models are *dynamic evolving tools* with a *history of their own*, which is often intertwined with the history of other scientific models. Having a history means that perspec-tival models are also often the battleground for scientific rivalries and questions about co-authorship. Why did Bethe celebrate Bohr's 1936 compound model without giving credit to Gamow's 1929 liquid drop model? And what about Wheeler, who missed out on the opportunity of sharing the 1975 Nobel Prize due to a delay in the publication of his insight?[11] This is without mentioning Lise Meitner, who was not given the Nobel Prize for her crucial work on nuclear fission.

This inevitable aspect of perspectival models' authorship should not, how-ever, detract from appreciating their by and large *social and collaborative func-tion*. Models make it possible for teams of scientists to work together over time, make changes to and tweak an original model and eventually deliver on the task of advancing scientific knowledge about the phenomena of interest by *making relevant and appropriate inferences*. This is evident in the history of nuclear models around the 1930s–1950s. For they offered perspectival representations for a number of phenomena (nuclear stability, nuclear fission, nuclear rotational spectra) not in the sense of ascribing inconsistent essential properties to the same target system (the nucleus as 'a given'). Instead, they enabled model-based *inferences offered by various authors over time*.

The 1949 Nobel Prize-winning shell model by Jensen and Goeppert Mayer is the final output of a long tradition of earlier models offered over time by Elsasser, Guggenheimer, Schmidt and Way. The 1975 Nobel Prize-winning 'uni-fied model' by Rainwater–Bohr–Mottelson is itself the final product of the long

---

[11] 'I learned a lesson. When one discovers something significant, it is best to publish it promptly and not wait to incorporate it into some grander scheme. Waiting to assemble all the pieces might be all right for a philosopher, but it is not wise for a physicist' (Wheeler quoted in Thorne 2019, p. 10).

historical intertwining of the Gamow–Bohr research on liquid drop models *and* the Elsasser–Guggenheimer–Schmidt–Way–Jensen–Goeppert Mayer studies on shell models.

But there is more. As outlined already in Chapter 1, I understand 'perspectival modelling' in a broad sense, rather than in an exclusively narrow one confined to the actual models. Perspectival modelling is an integral part of a scientific perspective in being embedded into historically and culturally situated scientific practices. This is evident in the history of nuclear models if one considers the complex historical intertwining of data-to-phenomena inferences behind plots of isotopic abundances—from Washington's petrological studies to Aston's mass spectrograph to Goldschmidt's cosmochemistry—that made it possible in the first instance to identify the phenomenon of nuclear stability that the shell models were designed to explain. It is through this plurality of intersecting scientific perspectives and ever-evolving perspectival₂ representations offered by perspectival models that knowledge claims about which nuclear phenomena *might be possible* (and which *might not*) were advanced.

To conclude, perspectival models are exploratory in offering *blueprints* with instructions that enable various epistemic communities over time to come together and *make relevant and appropriate inferences* for the phenomena under study. Much as each model bears someone's name and authorship, and Nobel Prizes are given on such a basis, the role of these models is in fact to facilitate knowledge production over time among very diverse epistemic communities. The nuclear models of the early 1930s were the arrival point of a number of perspectival data-to-phenomena inferences that saw petrologists, volcanologists, mineralogists, spectroscopists, meteorologists, and physical chemists robustly identify the phenomenon of nuclear stability across rocks, atmospheric gases, and meteorites.

The very same models were also the starting point for a number of further inferences from the identified phenomenon of nuclear stability to the underlying nuclear structure that *might be* responsible for it. The liquid drop model and the shell model provided the *inferential blueprints* that enabled physical chemists, atomic theorists, spectroscopists, and nuclear physicists to collaborate and make *relevant and appropriate inferences* from the phenomena of interest (e.g. nuclear fission, nuclear stability) to what nuclear structure *might be like*. The overall scientific knowledge delivered by them is knowledge ultimately produced by and shared among a great number of epistemic communities that are historically and culturally situated across scientific perspectives (I return to this point in Chapter 11).

The constraints within which this modelling exercise took place included, in this particular example, lawlike dependencies such as Coulomb's law of electrostatic repulsion at work among the protons; Pauli's principle, which guided the

analogy with the closure of electronic shells; and von Weiszäcker's semi-empirical formula, among others. And I will return to the role of these laws for perspectival models qua inferential blueprints in more detail in Chapter 5. But next
I turn my attention to two more case studies, which probe a little deeper into
the collaborative and social nature of the scientific knowledge produced via perspectival modelling, and shed light also on the semantic nature of the associated
inferences.

4.b

A tale from the ice, the sea, and the land: climate modelling

4.b.1. Model pluralism in climate science

In 2013, the Intergovernmental Panel on Climate Change (IPCC) published the Fifth Assessment Report, Working Group 1 (AR5-WG1), a systematic assessment of the physical basis of climate change.[1] In its Summary for Policymakers, the report concluded:

> Warming of the climate system is unequivocal, and since the 1950s, many of the observed changes are unprecedented over decades to millennia. The atmosphere and ocean have warmed, the amounts of snow and ice have diminished, sea level has risen, and the concentrations of greenhouse gases have increased. . . . In the Northern Hemisphere, 1983–2012 was *likely* the warmest 30-year period of the last 1400 years (*medium confidence*). (Stocker et al. 2013, pp. 4–5, emphasis in original)

The report identified a 'very likely' loss of ice sheets from glaciers around the world of the order of 275 [140 to 410] Gt yr^{-1} over the period 1993 to 2009. Being able to offer evidence-informed estimates of this nature (accompanied by statistical confidence levels) is key to informing policymakers. Countries around the world signed up to the 2015 United Nations Paris agreement with the goal of reducing greenhouse gases (GHG) emissions and keeping global mean surface temperature (henceforth 'global temperature') 'well below' 2 degree Celsius compared to pre-industrial levels (and 'endeavour to limit' temperature increase to 1.5 Celsius).[2]

To inform policy recommendations of this nature, climate scientists build a variety of models. Their purpose is not just to *reliably* establish *that* global

[1] I am very grateful to Roman Frigg, Benedikt Knüsel, Wendy Parker, Ken Rice, and Jon Turney for helpful comments on earlier versions of this chapter.
[2] See https://unfccc.int/process-and-meetings/the-paris-agreement/nationally-determined-contributions-ndcs.

Perspectival Realism. Michela Massimi, Oxford University Press. © Oxford University Press 2022.
DOI: 10.1093/oso/9780197555620.003.0006

warming is occurring, but also *how fast*. Most importantly, modelling allows to forecast *how global warming might change in the future.*

This is an example of what I called exploratory modelling. The task is to explore how global warming might change from now to, say, the year 2100 via what climate scientists call projections conditional on particular scenarios about GHG concentrations. Many complex factors have to be modelled to make such projections. This exploratory exercise is an example of perspectival modelling and it deserves a closer philosophical look.

A decisively pluralistic stance has been adopted on climate modelling since the Fourth Assessment Report (AR4) via what are called 'ensemble' methods 'used to explore the uncertainty in climate model simulations that arise from internal variability, boundary conditions, parameter values for a given model structure or structural uncertainty due to different model formulations' (IPCC AR5-WG1: Stocker et al. 2013, p. 754). The reasons for going pluralist are summarized by Reto Knutti et al.:

> [B]ecause of the complexity of the system, the computational cost, and the lack of direct confirmation of prediction, there is no single agreed-on 'best' model.... So while multiple models could be seen as ontologically incompatible (strictly speaking, they make conflicting assumptions about the real world), and one could argue that scientists have to assess how well they are supported by the data, the community seems happy with the model pluralism. The models are seen as complementary in the sense that they are all plausible (although not necessarily equally plausible) representations of the real system given the incomplete knowledge, data, and computational constraints; they are used pragmatically to investigate uncertainties. (2019, p. 840)

This model pluralism has since 2007 taken the form of so-called Coupled Model Intercomparison Projects (CMIP). The task is to bring together modelling groups across the globe and compare how a plurality of models fare vis-à-vis their respective climate projections when using the same forcings. CMIP model the evolution of climate up to the present and offer robust long-term projections taking into account various subsystems of the Earth's system. These include, among others, ocean temperature (which is part of the hydrosphere), sea ice (which is part of the cryosphere), land surface (which is part of the lithosphere), and carbon cycle (which is part of the biosphere).

The 2013 CMIP5, for example, included a number of so-called Atmosphere-Ocean General Circulation Models (AOGCMs) produced by modelling groups in 11 countries. The AOGCMs simulate the complex interactions among the atmosphere, land surface, ocean, and sea ice. A number of more fine-grained higher-resolution Earth System Models (ESMs) supplement AOGCMs by simulating in turn the effects of aerosols, atmospheric chemistry, land carbon, and ocean bio-geochemistry (BGC). Figure 4.b.1 lists Phase 3 models (CMIP3) used for the AR4 and expanded into Phase 5 (CMIP5) for the AR5 in 2013.

The main novelty of AR5 in 2013 compared to AR4 in 2007 is that ESMs included an interactive carbon cycle (i.e. how much CO_2 ocean and land can take up and emit) as well as an interactive representation of aerosols. Time-varying ozone was also included in some models. This was important because it allowed climate scientists to model in real time—with actual emissions as model input—how anthropogenic radiative forcings are affecting the atmosphere–ocean interaction.

More importantly for my purpose here, model intercomparison is an ongoing project. It started in 2007 with the IPCC AR4, continued with the 2013 AR5, and its current instalment is the AR6 in 2021.[3] One of the novelties of the IPCC AR6 is that it integrates the physical basis for global warming with considerations about possible socioeconomic assumptions on mitigation pathways—the so-called Shared Socioeconomic Pathways (SSPs).

Climate models evolve from one CMIP iteration to the next. They come with what Knutti et al. call a 'genealogy': '[M]odels in different centres sharing the same atmospheric model (even in different versions) were also closely related. . . . [M]odels evolve from their ancestors by modification and by exchange of ideas and code with other groups. Successful pieces are kept, improved, and shared, and less successful parts are replaced' (Knutti et al. 2013, p. 1194). Family trees or dendograms highlighting similarities among the models can be produced, showing in detail which model is most similar to which other, and how this similarity is due to their sharing codes, for example, or being an evolved version of an earlier model.

Model genealogy is important for two main reasons. From a philosophical point of view, it shows my general point that *models have a history* of their own. As I emphasized already in my account of nuclear models, perspectival models are not static entities in a one-on-one representation-as-mapping relation with relevant pieces of reality. They are dynamic and evolving, as different epistemic communities collaborate, share ideas, tools, codes, and datasets *over time*. Paying more attention to the *historical* dimension of perspectival modelling is helpful to better understand some methodological challenges. In the case of climate, models that historically share codes or fit to the same datasets are more likely to produce similar projections (something that goes under the name of 'model dependency'). This in turn raises questions about how to weigh the multi-model contributions to climate projections. Should all models be treated equally? (The so-called 'model democracy', in Knutti's words 2010). Or is it better to weigh

---

models differently, to avoid double counting when models share codes or fit to common datasets?

Model genealogy also highlights another important feature of perspectival modelling: namely that any modelling is a remodelling. Models function as inferential blueprints, as I shall explain in Chapter 5: they allow different groups and epistemic communities to work together over time in keeping the successful parts, tweaking, adjusting, and withdrawing the unsuccessful bits. The final output is never the work of one person or a single group, but only the latest instalment/version of often a very long genealogical chain of ancestor models. The evolution of model ensembles from CMIP4 to CMIP5 to the current CMIP6 is a perfect illustration of this historicity of perspectival modelling.

But what has model genealogy or historicity got to do with perspectivism and how perspectivism enters into discussions about model pluralism in climate science? In what follows, I unpack the peculiar (perspectival) kind of model pluralism here. First, I underline the multifactorial nature of the phenomenon 'global warming' and how it in itself is the result of what I call *perspectival* data-to-phenomena inferences that require a plurality of epistemic communities to work together. Such data-to-phenomena inferences are *perspectival* in bringing together a plurality of *scientific perspectives* qua historically and culturally situated scientific practices: from dendroclimatology to palaeoclimatology and glaciology, among many others in this example. This perspectival pluralism is key to improving the *reliability* of the inferences and the *justification* for the reliable knowledge claims advanced about global warming.

Second, I see climate modelling as an example of what I call *perspectival modelling* in being exploratory and delivering knowledge of what is *possible*. As with the first case study, I use the term 'perspectival modelling' in a broad sense and not to refer narrowly to CMIP models *exclusively*. Or better, I see the role of CMIP as an integral part of how a plurality of intersecting scientific perspectives (i.e. the aforementioned broader situated practices) deliver modal knowledge over time. I engage with recent trends in the so-called modal epistemology associated with climate modelling, namely the idea that climate projections represent *possibilities* (Betz 2009, 2015; Katzav 2014; Stainforth et al. 2007a, 2007b). I offer a novel way of thinking about climate possibilities and what is at stake in forecasting climate trajectories conditional on a number of scenarios about GHG concentrations. Climate models illustrate what I call 'physical conceivability' as a guide to possibility: they invite us to *physically conceive* of a number of GHG concentration scenarios so as to deliver knowledge of what *might be the case* about global warming in the year 2100.

4.b.2. Knowing the past global temperature: from tree rings, boreholes, and corals to 'climate signals', and from there to the phenomenon of global warming

It is necessary to establish *that* global warming is occurring, and *how fast* it has occurred since the pre-industrial era, to be able to make robust projections about the future. It is also necessary to raise public awareness about climate change, and to bring politicians around a table and persuade them to act to cut GHG emissions. To achieve these goals, large-scale reconstructions of past temperature have to be produced spanning centuries so as to be able to identifying warming trends.

A key variable in this complex process is so-called climate sensitivity, namely the warming resulting from doubling CO_2 concentrations. Anthropogenic forcings (e.g. GHG emissions, including CO_2 and water vapour produced by tropical deforestation, industrial activities, and extensive agricultural irrigation) affect the energy balance of the Earth (the balance between incoming and out-going radiative energy).

Over the past two decades, a number of global reconstructions of annual surface temperature patterns have been produced stretching back centuries. How do climate scientists go about reconstructing past temperature when instrumental climate data tend to be available only for the last 150 years? How do they quantify climate sensitivity to the extent of being able to make very specific knowledge claims as in the IPCC, for example: 'In the Northern Hemisphere, 1983–2012 was *likely* the warmest 30-year period of the last 1400 years (*medium confidence*)' (Stocker at al. 2013, p. 5, emphasis in original)? To be clear, I am not so much interested here in how to assign confidence levels to specific knowledge claims of this nature. My focus is instead on how scientists collect a great variety of datasets (including tree rings, borehole temperatures, corals, and ice cores) and are able to make *reliable inferences* from these data to the relevant 'climate signals' by filtering noise and confounding factors. Once climate signals have been reliably identified across a number of datasets, the next step consists in disentangling whether the signal is due to natural or anthropogenic forcing and reconstructing from the signals the large-scale global temperature for the planet.

A series of reliable inferences are needed in each step. Inferences from the dataset to the climate signal, which is different for different datasets: that is, the climate signal evinced from the isotopic composition of an ice core is very different from the climate signal emerging from the width and density of tree rings. And then inferences from a plurality of climate signals—laboriously harvested through sampling a variety of datasets across various regions of the globe—to the large-scale reconstruction of the global temperature. The phenomenon of global warming is the outcome of this plurality of data-to-phenomena inferences.

Such inferences are *perspectival* in bringing together a plurality of scientific perspectives, from dendroclimatology to palaeoclimatology and glaciology, among many others. Each scientific perspective does not *just* provide independent evidence for the same phenomenon of global warming. The perspectival lines of inquiry are interwoven, or better 'intersecting'. The geothermometrical data-to-phenomena inferences bear on the dendroclimatological data-to-phenomena inferences. This is necessary for improving reliability and for offering *cross-perspectival justification* of the individual methods adopted in each scientific perspective. For example, the justification for the reliably formed dendroclimatological data-to-phenomena inference cannot entirely lie with methodological-epistemic principles in dendroclimatology. It benefits instead from cross-perspectival justification coming from methodological-epistemic principles from other scientific perspectives such as geothermometry. Let us see why with some examples.

Palaeoclimatologists are able to provide large-scale reconstructions of climate variability over a long period via regional palaeoclimatic reconstructions (e.g. for the Northern Hemisphere). For example, Michael Mann et al. (1998) adopted this approach to offer a reconstruction of the Northern Hemisphere mean annual surface temperature stretching back to the year 1400 CE.[4] Having a representative sample of regional palaeoclimatic proxy data is key to this process and a plurality of 'proxy' climate indicators are typically used in the so-called 'multi-proxy' approach. Mann et al.'s proxies included dendroclimatic data (data from tree rings), corals, sediments, and ice cores, sampled from around the world and feeding in so-called 'multi-proxy networks' (Mann et al. 1998).

Tree rings are particularly interesting datasets with which to identify climate signals as it is possible to date them and measure their width and density. Tree growth is a seasonal/annual phenomenon that is affected by rain and changes in temperature year by year. Thus, the International Tree Ring Data Bank (ITRDB)[5] offers a valuable resource for this kind of climate proxy data. Proxy data need to be carefully chosen and standardized to be reliable indicators (see Bokulich 2018b). Moreover, they need to be calibrated (or 'trained') using instrumental climate data for land, sea, and air surface temperature (where these monthly data are available from meteorological stations around the world at least for the last century).

For the large-scale reconstruction of the Northern Hemisphere temperature to be empirically accurate, annual anomalies must be factored in. For example, there might be particularly cold or particularly warm years due to specific

---

[4] In a follow-up study, Mann et al. (1999) went even further back in time than 1400 CE, despite the sparser proxy data network available.

[5] See https://www.ncdc.noaa.gov/data-access/paleoclimatology-data/datasets/tree-ring

events, such as El Niño currents affecting the tropical west Pacific coast of South America. Correlation statistics were used by Mann et al. (1998) to identify which kind of forcing was responsible for the global warming trend of the last century emerging from this large-scale temperature reconstruction, with GHG emerging as the main culprit for it.

The palaeoclimatic multi-proxy approach is not the only method for reconstructing past global temperature. Indeed, despite the fact that some of the palaeoclimatic proxy data, such as tree rings, are relatively easy to obtain and spatiotemporally rather ubiquitous, they also have shortcomings (for a discussion, see Briffa and Osborn 1999). For example, younger trees and older trees display different patterns of rings (with the young ones having greater width). To reduce biases due to the different ages of the trees in the chosen sample, long tree-ring chronology performs what is called the 'standardization' of the sample. Standardization means removing from a tree-ring series possible trends related to the age of the trees (this is known as 'detrending').

Detrending a tree-ring width series is important for being able to assess the *synchronic growth* of a large number of sampled trees and their sensitivity to temperature changes that might have occurred *at one particular historical moment* (the sought-after 'climate signal'). But detrending also cancels out longer-term low-frequency trends that are in turn important to assess *diachronic growth* over a period and how it might be affected by a sustained trend of temperature changes (for a discussion of detrending in tree-ring chronology, see Sullivan et al. 2016). Moreover, tree growth is affected not just by warmer temperature but also by a higher percentage of nitrogen due to industrial pollution. Thus, the width and density of the tree rings may speak as evidence for increased mean temperature *or* as evidence for anthropogenic pollution. Disentangling them is a non-trivial matter when it comes to identifying the relevant 'climate signal'.

To obviate these specific methodological challenges and offer cross-perspectival justification for the reliability of the data-to-phenomena inferences, another scientific perspective might be deployed: for example, geothermometry (Pollack et al. 1998). Surface temperature changes affect the Earth's crust and propagate in the subsurface over a long period due to the low thermal diffusivity of rocks. Given a change in temperature, 'it takes about 100 years for perturbation to reach a depth of 150m and 1000 years to reach 500m depth' (Huang et al. 2000, p. 756). Measuring the subsurface temperature in boreholes (e.g. 200–600 m underground) in various parts of the world therefore provides another important geothermal dataset to complement paleoclimatic proxies and, ultimately, to identify climate signals over long periods.

In this case too, identifying a genuine climate signal (i.e. a borehole climate perturbation) is like extracting a needle from a haystack. There is background noise in borehole direct geothermal measurements due to idiosyncratic topographic features of each borehole (e.g. the nearby presence of water and plants). A database of borehole temperatures typically includes hundreds of samples across Northern and Southern hemispheres.

Geothermal data are very different in nature from palaeoclimatic multiproxy data in being direct temperature measurements. The theoretical-modelling basis is in geothermal physics. There are also significant methodological differences in the step that goes from the identification of relevant climate signals to the large-scale reconstructions of surface temperature. For example, Henry Pollack et al. (1998) made use of Bayesianism, by contrast with the regression method used by Mann et al. (1998) in palaeoclimatology. Bayesian analysis calculated the probability of the null hypothesis (i.e. no climatic variation) conditional on the evidence found from geothermal data. In this case too, the result indicated that global warming had occurred and that in the twentieth century the mean temperature went up by 0.5 ± 0.1 Celsius (Pollack et al. 1998, p. 280).

Direct measurements such as borehole geothermal data are not necessarily a better way of cross-perspectivally validating indirect measurements such as palaeoclimatic proxy data. In other situations, exactly the opposite happens. Consider a different data-to-phenomena inference, this time about ocean heat content as another powerful 'climate signal'. To measure ocean heat content, climate scientists today resort to direct measurements via, for example, ca. 3,800 free-drifting Argo floats,[6] among other types of probes. Argo floats monitor long-term warming trends in the upper volume (2,000 m deep) of oceans around the world and feed into computer models that allow measurements of the upper-surface ocean temperature over time.

However, to reconstruct past surface ocean variability, instrumental datasets like those from Argo floats are not sufficient. What is needed in addition are palaeoclimatic records stretching back a few centuries. Corals have provided important palaeoproxy data in this respect, because they can live for hundreds of years and have a fast annual growth rate that makes it possible to record ocean temperature changes over long periods of time. Fossil corals with their oxygen isotopic composition can offer detailed records of seasonal changes during the Holocene (the last 10,000 years). These records are particularly useful in the absence of instrumental data for past ocean temperature and also offer a more fine-grained picture of localized complex ocean cycles over multi-decadal timescales. For example, fossil corals from the Seychelles and other locations have been used to reconstruct the effects of El Niño on the Western and Central Indian Ocean (see Zinke et al. 2005).

Ice cores are a further source of data for reconstructing large-scale climate variability. The cryosphere includes Artic and Antarctic ice sheets, glaciers around

---

[6] See http://www.argo.ucsd.edu/

the world, but also snow, permafrost, sea ice, river ice, and lake ice. By drilling million-years-old ice cores (as much as 3–4 km thick) to extract ice samples, it is possible to analyse the ratio of stable oxygen isotopes (^{16}O and ^{18}O) over different epochs. This provides important information about climatic change over time.

As these examples indicate, establishing *that* global warming is occurring, and *how fast* it has occurred, involves identifying a number of clear and unequivocal 'climate signals'. These signals are to be found in the isotopic compositions of ice cores no less than in the width and density of tree rings; in the oxygen isotopes in fossil corals no less than in the subsurface rock temperatures from borehole data. Perspectival data-to-phenomena inferences are at work behind the phenomenon of global warming and its multifactorial nature, in three main ways:

1. there is a *multiplicity of factors* (e.g. sea, air, and land surface temperature, ice melt) that *jointly* answer the question how global surface mean temperature has changed;

2. these factors are each *indexed at a particular Earth's sphere* (hydrosphere, atmosphere, lithosphere, cryosphere, and biosphere);

3. they show variability *over time* from the so-called Little Ice Age (i.e. from the fourteenth to the mid-nineteenth century) to the industrial era, up to the present day.

The multifactorial nature of the phenomenon 'global warming' requires a plurality of perspectival data-to-phenomena inferences to be reliably established. But again, do not think of these inferences as perspectival$_1$ representations— they do not represent global warming as x when seen from the point of view of the sea surface temperature, or as y as seen from the glaciers, or as z when seen from land surface and dendrochronology. Each is perspectival$_2$ instead, in contributing to open a 'window on reality'. Jointly these perspectival data-to-phenomena inferences establish that the multifactorial phenomenon 'global warming' is real, it is occurring, and has been accelerating in the past 30 years. To establish the reality of this phenomenon, as we have just seen, it is necessary to identify a number of specific climate signals (e.g. ocean heat content, melting of glaciers, land surface temperature increase, among others).

Identifying each of these signals as 'stable events' involves an exceedingly subtle series of data collections and analyses. The stability of each climate signal is often underwritten by lawlike dependencies. For example, the stability of the climate signal identified in borehole data depends on a new equilibrium in the geothermal temperature gradient after the change in temperature at the surface has diffused through the underground rocks. Or, to give another example, it is the lawlike dependency between the ratio of different isotopes in the wood of each tree ring, and the rate at which photosynthesis must have taken place, which underpins the stability of the climate signal to be found in tree rings and how it reliably provides evidence for hot past summers or cold past winters.

To conclude, in order to provide reliable evidence that the multifactorial phenomenon 'global warming' is real, a plurality of lines of evidence and distinctively perspectival data-to-phenomena inferences are required. But climate scientists are not just interested in establishing *that* global warming is occurring, and *how fast*. They are also interested in investigating *how quickly* the phenomenon *could* continue to occur in the future and offer reliable evidence for the potentially devastating consequences. In other words, the phenomenon is not just happening. It *could* robustly continue to happen under *a range of conceivable future* GHG scenarios (from mild to more severe), unless interventions to cut GHG emissions are put in place. In my philosophical lingo, the phenomenon is *modally robust*. To explore the modal robustness of global warming, the IPCC put in place the distinctive model pluralism of CMIP5, to which I return next.

4.b.3. Future climate projections: RCPs, possibilities, and the role of perspectival modelling

Global warming *could* exceed 2°C over the next 100 years unless the GHG emission rate gets cut significantly by 2050. The Northern Hemisphere sea ice sheet *could* decrease dramatically by 2050 (see Figure 4.b.2). These are scientific knowledge claims that are modal in flavour. They are delivered by model ensembles such as those at play in CMIP5, which generate long-term projections. To make projections, the IPCC adopts 'ensemble approaches for model evaluation'. Philosophers of science have discussed and interpreted climate model evaluation in different ways.

For example, Elisabeth Lloyd (2010, 2015) understands model evaluation in terms of fit between the simulation offered by the climate models and the actual climate system. The model pluralism of CMIP5 is seen in this case as central for delivering *robust* projections (building on the notion of 'robustness' in Levins 1966; Weisberg 2006; Wimsatt 1981). If most models in an ensemble give the same (or roughly the same) outcome for the same computer simulation, the model projection can be regarded as 'robust'. On this view, model pluralism and inter-model comparison are important to evaluate and confirm the model outcome, and to produce robust projections such as 'Global warming *could* exceed 2°C max increase over the next 100 years'. Benedikt Knüsel and Christoph Baumberger (2020) have further argued that robustness in the agreement between, say, CMIP5 models and the energy-balance model in turn gives confidence in the 'representational accuracy' of the energy-balance model.

Other philosophers like Wendy Parker (2011) have downplayed inter-model agreement and robustness. Parker argues that robust model predictions neither increase scientists' confidence in the model outcomes nor provide evidence for the hypothesis about future climate change. Models often share common assumptions and even coding. Model robustness is not tantamount to model trustworthiness, Parker maintains: instead, climate models should be evaluated for their adequacy-for-purpose (Parker 2009; see Alexandrova 2010 for a discussion; Bokulich and Parker 2021; for a discussion of climate modelling see also Frigg, Thompson and Werndl 2015a and 2015b). Climate scientist Reto Knutti agrees: models 'represent reality well enough for a particular purpose. . . . Model evaluation for long-term climate predictions cannot be based on repeated confirmation of the predictions against observation-based data' (Knutti et al. 2019, p. 836). Along similar lines, Johannes Lenhard and Eric Winsberg (2010) see climate models as affected by a confirmational holism that would make the model pluralism endemic and not a transient phase towards some future convergence of results.

Even the best CMIP5 computer simulations have a spread in the projection of particular future events. Usually the multi-model mean in an ensemble is taken as the best estimate projection (with margins of uncertainty and confidence levels clearly stated—see, e.g., mean and associated uncertainty for all RCP scenarios as coloured vertical bars in Figure 4.b.2). The IPCC AR5 selected four possible Representative Concentration Pathways (RCPs)—RCP2.6, RCP4.5, RCP6.0, and RCP8.5—at an IPCC meeting in September 2007 following criteria established by the community of climate scientists.[7] RCP2.6 assumes GHG concentration peaks at mid-century and declines before 2100. At the high end, RCP8.5 envisages GHG concentration to rise steadily throughout the twenty-first century.[8]

The four RCPs represent four different ways in which one can imagine the balance between incoming and outgoing radiation being affected by changes in the concentration of GHGs over the next century. Each might be thought of as

---

[7] These included 'compatibility "with the full range of stabilization, mitigation, and reference emissions scenarios available in the current scientific literature" (...); a manageable and even number of scenarios (to avoid the inclination with an odd number of cases to select the central case as the "best estimate"); an adequate separation of the radiative forcing pathways in the long term in order to provide distinguishable forcing pathways for the climate models; and the availability of model outputs for all relevant forcing agents and land use. . . . The selection process relied on previous assessment of the literature conducted by IPCC Working Group III during development of the Fourth Assessment Report. Of the 324 scenarios considered, 32 met the selection criteria and were able to provide data in the required format. An individual scenario was then selected for each RCP' (Moss et al. 2010, p. 753).

[8] RCP8.5 has been at the centre of a debate in more recent times, which goes beyond my remit and scope to discuss here (see Hausfather and Peters 2020a, 2020b; Schwalm et al. 2020a, 2020b).

representing *conceivable concentrations* (rather than actual real-time emissions) of GHGs for the twenty-first century. They provide a set of different boundary conditions which are then used in climate modelling to simulate a range of long-term projections for some of the main markers of global warming.

Climate scientists model the relevant historical evolution (black line with grey shading in Figure 4.b.2) using historically reconstructed forcings from the pre-industrial age up to the present (ca. 1950–2005); and make a series of long-term projections that diverge depending on which RCP is factored in. For example, Figure 4.b.2 shows the time series of projections (and related uncertainty spread via shading) for GHG scenarios RCP2.6 (blue) and RCP8.5 (red). The coloured vertical bars for each RCP give the mean and associated uncertainties. The number of models used to calculate the multi-model mean (blue or red solid lines, respectively) is indicated in each case: e.g. in Figure 4.b.2(a), 32 models of the CIMP5 ensemble entered in the projection of the multi-model mean for the global average surface temperature (and related uncertainty range) associated with RCP2.6 and 39 for the RCP8.5.

Uncertainty in climate modelling crops up at three different levels: (1) in the very nature of the physical processes involved; (2) in the different conceivable GHG scenarios (called 'scenario uncertainty' under different RCPs) that depend on political choices, human behaviour, and socioeconomic factors at play in different countries; and (3) in the model itself with the particular choice of parameters, how they are calibrated to match large-scale observations, and so on. It is precisely with an eye to getting a handle on these three different sources of uncertainty that climate scientists embraced model pluralism a long time ago. Philosophers have dedicated their attention mostly to (1) and (3). In what follows, I concentrate on (2): 'scenario uncertainty'.

I see model pluralism in CMIP5 as an example of perspectival modelling in being distinctively exploratory: it delivers knowledge of what *might happen* under different GHG concentration scenarios (RCPs). The idea that ensemble models are a guide to possibility (rather than, say, to a probability distribution) is not new. Among philosophers of science, this has sometimes been called the 'possibilist view' (see Katzav 2014). The spread in the projections (as in Figure 4.b.2) can be interpreted as representing the range of future possibilities that cannot be discounted. Joel Katzav, for example, has argued that climate models do not just make predictions but 'describe real possibilities and . . . determine how remote the described possibilities are . . . which real possibilities obtain is a time-relative issue' (Katzav 2014, p. 236). He defines real possibilities relative to some time *t* in terms of states of affairs in a target domain whose '(a) . . . realisation is compatible with the basic way things are in the target domain over the period during which it might be realised and (b) our knowledge at *t* does not exclude its realisation over that period' (p. 236). Accordingly, he sees the function of climate models as that of showing 'that certain states of affairs are compatible with the basic way the climate system is over relevant periods of time' (p. 236).[9]

Historically, a drive towards the 'possibilist view' has been climate scientists's need to interpret climate models in a way that can inform policy, knowing all too well that while 'In the context of constant boundary conditions, and specifically no changes in atmospheric GHGs, . . . weather is chaotic. . . . Under changing concentrations of atmospheric GHGs, the behaviour is not chaotic but pandemonium (Spiegel 1987)', as D. A. Stainforth et al. (2007a, p. 2147) aptly characterized it. Communicating climate forecasts to policymakers and the public based on conceivable future scenarios of GHGs has its challenges. This kind of 'pandemonium' uncertainty cannot be eliminated. However, it is in some ways under 'our' control. Stainforth et al. compare it with more familiar varieties of uncertainties:

> The likelihood of drowning is low in the shower, higher if we choose to swim in a shallow children's swimming pool, higher still in an adult pool and even higher along a beach with a strong undertow. Given that the anthropogenic GHG emissions are considered to be the most significant drivers of changes in climatic forcing in the twenty-first century . . . , the future is therefore in 'our' control in the sense that we can choose if and where to swim. There is therefore no need to remove this uncertainty so long as reliable information can be given for the outcome of any particular choice. (Stainforth et al. 2007a, pp. 2149–2150)

Climate scientists continue to debate how to offer reliable information for the projected outcome of any particular choice (i.e. any particular RCP scenario). Multi-model ensembles have been regarded as exploratory models giving a wide range of possibilities fixing a 'lower bound on the maximum range of uncertainty. . . . The range of possibilities highlighted for future climate at all scales clearly demonstrates the urgency for climate change mitigation measures and provides non-discountable ranges which can be used by the impacts community' (Stainforth et al. 2007a, pp. 2155 and 2159).

Modelling the rate of climate change in the next 50–100 years requires canvassing a number of future scenarios of how CO_2 and other emissions might get capped and reduced over that time. And that depends on socioeconomic developments at regional and global scale, whose responsibility lies in the hands of policymakers.

---

[9] Gregor Betz (2015) has further discussed what he calls the *possibilistic challenge*. Building on Robert Sugden's (2009) and Till Grüne-Yanoff's (2009) work on modal modelling in economics, Betz explores the notion of 'credible worlds', the 'verified possibilistic hypothesis', and the 'falsified possibilistic hypothesis' to argue that climate models might verify possibilities despite idealizations.

Thus, in this take on climate modelling, model pluralism is endemic because of its exploratory nature. The purpose of multi-model ensembles is not so much to predict the future; or to reach an inter-model agreement and seek convergence in the models' outcomes. It is to explore instead the scenario uncertainties and associated range of possibilities so as to inform and reach effective policy decisions *now*. The point of displaying climate projections under a wide range of RCPs is to drive home that parts of the range of scenarios are exceptionally undesirable, should they come to pass, and therefore to offer model-based evidence to policymakers to ensure actions are taken *now* so that certain possible futures never eventuate. In the words of Moss et al. (2010, p. 747): 'The goal of working with scenarios is not to predict the future, but to better understand uncertainties in order to reach decisions that are robust under a wide range of possible futures'.

Each RCP specifies what I'd like to call a *conceivable* future concentration trajectory, rather than a real-time emission input. This is because each future concentration pathway can indeed be the result of more than one possible socioeconomic future scenario (and associated emissions scenarios). Why is this an example of perspectival modelling as modelling what is possible? While agreeing with Katzav and Betz on the need to pay attention to the role of modality in this modelling exercise, I part company with them for a number of reasons. I briefly outline some of them here; they will become clearer in Chapter 5.

I see the possibilities involved here neither in terms of *compatibility* of some *states of affairs* with the basic way a climate system is over a period of time (as in Katzav's 'real possibility'); nor in terms of *consistency* of some *statement P* (about future climate) with the relevant background knowledge *K* (as in Betz's notion of 'serious possibility'). I instead take ensemble models as an example of perspectival models: they invite us to *physically conceive* of a certain number of GHG concentration scenarios so as to deliver *modal knowledge* on what the future climate might be like. To be clear, all models in the CMIP5 run under all GHG scenarios according to highly standardised procedures shared across all the modelling groups. While each model gives information about all GHG scenarios, the extent to which their respective projections may converge is affected by many other aforementioned uncertainties entering climate modelling,

RCPs are an invitation to imagine certain GHG concentrations as the conditions under which the CMPI5 models deliver knowledge of what the phenomenon global warming *might be like* in the year 2100. Introducing a range of RCPs is a way of saying:

'Let us conceive that GHG concentration peaks by mid-century and decline before 2100 (as with RCP2.6). What might global warming look like in 2100?';

or:

'Let us conceive that GHG concentration rises steadily throughout the twenty-first century (as with RCP8.5). What might global warming look like in 2100?'.

To be *physically conceivable*, as I use the term here, the GHG concentration scenarios must comply with the state of knowledge and conceptual resources of the climate science community at time *t*, but they must also be consistent with the laws of nature known by the community at time *t*. Thus, a hypothetical GHG scenario that, say, unrealistically assumes a peak in 2023 and decline afterwards—while not violating any law of nature—would not count as physically conceivable. For it would not comply with the factual information known to climate scientists *as of today* (year 2021). Similarly, a hypothetical GHG scenario that assumes a steady rise in GHG escaping the troposphere and dissipating in outer space would not be physically conceivable as it would not be consistent with known laws of nature about the absorption and re-radiation at work in the greenhouse effect.

I see the relation between the physically conceivable RCP scenarios and the modal knowledge delivered by CMPI5 ensemble models as a *time-sensitive epistemic accessibility relation*. At any instant of time *t*, given the past historical evolution up to the present (black line with grey shading in Figure 4.b.2), the future climate ramifies and branches into a range of possible trajectories (e.g. the blue and red trajectories in Figure 4.b.2). Each of these trajectories is associated with a variety of modal knowledge claims about future climate, such as the following two emerging from Figure 4.b.2(b):

'the Northern Hemisphere sea ice extent could shrink to 2.50 (10^6 km²) by 2050' (under RCP2.6)

or:

'the Northern Hemisphere sea ice extent could shrink to 1 (10^6 km²) by 2050' (under RCP8.5).

Not all these possible future trajectories for the Northern Hemisphere sea ice extent will materialize as time goes by. But the purpose of CMPI5 models is not to predict the most likely ones, but to offer to policymakers the full array of possible futures. By taking action now to reduce GHG emissions, the international community has the power to eschew some of these possible future (extreme) trajectories and mitigate the severe ones. In other words, by taking action now and cutting GHG concentrations, as one moves from t_1 to t_2 to t_3 and onwards, the range of possible future trajectories changes too. Some previously possible trajectories might get ruled out and new ones might become live options, depending

on the extent to which international interventions on land use, pollution, deforestation, irrigation, farming, and crops can cut GHG emissions.

Year-by-year trends in GHG emissions allow climate scientists to revise in real
time the range of possible future trajectories. What at time t_1 was a modal claim
about, for example, loss of sea ice or a particularly warm year due to El Niño or
similar becomes at t_2 embedded into the state of knowledge of the climate science community. And at t_2 a new run of climate model simulations with updated
background information about GHG give epistemic access to another range of
refined modal claims. The process of projecting future global warming is itera-
tive and dynamic; it is responsive to real-time changes in anthropogenic forcings.

To conclude, global warming is a modally robust phenomenon. Not only *can*
it be robustly identified through a variety of perspectival data-to-phenomena
inferences, but it *can* also robustly continue to occur under a number of physically conceivable GHG concentration scenarios. Climate scientists identify and
study the complex multifactorial phenomenon of 'global warming' out of a vast
inferential space of what is *causally possible* thanks to the ability of perspectival
models to facilitate the network of inferences surrounding a plurality of climate
signals across the cryosphere, the atmosphere, the hydrosphere, the lithosphere,
and the biosphere.

But climate modelling is also perspectival modelling in a different sense: it
provides scientists and policymakers with evidence-based tools to monitor in almost real time how the branching future projections might change in response to
international efforts to curb anthropogenic radiative forcing. This is what multi-
model ensembles have achieved at IPCC AR5-WG1 and continue to achieve
in the current Sixth Assessment Report. Perspectival modelling allows climate
scientists to explore how global warming might pan out under different conceivable RCPs (and SSPs in AR6) so as to offer recommendations to policymakers
around the globe about which future branches we must avoid at all costs.

4.c

A tale from the development of language in children

4.c.1. Four scientific perspectives on dyslexia: behavioural, educational, neurobiological, and developmental

In the early 1970s, 10-year-old children on the Isle of Wight (UK) were part of a study in developmental psychology. The study was intended to measure 'under-achievement', defined as 'the ratio between the child's mental age and his achieve-ment age' (Rutter and Yule 1975, p. 183). It was common in the 1970s to use IQ tests as a measure of innate ability and a predictor of literacy. Underachievement was identified as a discrepancy between verbal and non-verbal IQ tests and age-appropriate reading and literacy attainment. The distribution of the latter in a population was assumed to be normal, with a Gaussian curve and under- and overachievers at each end.

The Isle of Wight study questioned the reliability of this method, which gave rise to 'misleading statistics'. It overestimated the number of underachievers in children with high IQ but underestimated it in other cases (Rutter and Yule 1975, p. 183; see also Yule et al. 1974).[1] The study went further in comparing groups of 10-year-old children on the Isle of Wight and in London. Comparing data about reading accuracy and comprehension, Rutter and Yule were hoping to find a statistically significant distinction between children with specific reading dif-ficulties and what might be called the 'garden variety' of children experiencing reading difficulties. They were looking for a 'hump' in the lower-end distribution of learners: a specific group of underachievers with specific reading difficulties.

The 'hump' was indeed evident from the data. But, crucially, the study was not able to establish any specific neurological pattern responsible for the 'hump' and the group it identified. Children with specific reading difficulties displayed no 'overt neurological disorder' and had 'delays in the development of speech and

---

[1] '10-yr-old children with a mental age of 9 yr should have an average attainment age of 9 yr and 10-yr-olds with a mental age of 13 yr should have an average attainment age of 13 yr. *But, neither in theory nor in practice, does this happen.* In fact, the mean reading age of 10-yr-olds with an average mental age of 13 yr will *not* be 13 yr, it will be more like 12 yr. Only in the middle of the distribution will the two be the same. The reason for this occurrence lies in the "regression effect"' (Rutter and Yule 1975, p. 183, emphasis in original).

Perspectival Realism. Michela Massimi, Oxford University Press. © Oxford University Press 2022.
DOI: 10.1093/oso/9780197555620.003.0007

language . . . *no* more frequent [than] in those from families of very low social status' (Rutter and Yule 1975, p. 190, emphasis in original). From the behavioural point of view, the study concluded that, yes, there was a statistically identifiable group of learners within the lower end of the normal distribution; but it was unable to identify any meaningful pattern behind it. The authors declared that there was 'no evidence for the validity of a single special syndrome of dyslexia. . . . Some kind of biological "marker" would be needed and so far none has been found' (p. 194).

At the behavioural level, traditionally dyslexia has been identified with an 'unexpected' gap between verbal and non-verbal IQ, on the one hand, and reading and literacy skills, on the other. Although there are clearly identifiable symptoms available from Wechsler tests (e.g. low scores on the Digit Span subtest) to help with diagnosis, this discrepancy approach to defining dyslexia has been heavily criticized since the Isle of Wight study (see Elliott and Grigorenko 2014, and Siegel 1992). First, IQ does not correlate with the specific subskills involved in reading and writing (e.g. phonological awareness, or word recognition—see Stanovich 2005). Second, discrepancies of this nature are developmental and tend to change depending on whether the tests take place at the age of, say, 7, 10, or 13. Third, the threshold for the discrepancy to count as 'unexpected' has to be set high to identify children with reading difficulties. But in so doing it excludes many children who might also experience reading difficulties.

Reliance on the 'unexpected discrepancy' approach highlights also the socioeconomic inequalities in access to early diagnosis and support for dyslexia. As the historian Philip Kirby underlines, '[D]yslexia (then as now) was being diagnosed in higher proportions in children from wealthier socio-economic groups. Differential access to dyslexia specialists and their tests was a reason for this, sparking accusations that dyslexia was curiously prevalent in Surrey. . . . Parents with higher educational levels were also more likely to be aware of the condition, and earlier' (Kirby 2018, p. 58). The 2019 All-Party Parliamentary Group report on dyslexia has underlined the ongoing high costs (over £1,000 extra per year) for families supporting children with dyslexia, which once again point to socioeconomic disparities in the ability to offer timely diagnoses and interventions available to all children (Hodgson 2019).

The Isle of Wight study raised awareness about the necessity for remedial education tailored to the specific needs of different groups of children with reading difficulties (see Vellutino et al. 1996). Later studies by educationalists like Marie Clay (1987) in New Zealand showed how the attainment gap for 7-year-old underachievers could be significantly improved by intensive weekly remedial teaching. And the trend among educationalists continues with some researchers studying what is now called 'child characteristic-by-instruction (C-I) interactions' (see Connor 2010). The idea is that 'the effect of literacy instruction strategies does indeed appear to depend on students' characteristics' (Connor 2010, p. 256).

New longitudinal studies and randomized control trials have helped in iden- tifying possible relevant C-I interactions. Students with word reading difficulties benefit from teaching that emphasizes word recognition (see Juel and Minden- Cupp 2000). Moreover, reading skills seem to improve visibly in teacher–child- managed instruction rather than child-managed instruction settings (see Connor et al. 2004b), preferably if the teacher reads to a small group rather than to the entire class (see Connor 2010, p. 259).

While these studies survey ways of catering for the educational needs associ- ated with dyslexia, they have also highlighted a methodological gap in teachers' training. A recent online survey of teachers in England and Wales revealed that the majority of them (79.5%) mentioned behavioural descriptors and 'visual stress' rather than cognitive descriptors such as 'phonological awareness deficit' (39.3%). This imbalance suggests 'a "stereotypical" view of dyslexia' as mainly attributed to the 'singular category of the behavioural level' (see Knight 2017, pp. 216 and 211). Even more problematic is the association with visual stress 'de- spite research being inconclusive about this relationship' (p. 216).

This gap between educational studies and the reality of classroom teaching shows the risks of a one-sided (mostly behavioural) understanding of dyslexia. It may prevent schools from identifying children with special educational needs and disabilities (SEND) at an early age. It may also reinforce socioeconomic inequalities in early diagnosis. The Children and Families Act 2014 in the UK has allowed parents to request funding from local councils to cater to the special needs of their children in specialized private schools. Yet there is still a long way to go in accessing timely diagnoses and interventions for children from socially disadvantaged backgrounds.

Going beyond behavioural and educational-psychological studies, significant research has also been conducted on the neurobiological basis of dyslexia. The failure to identify a 'single special syndrome of dyslexia' in the Isle of Wight study did not deter neuroscientists from looking into the possible neurobiological mechanisms behind it. And they suggest today that dyslexia is a neuro-develop- mental disorder of genetic origin with a neurobiological basis (see Frith 2002b, p. 51).

Neuroimaging studies using data from CT and fMRI scans started in the late 1970s. The goal was to find possible patterns of cerebral asymmetry or sym- metry that could be related to language development. These studies (as well as post-mortem anatomical studies) have revealed significant differences in areas of brain activation for patients with and without dyslexia, with the former some- times showing a more prominent activation of the right side of the brain (see Maisog et al. 2008; Richlan et al. 2009). Other studies using PET scans have related phonological short-term memory tasks to the concerted activation of the relevant areas of the brain (especially Broca's area, involved in segmented phonology, and the superior temporal and inferior parietal cortex—see Paulesu et al. 1996).

One fMRI study (Olulade et al. 2013) has suggested that dyslexia is associated with a deficit in the magnocellular system, which is involved in human vision and the ability to detect edges, positions, and orientations of objects. This fits with the suggestion that decoding difficulties might be the product of a magnocellular deficit. But this hypothesis is just one among others at the same neurobiological level. Among them, the so-called cerebellar abnormality traces decoding difficulties back to a disconnection between right and left hemispheres.

Yet an explanatory gap inevitably remains between neurobiology and observed behaviour. These studies have enhanced the understanding of the possible neurobiological basis for dyslexia, but no genetic test can be performed as of today to secure an early diagnosis. Nor has a genetic marker for dyslexia been found. Things get more complicated. Not only is there no genotype for dyslexia. There is no phenotype either.

Indeed, the absence of a phenotype was a significant stumbling block in early neuroimaging studies, which often assumed the existence of a 'dyslexic phenotype' vis-à-vis a control group. R. H. Haslam et al. (1981) criticized early neuroimaging studies for 'questionable dyslexic subtyping typologies in examining for possible interactions between subtypes and brain asymmetry' (Hynd and Semrud-Clikeman 1989, p. 463). Further problems arose from the choice of control groups. Doubts were raised about how representative of the typical population the control groups might be given the available psychometric data (Hynd and Semrud-Clikeman 1989, p. 449).

This was an instance of Simpson's paradox: namely, in order to find statistically relevant differences in brain morphology, one needs to know already who is dyslexic and who is not to partition the groups correctly. This in turn would require some control over a rather complex and wide-ranging set of behavioural, neurological, and cognitive variables that might lead to reliably identifying a prototypical control group. However, the authors of the study concluded, often 'one must accept on faith the notion that these control subjects were indeed free of other behavioural, neurological or psychiatric disorders' (Hynd and Semrud-Clikeman 1989, p. 449).

In addition to behavioural descriptions, educational studies, and neurobiological research, environment and culture play their role in understanding dyslexia. Learning how to read and write is an artificial skill that human beings *acquire over time*, across different cultures and languages, with huge variations among them. It takes years for any child to master. Developmental psychologists have been studying the stages in this process as a way of pinning down key junctures at which setbacks might take place. Expecting to understand dyslexia in light of a single one-size-fits-all approach cannot do justice to the great variety of cases.

Uta Frith at the University College London Institute of Cognitive Neurosciences has been a pioneer in the study of dyslexia and other neurodevelopmental disorders such as autism. Very early on, at a time when dyslexia was still being studied and understood primarily in terms of an information processing model—how fast the brain can process, store, and retrieve information concerning how letters represent phonemes—she pointed out the need to pay more attention to developmental change. She has identified four developmental stages—symbolic, logographic, alphabetic, and orthographic—that need to be mastered for a child to become fully literate (Frith 1986).

At the symbolic and then logographic stage, children acquire the ability to recognize symbols and then words on the basis of some salient graphic feature: for example, a child might be able to recognize the word 'McDonald's' from the yellow M symbol. At the alphabetic stage, the child goes beyond symbols, associates letters with sounds, and blends sounds into words. This is the most demanding stage, which usually children acquire over a period of time at the start of primary school. And the degree of automaticity and fluency in blending sounds varies considerably from one natural language to another.

In transparent languages like Italian, for example, where the association between letter and sound is fairly stable and there is not much variability in the pronunciation of the same sounds, children on average acquire this skill by the end of primary one. But in a non-transparent language like English, where the same letter (say, the letter *a*) is associated with different sounds depending on the word it is in (think of the sound *a* in the two words: n*a*ture vs n*a*tural), acquiring such a skill takes on average two years (longer in the case of dyslexia, see Frith 2002a).

This has led to some statistically surprising results as the prevalence of dyslexia (measured in behavioural terms by the aforementioned 'unexpected discrepancy') has been estimated to be half in Italy what it is in the United States, for example (see Lindgren et al. 1985). Studies of Italian-speaking and English-speaking dyslexics with carefully chosen control groups—sharing the same age, levels of tertiary education and so on—have revealed that the former tend to perform better than the latter in tests, even if they perform in fact as the English-speaking ones compared with their respective control groups. At the neurological level, PET scans reveal similarly reduced areas of activation for the language-related left hemisphere (Broca's area and Wernicke's area) in both the Italian-speaking and the English-speaking groups (see Paulesu et al. 2001). Studies like this have corroborated the view concerning the common neurobiological basis of dyslexia while also drawing attention to the remarkable variability in its manifestation across different natural languages (and associated degrees of compensatory strategies available in each one).

In the final orthographic stage, the child instantly recognizes morphemic parts of words or the whole word without the need for blending. The child who experiences difficulties with phonological awareness and sound blending at the alphabetic stage may adapt and compensate for the deficit by overdeveloping something similar to the orthographic strategy (e.g. guessing words from the initial morphemes). Children with dysgraphia, on the other hand, tend to master the alphabetic phase and produce accurate blending but find the orthographic phase more challenging. A supporting strategy that might work for dyslexia might therefore not work for dysgraphia. And a remedial learning strategy for Italian-speaking children may not necessarily work for English-speaking children.

The debate on the nature and definition of dyslexia remains highly contentious among specialists, parents, teachers, and policymakers (see, e.g., Elliott and Grigorenko 2014, Ch. 1).[2] Dyslexia is a life-long condition with early onset in pre-school or school years. There have been historical difficulties with defining dyslexia, with the term often used as an umbrella to describe a range of symptoms from learning disability to specific reading difficulties in relation to fluency, automaticity, and spelling accuracy. The *DSM-5* (*Diagnostic and Statistical Manual of Mental Disorders*), for example, treats dyslexia as part of a larger family of 'specific learning disabilities' (SLD).

A study for the British Dyslexia Association (Crisfield 1996) estimated that up to 10% of the population might have symptoms of dyslexia, ranging from mild to severe; while the National Institute of Child Health and Development (2007) estimated that up to 20% of the US population has some kind of language-based disability (for this and other statistics, see Elliott and Grigorenko 2014, p. 32). The figure of 20% appeared also in the Connecticut Longitudinal Study of children from kindergarten to secondary school run by Sally Shaywitz and colleagues at the Yale Centre for the Study of Learning and Attention (Shaywitz et al. 1999). And the aforementioned 2019 All-Party Parliamentary Group report (Hodgson 2019) gives a figure of 10%–15% for the UK population (i.e. affecting an estimated 6.6 to 9.9 million people, including up to 1.3 million of young people in education).

The absence of a clear cut-off point in these statistics shows the complexity of understanding the multifactorial nature of dyslexia and the challenge of timely diagnosis and effective educational interventions. A child undiagnosed during primary school—maybe because the symptoms are read as 'laziness' or 'daydreaming' or 'inattentiveness'—is likely to be adversely affected in secondary

---

[2] See this article in *The Guardian* for a taste of the ongoing controversy and ramifications in educational policy: https://www.theguardian.com/news/2020/sep/17/battle-over-dyslexia-warwickshire-staffordshire?CMP=Share_iOSApp_Other.

and higher education. This is a reminder of the importance of securing timely diagnosis, and school support for these children and their educational needs.

The philosophically interesting question on which cognitive psychologists, educationalists, neurobiologists, and developmental psychologists focus today is *not*, then, *whether* there is dyslexia,[3] or 'what dyslexia *really* is' (Morton 2004, p. 162, emphasis in original). As the 2012 Dyslexia Action Report says of the condition, '[T]here is no longer controversy about whether it exists and how to define it' (Dyslexia Action 2012, p. 7). The debate is on how to identify the symptoms for individual children, offer timely diagnoses, and put in place effective interventions so as to give better educational prospects.

4.c.2. The phenomenon of 'difficulties with reading' and perspectival data-to-phenomena inferences

Cognitive psychologists, educationalists, neurobiologists, and developmental psychologists face the need to understand the behavioural phenomenon of 'difficulties with reading'.[4]

That a child might experience difficulties with learning how to read is relatively easy to spot. Much more difficult is to ascertain whether the difficulties with reading are the tail end of a normal distribution or the symptom of a lifelong condition such as dyslexia. A number of data-to-phenomena inferences are required to tease out these very different conclusions. These data-to-phenomena inferences are perspectival in the same ways exemplified by my other case studies: (1) the data in each case are sourced from experimental, theoretical, and technological resources available to distinct epistemic communities to *reliably* advance their knowledge claims; and (2) the methodological-epistemic principles at play to *justify* the reliability of their knowledge claims also pertain to distinct epistemic communities.

For example, the educationalists' data may include reading and comprehension tests from sampled pupils of different ages and geographical locations. These statistical data may be used to identify school attainment gaps and underperformances in the student population (including the phenomenon

---

[3] Even an account (such as Elliott and Grigorenko's) that treats dyslexia as a 'construct' acknowledges that 'the primary issue is not whether biologically based reading difficulties exist (the answer is an unequivocal "yes"), but rather how we should best understand and address the literacy problems across clinical, educational, occupational and social policy contexts' (Elliott and Grigorenko 2014, p. 4).

[4] This phenomenon is often referred to in the psychological literature as 'poor reading' (see, among many others, Carroll et al. 2016; Lobier and Valdois 2015; Nation and Snowling 1998). I have chosen, however, to use the term 'difficulties with reading' here because it does not have infelicitous connotations.

of 'difficulties with reading'). Educationalists use resources at their disposal to monitor (often through longitudinal studies) how effective particular remedial strategies might be (e.g. teacher–child-managed instructions).

Neurobiologists use data from CT/fMRI/PET scans, as well as post-mortem anatomical studies, as evidence for a range of other phenomena such as symmetry/asymmetry in brain morphology, lesions in the brain, possible abnormalities in the cerebellum, or possible magnocellular dysfunction. This neurobiological evidence is in turn used to infer the possible comorbidity of the phenomenon of difficulties with reading with other phenomena such as difficulties with motor development, or difficulties with motion detection. 'Difficulties with reading' in this case is part of a wider spectrum of co-occurring phenomena for which a neurobiological basis is sought. Being able to tease out these data-to-phenomena inferences is diagnostically important to help children whose difficulties with reading might be downstream from slow processing speed, or may be a consequence of ADHD, for example.

Developmental and cognitive psychologists use data from cognitive tests (e.g. slow naming speed, difficulties in letter–sound decoding) as evidence for a phonological deficit in the ability to associate phonemes with graphemes. The consensus view these days is that dyslexia has to do with some kind of phonological deficit,[5] namely a defect in the representation of speech sounds which leads to difficulties with phonological awareness, slow naming speed, difficulties with letter–sound decoding, and hence non-fluent reading and difficulties with spelling.

The phenomenon 'difficulties with reading' in this case is the behavioural manifestation of a developmental-cognitive problem concerning the representation of phonemes and the ability to segment and blend them. This perspectival data-to-phenomena inference has been very important, among other things, in establishing the most effective pedagogical method for literacy in the so-called reading wars—whether it is a phonic approach (learning one letter–sound at a time, as is now believed to be the preferred method) or a whole-language approach (see Connor et al. 2004a).[6]

'Difficulties with reading' is, then, what I call a 'modally robust' phenomenon. For it robustly *can happen* in very many different ways. And, typically, it is the job of different epistemic communities to explore the network of perspectival inferences from specific data to the correct diagnostic profile in each case. Let us briefly take a closer look, first, at the *semantic nature* of those inferences, and, second, at the *perspectival modelling* that enables them.

---

[5] On phonological deficit, see Ramus (2001 and 2003). For an excellent introduction to the general topic, see Snowling (2019).
[6] For some political context on the 'reading wars', see, e.g., https://www.theatlantic.com/magazine/archive/1997/11/the-reading-wars/376990/.

Perspectival data-to-phenomena inferences have to be *reliable* to advance bona fide knowledge claims (rather than spurious claims). What is to be said about these knowledge claims? Consider, for example, the following claim:

(i) If a child has difficulties with reading, then they are dyslexic.

This claim rests on an indicative conditional 'if ... then' with present tense in the antecedent and consequent. It is clear from the discussion so far that an unqualified claim of this nature is useless in diagnosing children with dyslexia from their non-dyslexic peers who might still experience difficulties with reading. As the Isle of Wight study revealed, behavioural data about reading accuracy and comprehension are not—in and of themselves—unequivocally reliable evidence for dyslexia. A more reliable diagnosis depends on how one understands the antecedent of this indicative conditional.

This in turn involves uncovering a number of additional perspectival data-to-phenomena inferences behind the phenomenon 'difficulties with reading'. These inferences may again take the form of further indicative conditionals where the phenomenon features this time in the consequent. Here are two examples:

(ii)   If a child experiences difficulties with schooling, they will have difficulties with reading.

(iii) If a child has a phonological deficit, they will have difficulties with reading.

Reliably diagnosing children with dyslexia among children without dyslexia who also experience learning difficulties depends on teasing apart (iii) from (ii). But even after screening to rule out difficulties with schooling as a potential cause, the reliability of the diagnosis depends on telling apart more indicative conditionals, such as the following:

(iv)   If a child has a phonological deficit and an attention deficit, they will have difficulties with reading and with planning.

(v)   If a child has a timing/sequence deficit, they will have a phonological deficit and a motor control deficit, and as a result difficulties with reading and with balance.

(vi)   If a child has slow temporal processing, they will have a visual deficit as well as a phonological deficit, and as a result difficulties with reading and with motion detection.

(vii) If a child has a cerebellar abnormality, they will have a timing/sequence deficit, and as a result a phonological deficit and a motor control deficit with difficulties with reading and with balance.

These knowledge claims in the dress of indicative conditionals belong to different scientific perspectives. They are advanced by communities as diverse as cognitive psychologists (iv), vis-à-vis neuroscientists (e.g. vii). Each community relies on its own experimental, theoretical, and technological resources to source the relevant data and to reliably make these claims. Moreover, each community uses second-order (methodological-epistemic) principles to *justify* their reliability.

For example, from the neurobiological perspective, scans showing anomalies in brain morphology can be used as evidence for inferring the presence of both a visual deficit and a phonological deficit. From this perspective, the phenomenon of difficulties with reading goes hand in hand with others: say, difficulties with motion detection as in (vi). This could form the basis for a possible diagnosis of slow processing speed, for example. The validity of the diagnosis depends on the reliability of the relevant data-to-phenomena inference, which is in turn justified by methodological-epistemic principles adopted in neurobiology. Among them: that in screening brain images the relevant (non-biased) control group has been correctly identified; that there are functionally relevant pathways from the brain to the relevant behavioural phenomena; that there are 'dyslexia candidate susceptibility genes' (Fisher and Francks 2006) implicated in the relevant neurobiology. Each of these methodological-epistemic principles can of course be challenged and are typically called into question as new evidence and new studies come to the fore. Indeed, there are communities within communities where, for example, colleagues performing fMRI imaging do not typically make any assumption about possible 'susceptibility genes'.

Consider now the perspective of cognitive psychologists, who use data from cognitive tests and reading tests as evidence for inferring specific reading difficulties. The reliability of the inference and associated knowledge claims is also in this case justified by methodological-epistemic principles internal to the discipline. One of these, as already mentioned, is, for example, the IQ-achievement test often used as an indicator of learning potential, with unexpected discrepancy from it being used as a diagnostic tool.

Sometimes the phenomenon inferred is the same (difficulties with reading in my example). But the perspectival nature of the inferences means that the phenomenon in question is each time *differently located* in a space of possibilities. Sometimes it is comorbid with other phenomena and symptomatic of a broader phenomenology (e.g. slow processing speed). At other times, it is continuous with garden variety reading difficulties that call for remedial teaching strategies of wider benefit for larger portions of children, as Marie Clay (1987) originally argued for.

It is in this specific sense that the modally robust phenomenon of 'difficulties with reading' lies at the intersection of a plurality of scientific perspectives. Evidence for it does not accrue by accumulation of *more data of the same type* (more reading tests, or more brain images). Nor is the phenomenon the manifestation of some hidden dispositional property. The reliability of the inference from data to the phenomenon cannot be justified by appealing to a genetic marker in neurobiology. Nor can it be justified by generically invoking underachievement, because there is no prototypical phenotype either.

This does not make the phenomenon any less real. On the contrary. It is very much real and can happen in a variety of different ways. But the reliability of the knowledge claims advanced through data-to-phenomena inferences *within each scientific perspective* needs to be cross-checked and cross-validated. This is usually done by bringing one perspective to bear on the other and vice versa. For example, one can bring the cognitive perspective to bear on the neurobiological one and the educationalist perspective to bear on the cognitive one. Moreover, one needs to take into account a number of other considerations about the environment and the transparent/non-transparent nature of the language in question.

This is perspectival pluralism in action. The pluralism of scientific perspectives is not just a desirable methodological feature of science. It is not just a way of offering a menu of different explanations for the same phenomenon. It is a way of checking the *reliability* of each data-to-phenomena inference *within* its own perspective *in light of other scientific perspectives*. Only in this way can key justificatory principles of each perspective be monitored, cross-checked, and held accountable.

No scientific perspective can sanction the reliability of its own inferences alone. One needs to ask: reliable *with respect to what* and *to what degree?* What is it that researchers are trying to achieve each in their own scientific perspective? And how successful are they in their inferences? Perspectival pluralism is required to improve the open-ended network of inferences (i)–(vii) for a variety of purposes (diagnostic, educational, screening, etc.). Most importantly, perspectival pluralism is required to maintain checks on the *justificatory* (methodological-epistemic) *principles* of each scientific perspective. In so doing, it allows different epistemic communities to have a debate about dyslexia.

Being located in a space of possibilities means that there is an element of *contingency* in what the data may *reliably* provide evidence for each time. It is not a necessary truth that if someone has a phonological deficit, they will also display difficulties with reading. They may in fact successfully develop compensatory strategies during the early years, maybe thanks to timely interventions and appropriate teaching support, or thanks to the transparent nature of their native language. Therefore the consequents of the indicative conditionals are best read as hiding a modal verb:

(iii*) If a child has a phonological deficit, they *may* have difficulties with reading.

(iv*) If a child has a phonological deficit and an attention deficit, they *may* have difficulties with reading and with planning.

I will have more to say about the semantic nature of these conditionals in Chapter 5. But, next, I want to illustrate how perspectival modelling understood as exploring the space of possibilities finds a natural expression in what is known as developmental contingency modelling for dyslexia, championed by Uta Frith and John Morton.

4.c.3. Developmental Contingency Modelling (DCM) as perspectival modelling

The Developmental Contingency Modelling (DCM) of Uta Frith and John Morton (Morton 1986, 2004, Ch. 8; Morton and Frith 1995) perfectly illustrates the cross-perspectival process of refining the reliability of knowledge claims in perspectival modelling. In continuity with the other two case studies, the term 'perspectival modelling' is again used in a broad sense and not to refer *exclusively* to the causal models within DCM. Or better, the individual causal models within DCM are an integral part of how the intersecting scientific perspectives of the educationalists, developmental psychologists and neurobiologists jointly deliver modal knowledge of the relevant phenomena over time. What is unique and particularly interesting about DCM is that it builds in enough modularity and contingencies[7] to allow a variety of researchers—neurobiologists, cognitive psychologists, educationalists, among others—to differentiate similar learning difficulties by tracing and retracing them back to specific contingent points where breakdown might have occurred. It includes two main components: (1) a number of distinctive levels (biological, cognitive, behavioural, and environmental); and (2) a number of distinctive temporal stages in the acquisition of reading and writing skills.

In the words of Uta Frith:

A great challenge for cognitive theories is that they have to explain the diversity of dyslexia as it manifests itself in different people. Most cognitive theories are not designed to cope with individual variation. They address the prototypical

---

[7] '[O]ne can imagine a skill whose emergence is a function of a late maturing structure but which also depends on the prior existence of other processes or knowledge. We would want to be able to represent all such contingencies. The general form of the contingency model is that of elements connected in a directed graph. The elements can be of a variety of kinds—processes, structures, knowledge, perceptual or other experiences, or biological elements. The symbols on the connecting lines have temporal/ causal implications' (Morton and Frith 1995, p. 377).

case instead. The behaviour patterns characteristic of the prototypical case are distilled from many individual cases, and it is this distilled information that is usually the target of explanation. (Frith 2002b, p. 53)

DCM accommodates a plurality of causal models that aim to disentangle similar phenomena at the behavioural level and explore the variety of contingent pathways that might be at play in each case, as the following rival causal models show (Figures 4.c.1, 4.c.2, 4.c.3, and 4.c.4):

Prima facie similar behavioural phenomena (e.g. difficulties with phonological awareness and in general difficulties with reading) can hide diverse cognitive deficits. In some cases, the phonological deficit that is primarily responsible for difficulties with reading is the direct consequence of some neurobiological anomaly (like the left-hemisphere disconnection in Figure 4.c.3). In other

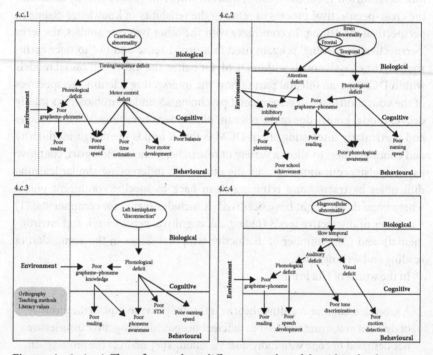

Figures 4.c.1–4.c.4 These figures show different causal models within the three-level developmental framework for dyslexia associated with different hypotheses about the neurobiological basis and different causal graphs across the three levels. Copyright © Fig 3.5, 3.6, 3.7, and 3.8 from Uta Frith (2002b) 'Resolving the Paradox of Dyslexia', in G. Reid and J. Wearmouth (eds), *Dyslexia and Literacy: Theory and Practice*, John Wiley & Sons, pp. 56–60. Reproduced with permission of the Licensor through PLSclear.

cases, the phonological deficit is the joint effect of a common cause such as slow temporal processing in Figure 4.c.4 that manifests itself in a number of other symptoms such as visual deficit and difficulties with motion detection. Or a timing/sequence deficit, as in Figure 4.c.1, that results in motor control deficit and difficulties with motor balance. Once again, the search for a one-size-fits-all phenotype would be misguided.

As Frith and Morton have long been arguing, dyslexia is about individuals. Effective remedial strategies should address specific individual needs. The dyslexic child who also has an attention deficit problem (inattentive ADHD) needs learning support strategies different from those of a child with slow processing speed. Identifying the possible pathways within the developmental framework is key to go from 'won't read' to 'can't read' (because of visual deficit, motor control deficit, slow processing speed, . . .). And this in turn would allow appropriate educational support, which is necessary to transform 'can't read' into 'can read'.

Causal models within DCM take different forms (X-shape, V-shape, A-shape) depending on the number of factors and their causal relations across the three levels. In every case, a single causal nexus is implied. This can be at the cognitive level with multiple causes and multiple behavioural manifestations (X-shape, e.g. dyslexia). It can be at the behavioural level with multiple causes (V shape). It can be at the brain level, with a single known cause and a variety of behavioural consequences (A-shape, e.g. a rare single-gene defect). Thinking in terms of these shapes for different conditions allows practitioners to identify robust (almost lawlike) dependencies among relevant features of the phenomena of interest across different levels. For example, the ability to decode letter strings causally depends on both unimpaired vision that allows the child to discriminate visual features of letters at the alphabetic stage *and* the phonological ability to sequence phonemes in a particular order.

In turn, the phonological ability that is critical to identifying and sequencing phonemes may also causally determine the speed of object naming. Object naming tests are thought to be sensitive tests for diagnosing dyslexia, regardless of impairments in other phonological tests. Likewise, there might be children who do not have any problem with object naming and yet experience difficulties with decoding skills. All else being equal, causal graphs like Figure 4.c.5 allow practitioners to conclude (e.g. via object naming tests) that the setback might be at the alphabetic level of knowledge of the letters (maybe because of some vision deficit or delay in the transition from the logographic to the alphabetic phase) and 'tell the difference between a child who cannot decode simply because letter knowledge is absent and a child who lacks the requisite phonological skills' (Morton and Frith 1995, p. 378).

Let us draw some philosophical conclusions. DCM is a perfect illustration of what I call 'perspectival modelling' in that the model pluralism here at stake

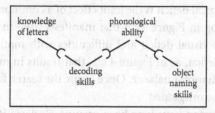

Figure 4.c.5 A zoomed-in detail of a causal graph within the DCM for dyslexia, where decoding skills is the joint effect of two different abilities (with phonological ability having object naming skills too as a secondary effect). Copyright © Fig. 13.41 from J. Morton and U. Frith (1995) 'Causal Modelling: A Structural Approach to Developmental Psychopathology', in D. Cicchetti and D.J. Cohen (eds), *Manual of Developmental Psychopathology, Vol. 1: Theory and Methods*, New York: Wiley, p. 378. Reproduced with permission of the Licensor through PLSclear.

is exploratory. It enables a particular kind of inferential reasoning necessary to explore how learning to read and write *might be affected* by a number of contingent setbacks during the early years. Charting this space of possibilities and being able to correctly locate the modally robust phenomenon of 'difficulties with reading' is key to understanding all the possible routes through which a child becomes fully literate, and the adaptive strategies for each possible setback. Thus, the causal models are perspectival in that they chart the possible routes through which language skills can be learnt and relearnt over a period of time in response to a variety of conceivable neurobiological, cognitive, or environmental stumbling-blocks. Perspectival modelling in this case is pivotal to effective diagnostics and educational interventions tailored to the specific needs of individual children. Because every child is unique, so is every developmental setback.

5

Inferential blueprints and windows on reality

> Kublai interrupted him: 'From now on I shall describe the cities and you will tell me if they exist and are as I have conceived them. I shall begin by asking you about a city of stairs, exposed to the sirocco, on a half-moon bay . . .'
> 'Sire, your mind has been wandering. This is precisely the city I was telling you about when you interrupted me'.
> 'You know it? Where is it? What is its name?'
> 'It has neither name nor place. I shall repeat the reason why I was describing it to you: from the number of imaginable cities we must exclude those whose elements are assembled without a connecting thread, an inner rule, a perspective, a discourse'.
>
> Italo Calvino (1972/1997) *Invisible Cities*, p. 37[1]

5.1. What blueprints are for

In 1842, the astronomer Frederick William Herschel stumbled upon a peculiar phenomenon during his experiments with colours derived from flower petals and plant leaves. He reported in the *Philosophical Transactions of the Royal Society* that if a drawing on paper were washed over with a solution of 'ammonio-citrate of iron, dried and then washed over with a solution of ferro-sesquieyanuret of potassium [potassium ferrocyanide]', the paper received an image which 'from being originally faint and sometimes scarcely perceptible, is immediately called forth on being washed over with a neutral solution of gold' (Herschel 1842, p. 394).

Perspectival Realism. Michela Massimi, Oxford University Press. © Oxford University Press 2022.
DOI: 10.1093/oso/9780197555620.003.0008

Herschel had invented 'blueprinting'—one of the oldest methods of reproducing large-scale drawings and maps. The method was simple and cheap and did not require copyists and engravers. The original master drawing had to be on paper or a fabric which light could pass through. Another piece of paper or fabric had to be pre-washed with the ferric salts, put in direct contact with the master, and then exposed to light. The coating of ferric salts turned dark Prussian blue, leaving white lines where the master drawing blocked exposure.

Blueprints mark a turning point in the history of architecture and cartography (see Murray 2009). Anyone could now produce copies of the master drawing at roughly the same scale as the original (the pre-washed paper tended to shrink while drying). The original method spread widely through Europe and North America in the second half of the nineteenth century. Electric arc lamps proved more reliable than sunlight in the overcast sky of Northern Europe, and were introduced for blueprinted maps of trenches in World War I. After the war came blueprinting machines that were able to deliver large-scale, better-quality blueprints in three minutes.

In architecture, blueprints made it easy to go through various iterations of the same design, introduce changes to the original, delete original features, tweak roof height, enlarge chimneys, subdivide internal spaces, and so on (see Figure 5.1). They also made it possible to reproduce any sequence of perspectival

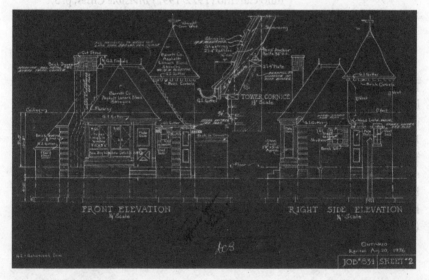

Figure 5.1 Blueprint of Joy Gas service station elevation drawings, 910 Lake Shore Road Blvd W., Toronto, Canada. 17 November 1936. City of Toronto Archive, *Series 410, File 505, Item 8*. (For interpretation of the references to colours in this figure, the reader is kindly referred to the Open Access eBook version). Public Domain, Wikimedia Commons: https://commons.wikimedia.org/wiki/File:Joy_Oil_gas_station_blueprints.jpg

representations of the same building. The blueprints of Frank Lloyd Wright's Shampai House, for example, represent the house from a plurality of points of view: from below, above, from the left side, and so on.

A sequence of blueprints through an architectural project made it possible to see (via overdrawing) *who* was responsible for the changes over time and *who* could legitimately claim ownership of the outcome. Blueprints became the battleground of authorship disputes, such as the one between Lloyd Wright and one of the architects in his team, Schindler, in the final design of the Shampai House in 1919 (for details, see Park and March 2002).

I suggest that perspectival models are a bit like architectural blueprints, and allow teams of scientists to work together over time, make changes, and eventually advance scientific knowledge about the target system. In a way, the analogy continues the one in Chapter 2 about perspectival representations in art. Perspectival representations are not just a bunch of coplanar lines directed towards vanishing points. They require a medium. That is what makes it possible to apply the perspectival techniques in a number of contexts (paintings, architectural designs, maps), with different effects and results. In the case of architectural blueprints, the medium enables the production of multiple perspectival drawings of the same building on which teams of architects can work together over time, make changes, and finally implement the original project.

So far I have not said much about what a perspectival model is; or how it works. I attend to this task now. Here is the main philosophical idea I will unpack:

(I) Perspectival models model possibilities by *acting as inferential blueprints* to support a particular kind of conditionals, namely *indicative conditionals with suppositional antecedents*.

Scientific models have long been associated with surrogative reasoning. My view of the distinctive way in which perspectival modelling is a particularly interesting example of inferentialism about models connects with the literature on conceivability and possibility in modal epistemology. I see perspectival models as inviting us to physically conceive a number of scenarios within the constraints of laws of nature as a guide to modelling what is possible. Perspectival models support what logicians call indicative conditionals with a suppositional antecedent (where the suppositional antecedent captures the physically conceivable scenario).[2] I dig into modality, laws of nature, and these indicative conditionals

---

[2] It is important once again to bear in mind the distinction between 'perspectival modelling' and 'perspectival models'. The latter are a variety of scientific models that are an integral part of 'perspectival modelling' understood broadly as the situated practice of modelling. As the three case studies in Chapter 4.a, 4.b, and 4.c illustrate, typically several epistemic communities jointly contribute to perspectival modelling. Perspectival modelling, in its broad sense, include data collections and a variety of data-to-phenomena inferences extending well beyond the specific task of modellers. Perspectival

in the rest of the chapter. But first, in what sense do perspectival models act as *inferential blueprints?*

5.2. Perspectival models qua inferential blueprints

The word 'blueprint' is sometimes found in connection with scientific modelling. Most of the time it is used to mean 'plan', rather than in its original sense. For example, Nancy Cartwright (1997) has argued that models are blueprints for what she calls nomological machines (see also Cartwright 2019 for an insightful recent discussion of how models deliver knowledge). Rachel Ankeny and Sabina Leonelli (2016 and 2020, pp. 40–41) define the notion of *repertoire* as 'a general framework for analysing the emergence, development, and evolution of particular ways of doing science. In a repertoire, the successful alignment of conceptual, material, logistical, and institutional components (including specific skills and behaviours by participants in scientific efforts) results in a blueprint for how to effectively conduct, finance, and support research in the longer term'. Tarja Knuuttila and Andrea Loettgers (2013) have also seen mathematical models as 'blueprints' for the engineering of biological systems.

I am using the term differently. I contend that perspectival models act as inferential blueprints. There are a number of elements at play in this analogy:

1. Blueprints are perspectival$_1$ representations. Each represents the target system *from* a specific vantage point. But when taken together, a collection of blueprints offers a plurality of points of view (e.g. from above, from below, from the rear) for opening up a 'window on reality', namely how the final building is going to look. No individual blueprint supplies a mirror image of the final target system because—as with Lloyd Wright's Shampai House—the final building is often the end point of a long process of changes made by different architects.

2. The perspectival$_1$ representation of each blueprint is distorted. A blueprint gives a representation that is not a mirror image or a perfect copy of the original master document. It is not a perfect copy of the original because the paper shrinks in the process of making the blueprint and scale distortion might result.

3. Blueprints are the starting point of a collaborative effort to implement the original design. They show what different communities *can do* with perspectival$_1$ representations, how they can *use* them. As such, blueprints are

models, by contrast, refer to a subset of scientific models that are at play in perspectival modelling (be they causal models within DCM for dyslexia, CMIP5 in climate science, or nuclear models around 1930–1950, among other examples).

the authorship of multiple agents; they have a history of their own; and are often the battleground of copyright ownership disputes.

4. Blueprints act as inferential tools. A blueprint is a medium for perspectival representations of a given building. It facilitates exchange of instructions among different communities *over time* for *making relevant and appropriate inferences.*

Before the development of xerography, blueprints facilitated the exchange of instructions among architects, draughtsmen, builders, carpenters, artisans, joiners, and masons. They allowed them to work in parallel on multiple aspects of the same project. Blueprints offered instructions (often marked by a number of architects over time) for builders, joiners, masons, and carpenters to make inferences: how deep in the ground should the foundations lay? How high will the roof be? Will this wall include a window or a French door?

These instructions have to be *relevant*. A blueprint of a two-storey house does not specify the colour of the sofa or the shape of the dining table. Blueprints are for the architect to design the house, not for the interior designer to choose the furniture. Information also has to be *appropriate* to the task. A blueprint recommending a roof 'higher than lower roof by 20m' or 'chimney 15m wide' would clearly be inappropriate for building a two-storey house. The house would be structurally unstable or would no longer be a two-storey house.

Thus the representational value of a blueprint consists in its ability to enable the relevant users to make *relevant and appropriate inferences* over time. Appropriate rather than correct, because correctness implies that there can only be one set of inferences. The reality is that—within broad constraints—various sets of instructions are conceivable and can be appropriate for the assigned task.

Blueprints do more than help particular communities of users to make inferences. They are a natural medium for perspectival representations. And a sequence of blueprints can offer different perspectival$_1$ representations of the same object: for example, the house as seen from a front view; as seen from a so-called worm's-eye perspective using three vanishing points; or as seen from the rear garden.

This plurality of perspectival$_1$ representations for the same object operates within the constraints of the original master design. If the original plan included a wall-faced living room, it is possible to amend the design in the blueprint and open up the living room to a rear garden. But it would not be possible to transform the living room into a tennis court. Within broad constraints, the novelty of blueprints in architecture was that they allowed users to *track changes by various authors over time*. The final outcome is the product of various iterations. Blueprints are the tangible evidence of the *collaborative nature* of architectural drawing. It is through this plurality of ever-evolving perspectival$_1$ representations that the Shampai House was conceived, and eventually realized. The same, I suggest, is true for perspectival models:

(I) Like blueprints, each perspectival model represents the target system from a specific vantage point. But it is only the plurality of perspectival models that can open a 'window on reality' about the object under study.

(II) The perspectival representations of each model are inevitably distorted and idealized as with any scientific model (see Bokulich 2011; Potochnik 2017; Rice 2018).

(III) Perspectival models have a history of their own; they are a collaborative effort, authored by multiple agents and the battleground for ownership.

(IV) Perspectival models act as inferential blueprints in making it possible for different epistemic communities to come together, revise, and refine the reliability of each other's claims and advance scientific knowledge over time.

The multi-model ensembles at work in CIMP5 are the outcome of large worldwide collaborations among teams of scientists involved in integrating models and *making inferences* about the phenomenon of global warming and its longterm projections. These models have a *genealogy* of their own (Knutti et al. 2013), building on the earlier generation of CMIP4.

The shell model has enjoyed a long history from Elsasser and Guggenheimer to Goeppert Mayer and Jensen and the physicists (Bohr, Mottelson, and Rainwater) who subsequently developed the model and its relation to the liquid drop model and made new inferences concerning nuclear structure. Perspectival models are an integral element of the broader practice I called 'perspectival modelling,' in that nuclear physicists, atomic theorists, and physical chemists could all avail themselves of the relevant and appropriate instructions incorporated in these models to explain the phenomena of interest.

Likewise, the causal models within developmental contingency modelling (DCM) for dyslexia enable different users (behavioural psychologists, educationalists, neurobiologists, and developmental psychologists) to make relevant and appropriate inferences. For example, they are used by developmental psychologists to explore developmental pathways behind the phonological deficit hypothesis. Neurobiologists can use them in studying possible neurobiological bases for a range of specific learning disabilities (SLD). And educationalists can avail themselves of these models to make diagnoses and put in place suitable educational interventions.

In each example, the perspectival models offer instructions to diverse epistemic communities for making relevant and appropriate inferences about the phenomenon of interest within broad constraints. These constraints are different in each case. The lawlike dependency between, for example, decline in land snow cover and surface energy flux[3] is one among many for making inferences about global warming. The lawlike dependency between the ability to decode letter strings and sequencing phonemes is an important constraint in making diagnostic inferences from the phenomenon of difficulties with reading to dyslexia. And Coulomb's law and Pauli's principle, among others, prove important to make relevant and appropriate inferences about magic numbers, nuclides' stability, and chemical properties of elements when modelling nucleon–nucleon behaviour.

In all these cases the task of *modelling what is possible* requires perspectival models acting as inferential blueprints. Each model offers a partial, incomplete, and perspectival₁ representation of the intended target system just as each blueprint offers only a specific vantage point on the target. But the purpose is ultimately to jointly make possible perspectival₂ representations of the intended target system— to open a 'window on reality' when it comes to global warming, nuclear stability, and difficulties with reading. In the next sections, I tease out how perspectival models deliver on this task.

5.3. Conceivability and possibility for perspectival models

That scientific models facilitate or support inferences (or surrogative reasoning, as it is often called) has been well established in the literature on modelling, in the work of Frigg (2010a, 2010b), Godfrey-Smith (2006), Magnani and Nersessian (2002), Mäki (2011), and Nancy Nersessian (2010), among others (see also Frigg and Nguyen 2020; Ippoliti et al. 2016). Suárez has put forward the inferentialist view of scientific representation, whereby for something to represent something else is for it to allow 'competent and informed agents to draw specific inferences about' it (Suárez 2004, p. 773; 2009). Contessa (2007) has in turn stressed how scientific models provide epistemic representations of a certain target for a certain user if the user adopts an interpretation of the model in terms of the target T.

By contrast, that scientific models can be used to make modal claims—for example, claims about what is possible—is a relatively new idea (see Bokulich 2014; Currie 2020; Godfrey-Smith 2020; Grüne-Yanoff 2009; Grüne-Yanoff and Marchionni 2018; Rice 2018, 2019; Verreault-Julien 2019; Sjölin Wirling and Grüne-Yanoff 2021, and forthcoming).[4] The growing interest in model-based

---

[3] Energy flux, namely, how incoming net radiation is balanced by outgoing radiation, is affected by anthropogenic greenhouse gases (GHGs) forcings. A number of lawlike dependencies enter in these processes, including, for example, the Penman–Monteith equation (see Bonan 2015).

[4] Sjölin Wirling and Grüne-Yanoff (2021, and forthcoming) refer to Massimi (2019a) but not to the wider context in which that paper sits (i.e. my perspectival realist view already in Massimi 2018a, 2018b, 2018e, 2018f, and 2019c). As such they mischaracterise my analysis of physical conceivability and misrepresent my view as blurring the epistemic possibilities vs. objective possibilities distinction.

modal reasoning among philosophers of science originates in part from an interest in the literature on the epistemology of modality in other areas of philosophy.

I want to bring together the literature on inferentialism about scientific representation and the more recent work on model-based modality. My task is to articulate how I see perspectival models delivering modal knowledge about what is possible by enabling a particular kind of inferences. I will clarify where the modal traction of the models resides and anticipate an objection that will be the focus of Part II of the book: namely *where is realism in all this?*

Let me begin with the epistemology of modality. Conceivability and possibility have featured prominently in the literature here (Berto and Schoonen 2018; Chalmers 2002; Fischer and Leon 2017; Gendler and Hawthorne 2002; Roca-Royes 2011; Yablo 1993). A key issue is whether conceivability can act as a guide to possibility. In the epistemology literature, conceivability is often honed by specific accounts of imagination, mental imagery, and mental stipulation (picturing something in one's own mind). Most of the discussions (e.g. Hawke 2011; Kung 2010; van Inwagen 1998)[5] analyse the role of intuitions and mental imagery in thinking of possible worlds that might verify a proposition p as a possibility.[6]

A classical argument goes as follows. An *epistemic* notion of conceivability (along the lines of Worley 2003 and Yablo 1993) is defined thus:

> p is epistemically conceivable (conceivable$_{E}$) if it is imaginable for human agents with a given state of knowledge and conceptual resources.

This is in contrast with a non-epistemic notion of conceivability (along the lines of Chalmers 2002), where to be conceivable is to be imaginable for an idealized conceiver with unlimited conceptual resources and knowledge of non-modal

As the discussion in the past chapters have made it clear, and this chapter further explains, this distinction is in fact key to my view of perspectival models as exploratory models that deliver modal knowledge. Philosophical distinctions are never universal truths from nowhere, especially when it comes to philosophy of science, where contextual details of modelling practices matter. Moreover, philosophical distinctions are useful only as long as they do make a difference. In the case of perspectival realism, the epistemic-objective distinction maps onto the division of modal labour that indicative conditionals and subjunctive conditionals are respectively entrusted with (see Section 5.7, building on Massimi 2019c) in addition to marking the ontological distinction between modally robust phenomena and stable events (see Chapter 6, Section 6.7, expanding on Massimi 2007, 2008, 2011a, and 2014) on which my phenomena-first ontology rests.

[5] I am very grateful to Franz Berto and Tom Schoonen for helpful discussions on this topic.
[6] More recently there has been an increasing interest in the role of scientific imagination among philosophers of science and physicists too: see, for example, Ivanova and French (2020), Levy and Godfrey-Smith (2020), and McLeish (2019). See in particular Salis and Frigg (2020), where imagination is understood along fictionalist lines as a make-believe game.

empirical facts about the actual world.[7] The ideal conceiver reminds me of what in a different context Ned Hall (2015) has called a Limited Oracular Perfect Physicist (LOPP), whose omniscient knowledge of non-modal empirical facts about nature could act as the basis for selecting the Best System of laws, according to David Lewis's influential view.

Let us assume an *epistemic* notion of conceivability as just defined. Consider the following example. It was conceivable$_E$ for the ancient Greeks that water consisted of something different from H_2O. For all they knew at the time given their informational state and conceptual resources, water was in fact an elementary substance. However, despite what was conceivable$_E$, it is not possible for water to consist of anything but H_2O since consisting of H_2O is an essential property of water. Or so goes the argument for why conceivability$_E$ is not a guide to possibility.

The argument assumes Kripkean essentialism to show that epistemic conceivability does not reliably deliver *de re* modal knowledge. If consisting of H_2O is an essential property of water, it follows that even if someone conceives$_E$ water to be different from H_2O, conceivability does not offer any handle on what is possible, unless a modal error strategy can be devised (see Kung 2016 for a discussion).[8]

Understanding how modelling practices in the natural sciences deliver modal knowledge is, then, important to shed light on the role that essentialism might or might not play in that delivery. I maintain that perspectival models resort to a particular version of conceivability that I call 'physical conceivability' with the goal of delivering knowledge of possibilities. Preliminary caveat: the adverb 'physically' does not mean that the conceivability here at issue is confined to the physical sciences. On the contrary. I see it equally applying to developmental psychology and climate science, to refer to my case studies in Chapter 4.b and

---

[7] A classical problem with this alternative notion is that conceivability facts might not necessarily be epistemically accessible to us (see Balcerak Jackson 2016).

[8] Yablo (1993), for example, traces modal error back to some lack of knowledge about actual-world facts (e.g. the ancient Greeks' lack of knowledge about hydrogen or oxygen); and to the role of such actual-world facts qua modal defeater in making it impossible for water to be anything but H_2O. Sonia Roca-Royes (2011) suggests that epistemic conceivability faces a metaphysical contradiction: the epistemic conceivability of *p* would depend on the subject not being aware of any metaphysical contradiction in *p*. An epistemic agent who *knows* that water is H_2O will find no conceptual contradiction in conceiving of water as different from H_2O. But they will have to concede that 'H_2O is not H_2O' is metaphysically contradictory. I am interested in exploring how these notions play out in the context of model-based natural sciences. For example, Daniel Nolan (2017) has advocated naturalizing modal epistemology and taking model-based sciences as a way of investigating how modal information is gained in science. Timothy Williamson (2016, 2020, Ch. 15) has drawn attention to what he calls 'objective modalities', the ways things could have been in a non-epistemic, non-psychological, and non-intentional sense. The notion proves interesting to tease out varieties, including nomic modality in the natural sciences.

4.c. More broadly, I see the notion as applying to any scientific area where perspectival models are at work.

Whatever its intellectual merits, a non-epistemic notion of conceivability is simply not available to perspectivism. Perspectivism is at odds with the very idea of a LOPP with unlimited conceptual resources and access to non-modal empirical facts about the actual world. Although scientists *do have* reliable access to empirical facts and reliable methods to make data-to-phenomena inferences, as I argue in Chapter 6, they do not have a LOPP-ian kind of access to them. Thus, my starting point will be the epistemic notion of conceivability.

However, conceivability$_E$ will not do when it comes to discussing the kind of imagining that goes on in scientific modelling, and in perspectival models in particular. The main problem, as I see it, is its unconstrained nature.[9] Let me, then, focus here on the kind of imagining that goes on in perspectival models. These models are neither unbridled nor the outcome of philosophical intuitions and mental imagery. Of course, there is a good dose of creativity and imagination at play in scientific modelling. However, that *p* is possible is not justified on the grounds that a scientist *S* (or an epistemic community *C*) has imagined/conceived$_E$ a scenario that *they* take to verify *p*. No matter how fine-grained the details of such a conceived$_E$ scenario, in science no one would take seriously a possibility claim that *p* advanced on such thinly conceived$_E$ ground.

Scientific imagining has to respond to the tribunal of experience in the form of *factual information* available to an epistemic community *C* at any given time, *and* to well-defined lawlike constraints that fix the boundaries of what is 'physically conceivable'. Hence my notion of *physical conceivability* departs from the more familiar philosophical notion of epistemic conceivability. Here is my working definition:[10]

> *p* is physically conceivable for an epistemic subject *S* (or an epistemic community *C*) if *S*'s (or *C*'s) imagining that *p* not only complies with the state of

[9] For example, in Yablo's influential account, some *p* is conceivable if 'I can imagine a world that I take to verify *p*' (Yablo 1993, p. 29). In other words, epistemic conceivability can act as a guide to possibility (albeit a fallible one) as long as the imagined world is taken to verify *p* and hence is incompatible with the falsity of *p*. Van Inwagen (1998) has urged caution against extending this to 'far-out' modal claims, like imagining a purple cow or transparent iron. This imagining would fall short of Yablo's conceivability$_E$ because the imagined scenario is compatible with the falsity of a naturally occurring purple cow, unless in my imagined world I can also imagine a purple pigment that becomes part of the DNA of the cow. But such imagining would stretch far beyond the usual remit of (everyday or scientific) possibility claims (see Hawke 2011).

[10] Here and in Sections 5.3, 5.4, and 5.5 I build and expand upon Massimi (2019a) 'Two kinds of exploratory models', *Philosophy of Science* 86, 869–881. Reproduced with permission from University of Chicago Press, copyright (2018) by the Philosophy of Science Association. I am very grateful to the audience at the British Society for Philosophy of Science 2016 Annual Conference in Edinburgh, the Scottish Philosophical Association 2017 keynote lecture and the Barcelona Logos Research Group for helpful comments on earlier drafts of the material here presented.

knowledge and conceptual resources of S (or C) but is also consistent with the laws of nature known by S (or C).

I understand the exploratory function of perspectival models in terms of the physical conceivability they afford, and the modal knowledge they accordingly deliver. Far from being mere instruments, perspectival models explore what is physically conceivable to deliver scientific knowledge about what *might be the case* in nature.

I will ultimately argue that the modal gap between physical conceivability and knowledge of possibilities is *not* filled through essential-property ascription. I part ways with Kripkean essentialism, which has been built from the ground up in the epistemological debate on conceivability and possibility. Perspectival models should not be entrusted with the task of ascribing essential properties to the target system. This anti-essentialist line inherent in my perspectival realism will become more transparent in Part II of the book (especially Chapters 7 to 9), where I discuss why I am anti-essentialist about natural kinds. I will defend realism about modally robust phenomena instead. Such realism is born out of historically and culturally situated scientific perspectives whose practices include perspectival modelling, as illustrated in Chapter 4.a, 4.b, and 4.c.

An important clarification is in order. I see physical conceivability at work not just in perspectival models but more broadly across a large number of other models. I see it also in Maxwell's honeycomb model of the ether as a classical example of a fictional model that I would not qualify as perspectival (see Massimi 2019a). There were several models of the ether available at the time, but the pluralism there in play resembles more the non-perspectival varieties described in Chapter 4. In what follows, my discussion of physical conceivability focuses on perspectival models even if the notion extends to other relevant classes of scientific models too.

5.4. Physical conceivability

Let us unpack the definition of physical conceivability. *Factual information* available to a community C at any time is encapsulated in the 'the state of knowledge of S (or C)'. Imagining is always sensitive to the informational state of a community. It is physically conceivable to the high-energy physics community in 2020 that the lightest supersymmetric particle is a candidate for dark matter. But this was not conceivable to physicists in 1920 because neither supersymmetry (SUSY) nor the idea of dark matter was part of their state of knowledge at the time. Likewise, it was physically conceivable to Maria Goeppert Mayer in 1948 that nucleons could be arranged in orbitals/shells given the information she had about the analogy with electronic shells and spin–orbit coupling for electrons. But it was not physically conceivable at the time of Rutherford and his alpha particle scattering experiments in the early 1910s, when the community had barely started contemplating the idea of a nucleus. What is physically conceivable depends on the historically and culturally situated scientific perspectives in which an epistemic subject S (an epistemic community C) operates. It depends on the experimental, technological, and theoretical resources that are an intrinsic part of their scientific perspective, and ability to perform data-to-phenomena inferences.

Thus, in my aforementioned definition, I mean something stronger than p has to be consistent with the set of beliefs held by S (or C). First, beliefs are not a 'state of knowledge' (unless they are both true and justified). Second, consistency is a logical property of a set of beliefs. But consistency is not time-sensitive. It does not track how the state of knowledge and conceptual resources evolve and change over time.[11]

Also, the verb 'complies' is there for a reason: to mark the difference between the broader epistemic notion of conceivability and the more specific variety of it I am operating with. To 'comply' conveys that for p to be physically conceivable (rather than just conceivable$_E$), imagining that p has to *be in accordance with the command* of factual information known by S (or C) at time t. For example, think of the role of RCPs in climate models. A concentration scenario that, say, peaks in the year 2022 and declines afterwards is conceivable$_E$ as I edit this chapter (in 2021) because we can imagine a world that would verify it. This could be, for example, a world where President Trump did not withdraw from the 2015 Paris agreement, among other things happening. I can close my eyes and imagine such a world that would verify p as the proposition <in the RCP$_{\text{conceivable}_E}$, greenhouse gases concentrations peak in the year 2022 and decline afterwards>.

However, such RCP$_{\text{conceivable}_E}$ is not physically conceivable because to the best of our state of knowledge as of December 2021, a lot of steps should have already happened for this imagining that p to be in accordance with the command of known factual information. For example, steps that should have already happened by now (December 2021) include all countries keeping all along the 2015 Paris agreement, and having cut GHG emissions dramatically by now. This is an important difference between physical conceivability and the broader notion of epistemic conceivability: the former ought to 'comply', that is, to be *in*

---

[11] Thus, I take it that if something is physically conceivable at time t_c, it has to comply with the state of knowledge at t_c and stretch beyond it by using the available experimental, technological and modelling resources at t_c. Claims of knowledge that are advanced in this way may of course turn out to be either true or false. Assessing their truth or falsity requires other scientific perspectives either at t_c or at later times $t_m, \ldots, t_s$, as I explain in Section 5.7.

accordance with the command of factual information known by S (or C) at time *t*. There is more.

Physical conceivability is not just sensitive to the factual information available to a scientific community. It must also be *consistent with the laws of nature* known by S (or C). This requirement makes the link between imagining and scientific modelling evident. Physical conceivability has to be embedded in a suitable scientific model *built in a way that is consistent with known laws of nature* for it to do any work at all.

This is another difference between epistemic conceivability in general and physical conceivability. Physical conceivability is not 'close-your-eyes-and-imagine' a world that can verify *p* by ruling out its compatibility with not-*p*. It is instead the *embedded imagining* of a given scenario which is (1) consistent *with known laws of nature* (2) *through the models* available within a scientific perspective at time *t*. Let us see why, starting with point (2), there is no physical conceivability in the absence of a scientific model.

If not embedded in a scientific model, physical conceivability qua consistency with the known laws of nature is vacuously true. For example, let us conceive two yet to be discovered half-integral spin particles (let us call them Castor and Pollux), one with spin up and one with spin down. This conceivable scenario would be consistent with known laws of nature (the Pauli principle says that no two half-integral spin particles can be in the same state). But it would be vacuously true—a mere re-statement of what Pauli's principle says camouflaged by the expression 'let us conceive of two yet to be discovered half-integral spin particles'.

Even for figuring out how such hypothetical half-integral spin particles would behave—were they to enter an electrical or magnetic field, for example, or be part of a large statistical collection of similar particles—a model is required to tease out how having half-integral spin is going to affect their behaviour. Without a scientific model for such particles, the physical conceivability exercise cannot deliver any scientifically interesting claims of knowledge.

Turning to (1), why does consistency with known laws of nature matter? That physical conceivability should refer to the *known* laws of nature, rather than laws of nature *simpliciter* (be they known or not), is a consequence of it being a variant of epistemic conceivability. It would be contradictory to allow for physical conceivability to be conceivability with respect to the state of knowledge and conceptual resources of an epistemic agent S (or community C) while also making reference to some atemporal set of laws of nature that might be epistemically inaccessible to the agent or community.

A worry here is that physical conceivability so understood can accrue cheaply given the reference to the *known* laws of nature. For example, a critic might argue, the alchemists of past centuries thought that it was possible that lead transmutes into gold. The modal knowledge claim p <lead can be transmuted into gold> is not available to us today because we now know that lead is one of those stable nuclei with magic number 82 and that nuclei formation via slow neutron capture and beta decay goes in the direction of elements with higher (rather than lower) atomic number Z (lead has $Z = 82$ compared to gold with $Z = 79$).

But for the medieval alchemists, whose state of knowledge did not include knowledge of nuclear structure, atomic numbers, or the shell model, it was in a way (epistemically) possible that lead transmutes into gold, based on their own physical conceivability exercises at time t. These counterexamples are ubiquitous in the history of science. They urge us to tread with caution and not trivialize physical conceivability as a guide to modal knowledge. How to proceed, then?

Recall that physical conceivability is a kind of imagining that complies with the state of knowledge and conceptual resources of S (or C). The state of knowledge varies over time as our epistemic access to factual information evolves and changes. Thus, although the alchemists' claim that lead can be transmuted into gold was based on their own imaginary scenarios about lead in accordance with the factual information and laws of nature known at the time, for physical conceivability not to be trivial, something ought to be said about how it has a purchase on modal knowledge claims.

Here is an idea, loosely inspired by work by Ippolito (2013) on past modality.[12] Think of the relation between physical conceivability and modal claims in terms of a time-sensitive epistemic accessibility relation that at any point in time branches as in Figure 5.2.

At some early time t_c, physical conceivability$_{t_c}$ gives epistemic access to a range of modal claims$_{t_c}$ (the branching paths in Figure 5.2), some of which include false claims such as the alchemist's p (i.e. <lead can be transmuted into gold>). At time t_m, as the state of knowledge changes, physical conceivability$_{t_m}$ gives epistemic access to another range of modal claims$_{t_m}$ which is a subset of the originally expanded range of modal claims$_{t_c}$. The modal claims that at some later stage t_m turn out to be true—let us call them *bona fide* modal knowledge claims$_{t_m}$—become embedded in the state of knowledge of the community C so that at time t_s a new physical conceivability$_{t_s}$ exercise gives epistemic access to another range of modal claims$_{t_s}$ again branching as a subset of the *bona fide* modal knowledge claims$_{t_m}$. And again those that prove *bona fide* modal knowledge claims$_{t_s}$ become embedded in the state of knowledge at t_z. And so forth.

---

[12] Ippolito's main concern is with articulating a semantic theory of subjunctive conditionals where the Lewisian relation of similarity among possible worlds embeds some kind of historical dimension. In particular, Ippolito sees the accessibility relation to possible worlds as historical and time-dependent: 'given a world w at time t, the worlds historically accessible from w at t are those worlds that share the same history at w up to t' (Ippolito 2013, p. 3). Obviously here I am not discussing Lewisian possible worlds, but a similarly time-dependent accessibility relation can be seen at work between the physically conceived scenarios and modal knowledge claims.

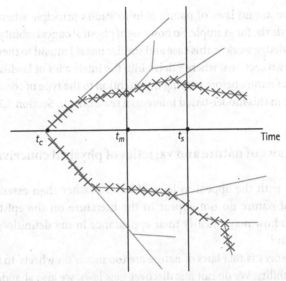

Figure 5.2 The time-sensitive epistemic accessibility relation between what is physically conceivable at any point in time $(t_c, \ldots, t_m, \ldots, t_s, \ldots)$ and what is possible. The branching paths in grey are physically conceivable scenarios consistent with the known laws of nature at any point in time $(t_c, \ldots, t_m, \ldots, t_s, \ldots)$. The paths marked with blue crosses are the ones that track what is possible as conceptual resources and state of knowledge evolve over time (e.g. new experimental practices become available, new laws become known, new models are devised, etc.). Not every physically conceivable path tracks modal knowledge—only the ones with blue crosses that epistemic communities learn to navigate and find out across a plurality of scientific perspectives. (For interpretation of the references to colours in this figure, the reader is kindly referred to the Open Access eBook version).

Instead of a linear or cumulative acquisition of modal knowledge, this is a branching one where some claims of modal flavour at t_c are either true or false, and accordingly either retained (blue crosses) or abandoned (grey paths) at t_m. This procedure gets reiterated at t_s, t_z, and so forth. This is what exploring the space of possibilities looks like from a perspectival realist standpoint: zig-zagging across the branching paths of what is physically conceivable for epistemic communities with the experimental, technological, and theoretical resources available at any historical time to model particular phenomena of interest.

Two more objections await. First, if the known laws change over time, how can they provide any solid ground for the notion of physical conceivability? If the known laws of nature are 'on wheels', would it not be better to appeal to essential properties to offer a stronger tether for physical conceivability?

Second, as the term 'physical' conceivability might suggest, this account seems to serve well for physics, but less well for sciences that do not trade in laws of nature. There are no laws of nature akin to Pauli's principle when it comes to DCM for dyslexia, for example. So how might physical conceivability as a guide to modal knowledge work in this case and similar ones? I attend to these objections in the next two sections, where I delve into the intricacies of lawlike dependencies vs laws of nature before turning my attention to the type of conditionals that I see at place in this model-based inferential reasoning in Section 5.7.

5.5. Laws of nature and varieties of physical conceivability

Let us start with the appeal to laws of nature rather than essential properties. Laws of nature do not appear in the literature on the epistemology of modality. So how may I justify their appearance in my definition of physical conceivability?

The first worry is that laws of nature are too much 'on wheels' to tether physical conceivability. We do not just discover new laws, we also abandon old ones. We abandoned the laws and principles of Aristotelian physics about free fall with Galileo. Indeed a lot of literature on thought experiments has stressed how violating known laws of nature is key to make progress (think of Galileo's thought experiment about weights falling from the leaning Tower of Pisa). Similarly, we gave up on the law of conservation of caloric with the advent of thermodynamics and its second law. Why trust laws of nature, then? Or why trust laws more than essential properties?

The answer is that perspectival modelling is modelling phenomena that show lawlike dependencies. Indeed the whole process of coming to know those phenomena (be they nuclear stability, global warming, or difficulties with reading, to cite the examples of Chapters 4.a, 4.b, and 4.c, among others) requires inferring them from a plurality of datasets with perspectival models acting as inferential blueprints. Like architectural blueprints specify the instructions that carpenters, masons, and joiners have to follow, perspectival models specify the nomic boundaries within which the relevant and appropriate inferences have to be made by various epistemic communities at work on the same task.

Consistency with laws does not, then, imply that laws cannot be violated in the modelling exercise. They *can be violated* and historically often *are* violated, as Galileo did with the Aristotelian laws of physics with respect to the phenomenon of free fall. Interestingly, these violations typically happen at key historical junctures when a system of laws is replaced by another one—something which is perfectly compatible with my novel perspectival Best System Analysis (*npBSA*) discussed in Section 5.6. More to the point, these historical violations have to pass a high bar: that of improving on the relevant and appropriate inferences about the phenomenon of interest (as did happen with Galileo's example). Violations that result in pathological solutions in the models are typically discarded, as are blueprints with instructions that would make the building structurally unstable.

Thus we trust the laws of nature because we need to buttress the exercise of physical conceivability. In some case, the buttresses can be disposed of once the modelling is complete, as the principle of conservation of caloric was expunged from the Carnot cycle when its inconsistency with Joule's paddle-wheel experiment became evident. In other cases, the buttresses stay in place until they continue to serve well communities of modellers over time and across scientific perspectives.

Laws play different roles in different conceivability exercises inherent in different models. Laws can *drive* the identification of causal mechanisms behind given phenomena. Or they can *enable non-causal explanations* of some phenomena. In yet other cases, laws *fix broad nomological boundaries* for making inferences about the very existence of some hypothetical entity. Elsewhere (Massimi 2019a), I have called the first *law-driven physical conceivability* and the third *law-bounded physical conceivability*. Let me expand on each of these distinctions in what follows.

5.5.1. Law-driven physical conceivability

In *law-driven conceivability*—physical conceivability$_{LD}$—I see laws of nature as driving the analogical reasoning between one modelling practice and another.[13] Think, for example, of fictional models such as Maxwell's (1861–2/1890) honeycomb model of the ether (to use the example from Massimi 2019a). The goal was to identify the possible causal mechanism for electromagnetic induction. Maxwell physically conceived of an ether with a honeycomb structure where the ether vortices produced a magnetic field which in turn put in motion what the model described as idle wheels among vortices—little particles in between vortices corresponding to electric displacement.

The model worked by analogy with fluid dynamics. The relevant laws of nature were Faraday's law of electromagnetic induction and Helmholtz's equations for fluid dynamics. This law-driven exercise of imagining perfectly spherical cells moving among hexagonal cylinders in an elastic ether led Maxwell to his discovery that the elasticity of the medium for electromagnetic induction (magnetic ether) was the same as the elasticity of the medium for the transmission of light (luminiferous ether).[14] The conclusion that light was nothing but an

---

[13] For the key role that analogies and analogical reasoning plays in science and scientific models, see Hesse (1966).

[14] This is how Maxwell (1861–2/1890, pp. 13–14 and 21, emphasis in original) presented his model: '[W]e may conceive that the electricity in each molecule is so displaced that one side is rendered positively, and the other negatively electrical, but that the electricity remains entirely

electromagnetic wave followed. In the following decade, Maxwell abandoned the model of the ether and codified the equations for the electromagnetic field that bear his name.

Physical conceivability$_{LD}$ can be found in many modelling practices. Think again of climate modelling. Modelling radiative forcings is an exercise in physical conceivability$_{LD}$ because we want to find out how the overall energy balance of planet Earth is affected by changing concentrations of greenhouse gases. And here too laws of nature are involved.

Analogical reasoning treats planet Earth as if it was a spherical black body that obeys the Stefan–Boltzmann law. The ensuing outgoing radiative energy E_{out} is set to be equal to the incoming radiative energy E_{in} in the so-called zero-dimensional energy-balance model of the Earth. Eric Winsberg has given a clear analysis of how we use models such as the zero-dimensional energy-balance model of the Earth to make inferences about climate:

> [L]et's re-examine our notion of a mediating model with the example of a zero-dimensional energy balance model. The first thing we said was that a mediating model is 'used to characterize a target system or phenomenon in such a way that various salient bits of theory and other mathematical regularities can be applied'. That's what we did. We characterized the earth as a homogeneous black body being bathed in a uniform field of radiation with a uniform degree of reflectivity and a single-parameter greenhouse coefficient, etc. We did this so that we could apply bits of theory like the Stefan–Boltzmann law, the conservation of energy, and our little side-model that told us about the effect of carbon dioxide and water vapor on the passage of long-wavelength radiation. We used the model to mediate between those abstract bits of theory and our target system: the sun, the earth, and its atmosphere, etc. . . . What's the point of these (energy balance) models, after all, with all of their crazy distortions? Arguably, the point is to get at a basic, but obviously *ceteris paribus*, causal effect: long-wavelength-radiation-trapping molecules in the gases of the atmosphere can, *ceteris paribus* (more on this in a bit), make a planet hotter. But also arguably, the point is to be able to write down an analytically tractable model that can be used to make calculations without a supercomputer. (Winsberg 2018, p. 31)

connected with the molecules. . . . I have deduced from this result . . . that the elasticity of the magnetic medium in air is the same as that of the luminiferous medium, if these two coexistent, coextensive, and equally elastic media are not rather one medium. . . . [W]e can scarcely avoid the inference that *light consists in the transverse undulations of the same medium which is the cause of electric and magnetic phenomena*'.

Another way of expressing this point (without using the language of models as mediators) is to say that physical conceivability$_{LD}$ is a guide to possibilities, where in this particular case the possibilities in question are *causal*. For example, we want to find out what is causally possible about global warming under a number of physically conceivable$_{LD}$ GHG scenarios. That is how the CMIP5 models of Ch. 4.b act as inferential blueprints in delivering robust climate projections under a number of physically conceivable$_{LD}$ RCPs and based on reconstructions of past climate and global mean surface temperature to date.

5.5.2. Law-enabled physical conceivability

A second variety of physical conceivability can be described as *law-enabled* conceivability—or physical conceivability$_{LE}$. This is conceiving something about a target system with the goal of *explaining* some aspects of it. What is the shell model good for? As mentioned in Chapter 4.a, the model explains why, for example, certain nuclear isotopes are more stable than others. The model delivers such *explanation* without necessarily providing the physical causes for the phenomena. For it would be mistaken to take the shell model as a *bona fide* metaphysical-causal description of the atomic nucleus. The specific orbitals may be referred to as 'shells', but the term 'shell' is not to be understood to imply confinement to a narrow radial range. The shell model has been adapted to include the effects of the strong short-range forces between nucleons so that, nuclei being quantum systems, no nucleon is completely in just one orbital and modern applications of the shell model test modern computers to the limit.

Thus, a philosophically more accurate way of thinking about the shell model is that by physically conceiving the nucleus as being thus and so, scientists can *explain* nuclear structure and the stability of some specific nuclides. This is because the Pauli principle dictates that there cannot be more than two identical fermions (neutrons and protons being fermions) in the same dynamic state, and because of the particular spin-orbit coupling among nucleons that the shell model physically conceives. This is an exercise in physical conceivability$_{LE}$. The Pauli principle enables such non-causal explanations within the conceivability exercise of the shell model. A brief clarificatory remark on this point.

Sometime laws of nature do deliver causal explanations. Phenomenological laws such as Snell's law can be regarded as delivering causal explanations about relevant phenomena (how light gets refracted in different media). The Pauli principle exemplifies a kind of non-causal explanation. The principle does not tell a causal story as to why half-integral spin particles cannot be in the same dynamic state, but it simply imposes a constraint on the states that an ensemble of such particles is allowed to be in.

There are similarities here with Marc Lange's (2017) account of non-causal explanations 'by constraint'.[15] I take this 'non-causal explanation by constraint' to be the role of the Pauli principle within the shell model: it fixes constraints on *possible* states allowed for nucleons regardless of causal relations. Indeed, it is precisely for this reason that the principle *does* feature across models for very different entities (nucleons, electrons, quarks, etc.).

But scientists can equally physically conceive$_{LE}$ the atomic nucleus as if it were a drop of incompressible nuclear fluid, among other nuclear models. The rival liquid drop model explains other phenomena than the stability of nuclides, such as nuclear fission, for example. The possibilities in question are *explanatory* (rather than causal): *possible explanations* for given phenomena. Laws here do not drive the identification of a causal mechanism. Instead they enable our physically conceiving$_{LE}$ scenarios with the goal of delivering knowledge of what is explanatorily possible. In this respect, I see physical conceivability$_{LE}$ as akin to what Robert Brandon (1990) and Alisa Bokulich (2014) have called how-possibly model explanations. I share with them the idea that how-possibly explanations are highly contextual and in Brandon's vocabulary they involve some kind of 'speculative' explanatory conditions. I see these contextual and 'speculative' explanatory conditions at work in the notion of physical conceivability. Moreover, I give a more substantive role to laws of nature than I would imagine either Brandon or Bokulich might be willing to give. Most of all, I see the physical conceivability exercise as feeding into an inferentialist story about how a plurality of epistemic communities use perspectival models to garner and refine the reliability of the explanations for relevant phenomena (e.g. think again of nuclear stability and the story behind it as recounted in Chapter 4.a).

[15] Lange considers ways in which the Pauli principle might indeed act as a constraint on non-causal explanations similar to mathematical explanations. Take Lewis's example of the collapse of a star, which stopped because of the Pauli principle. Lewis thought this was a case of causal explanation because the Pauli principle would act by preventing the star from collapsing all the way down (as sometimes absence of forces act by preventing bodies from accelerating). But Lange takes a different stance on the example and argues that the Pauli principle explains non-causally in this case: '[T]he explanation works by showing that even if the world's network of causal relations had been different, with different force laws (or even different laws about what happens in the absence of forces or other causes), the explanandum [the stopping of the star's collapse] would still have obtained, since the explanandum arises from constraints on all *possible* causal relations—where these possibilities range far beyond the causal relations allowed by the various particular force laws. Indeed, these constraints are not constraints on *causal* relations per se. They constrain *all* states and events, regardless of whether or not those states and events are caught up in causal relations or have causal explanations' (Lange 2017, p. 184).

5.5.3. Law-bounded physical conceivability

In my third case, physically conceivable scenarios guide scientists to infer—from particular lawlike arrangements—what *might be objectively* the case about a putative target system. This can be described as *law-bounded* conceivability or physical conceivability$_{LB}$. It is an exercise in conceiving something about a target system which we do not know is real. The goal is not to find out what is causally possible or explanatorily possible but to make inferences about the very *existence* of the relevant phenomena of interest. We are interested in what is *objectively possible*.

In the literature, one finds a specific notion of objective modality, namely how things could have been otherwise. There could have been no inhabited planets. There could have been no nuclei, and so on. How things could have been otherwise invites interesting questions about modality and the metaphysics of possible worlds that Lewis (1986), Stalnaker (2012), and Williamson (2016)[16] have brought to general philosophical attention. But this falls outside the scope of my analysis.

Perspectivism is by its own nature confined to epistemic modality. What is possible is always and only what is possible *for historically and culturally situated epistemic communities*. This is the case for claims of causal possibilities and explanatory possibilities, as well as for what I call here 'objective possibilities'. In asking questions about objective possibilities such as why there is not in nature a nucleus like ^{100}H, or why heat could not consist in caloric, the objective possibilities are themselves related to our models and associated empirical evidence. In other words, I see objective modality as emerging from the social and cooperative nature of scientific knowledge, a feature that resonates with Helen Longino's (1990) account of objectivity in science.

Within perspectivism, the distinction between epistemic modality, on the one hand, and causal, explanatory, and objective modality, on the other hand, is not

---

[16] Timothy Williamson has characterized the notion of objective modality as follows: 'Let "n" name the actual number of inhabited planets. There are exactly n inhabited planets, as our stipulation guarantees. Since our planet is inhabited, we know that $n \geq 1$. However, even though we know for sure that there are no fewer than n inhabited planets, there could have been fewer than n, because there could have been *no* inhabited planets. Such a sense in which things could have been otherwise is *objective* rather than *epistemic*. It is not a matter of what any actual or hypothetical agent knows, or believes, or has some other psychological attitude to; nor is it a matter of what any actual or hypothetical agent *ought* to be or do, either morally or in order to achieve a given purpose. Conversely, some epistemic possibilities are not objective possibilities of any kind. For instance, since we do not know whether other planets are inhabited, it is in some sense both epistemically possible for us that $n \geq 2$ and epistemically possible for us that $n < 2$. . . . Objective modalities are non-epistemic, non-psychological, non-intentional. Thus they are not sensitive to the guises under which the objects, properties, relations and states of affairs at issue are presented. . . . The present category of objective modality corresponds roughly to Angélika Kratzer's "root" or "circumstantial" modals and to Paul Portner's "dynamic" modals' (Williamson 2016, p. 454).

as clear-cut as it is *outside perspectivism*. For in the latter case, a whole repertoire of metaphysical and semantic tools are available to talk about objective modalities that are not available to perspectivists.[17] One might reply at this point: so much the worse for perspectivism if it cannot talk about non-epistemic non-agential objective possibilities.

But there is still a lot that a perspectival realist can say about objective possibilities as long as one accepts that they are ultimately within the domain of epistemic modality. There is no requirement to go down some lengthy metaphysical route to be able *to talk* and *think* about what is objectively possible within the resources of perspectival modelling.

For example, a perspectivist—no less than a Lewisian—may want to talk about objective possibilities and distinguish them from more distinctively epistemic varieties. The latter are captured by sentences of the form '*Lavoisier thought that* oxygen was at work in combustion', or '*Goeppert Mayer believed that* a strong spin–orbit coupling could be at play in magic numbers'. That person X thought that *p*, or person Y believed that *q*, is a familiar way in which epistemic possibilities are cashed out.

But a perspectivist may also want to speak about objective possibilities as being *de re* possible, and not just possible in the eyes of person X; or in the light of what person Y thought, believed, and so on. We might be interested in which kind of nuclides can naturally occur, whether any SuperSymmetric (SUSY) particles might be found at the Large Hadron Collider, or whether a DNA with eight nucleotides could be engineered. The modal verbs 'can', 'might', and 'could' in these examples are not embedded in sentences of the form 'Person X thought that *p*' or 'Person Y believed that *q*'.

Instead they express what I'd like to call 'objective possibilities'. However, and here comes the difference from other accounts, such claims are neither free-floating, nor do they live in a Lewisian metaphysical realm of possible worlds, according to perspectivism. They are always claims of modal knowledge available to historically and culturally situated epistemic communities that have the technological, experimental, and modelling practices to explore them.

Exploring objective possibilities requires a *law-bounded* physical conceivability$_{LB}$, where the role of the laws is more broadly to fix the nomological boundaries for imagining. I have discussed this elsewhere in the case of searches for supersymmetric particles (Massimi 2018b, based on models from ATLAS Collaboration 2015). How could $x_1 \ldots x_z$ be possible, were $y_1, \ldots y_n$ physically conceived$_{LB}$ thus and so? For example, how could the particle called the Higgsino decay were a certain model point (in the phenomenological minimal

[17] I am very grateful to Franz Berto, Peter Hawke, and Tom Schoonen for helpful conversations on this topic.

SuperSymmetric model pMSSM-19) fine-tuned in this way? Any conceivable$_{LB}$ model point featuring the relevant particle that has not been ruled out (at a 95% confidence level) by experimental evidence from proton–proton collisions at the LHC remains a live objective possibility. As new evidence becomes available, it refines those inferences by excluding more and more possibilities $x_1, \ldots x_z$. This is because new evidence gradually increases constraints on what is physically conceivable$_{LB}$. Via bootstrapping, more and more stringent how-possible inferences are obtained, so that more and more possible SUSY particles get excluded with a high confidence level as new energy regions at the LHC are probed.

But one could similarly ask 'How could difficulties with reading arise were language development to be physically conceived$_{LB}$ thus and so?' (i.e. according to one of the causal models in the DCM framework discussed in Chapter 4.c). And here too, given the comorbidity of the phenomenon with others, it is through more and more stringent inferences supported by various pieces of empirical evidence—psychological, behavioural, environmental and neurobiological—that we can give the answer to the question.

Let us take stock. We have seen how laws of nature can play different roles: driving the analogical reasoning behind some modelling practices; enabling non-causal explanations of some relevant phenomena via models; or fixing the nomological boundaries within which the modelling exercise can answer questions about what is objectively possible. However, some pressing questions remain.

The first worry is that appeal to laws of nature is both too strong and too weak for a perspectival realist to make. It is too strong because one may suspect that appeal to the known laws of nature does the metaphysical heavy-lifting for the perspectival realist, that it sneaks in realism through the back door. It is too weak too because our laws of nature seem themselves susceptible to change over time. The second worry is that such a role for laws of nature cannot apply to areas of scientific inquiry that do not trade in laws like physics does (e.g. developmental psychology).

In the next section, I deflate both worries by drawing a distinction between lawlikeness and lawhood. I am a primitivist about the lawlikeness of stable events in nature (this is my realist tether, to which I return in Chapter 6). But I am a perspectivalist all the way down about lawhood, or better about how to understand the 'known laws of nature'. Thus, that laws of nature might change over time and across Kuhnian revolutions is grist to my perspectivalist mill. The burden of realism does not rest on the known laws of nature but on the lawlikeness (or, better, lawlike dependencies) that I see as a primitive relation among relevant features of phenomena (or, better, as a primitive relation for what I call a 'stable event'). And the good news is that lawlikeness cuts across the distinction between sciences that trade in laws and those that do not (be they psychology, biomedical sciences, or similar examples).

5.6. A perspectivalist take on laws of nature

Going back to the first worry that appeal to laws of nature is too strong, suppose a critic complains that perspectival modelling can deliver on realism only to the extent that the (shared) 'known laws of nature' enter in the various notions of physical conceivability and ensure that the inferences we draw from these models *converge*.

The complaint would carry weight if I had suggested a metaphysically robust notion of laws of nature. In other words, my envisaged critic would have a point, if I were a Necessitarian, believing that laws express necessitation relations between universals (following Armstrong and Tooley), or if I were defending some kind of dispositional essentialism about laws of nature whereby the nomic necessity of the laws is ultimately grounded in essential properties and their dispositions (recall the discussion in Sections 3.5 and 3.6 in Chapter 3).

But I have not said anything about realism yet. Part II of this book articulates the kind of empirically motivated realism that I see as emerging from scientific perspectives and the practice of perspectival modelling. Before then, let us take a closer look at what I mean by laws of nature. I will do so by drawing on some of my earlier work (see Massimi 2017d and 2018f).

I defend neither Necessitarianism nor dispositional essentialism about laws of nature. I advocate a thoroughgoingly perspectival account of laws of nature, which takes as its starting point David Lewis's (1973) Best System Account (BSA) of laws of nature and gives it a perspectival twist. In Lewis's BSA, laws are defined as axioms or theorems in the best deductive system. This is the system of knowledge that can be deductively organized into premises and conclusions, assuming one has complete knowledge of such things as a Limited Oracular Perfect Physicist.

In Lewis's account, such deductive organization follows principles of simplicity, strength, and balance. Simplicity means using few premises or axioms. Strength invites us to expand the informational content of the system. But simplicity and strength pull in opposite directions—the simpler a system, the less informational content. Hence the need for a 'balance' between the two.

Elsewhere (Massimi 2018f, p. 153), I have argued that simplicity, strength, and balance (SSB) are always contextual standards, defined by any given scientific perspective *sp* at a given historical time. Therefore, different combinations of standards (let us denote them as SSB_{sp1}, SSB_{sp2}, SSB_{sp3}, . . .) are in place across different scientific perspectives. I argued that SSB_{sp1}, SSB_{sp2}, SSB_{sp3}, . . . should not be seen as either defining the truth or providing the justification for the scientific knowledge claims within each scientific perspective. Instead, borrowing an apt expression from Jay Rosenberg (2002, Ch. 4), I see them as *standards of performance adequacy* monitoring how well those scientific knowledge claims have served the need of epistemic communities who advanced them and the need of any subsequent community that inherits them.[18] In the same paper, I proposed what I called a 'novel perspectival Best System Account' (*npBSA*) of lawhood whereby

> (*npBSA*) Given a scientific perspective *sp*, and given SSB_{sp} qua standards of performance adequacy, laws of nature are axioms or theorems of the *perspectival series* of Best Systems, which satisfy SSB_{sp1}, SSB_{sp2}, SSB_{sp3}, and so forth.

This was my response to criticisms of David Lewis's view, in particular the objection that simplicity and strength are vague and subjective. But *npBSA* can also respond to my own envisaged critic's wariness of my appeal to the 'known' laws of nature. Laws of nature, as I understand them, do not provide any metaphysical tether. The known laws of nature are nothing beyond axioms or theorems of the perspectival series of Best Systems historically developed across many scientific perspectives.[19] Thus, I am not suggesting laws of nature as foundations of some sort to the conceivability exercise of perspectival modelling. My account is thoroughly perspectival all the way down.

Perhaps I am conceding too much now? If the known laws of nature are axioms or theorems in the perspectival series of Best Systems, is not there a risk of them coming and going over time? One does not have to invoke here the 'lunacy of the ratbag idealist' that Lewis himself was at pains to avoid (for a discussion,

[18] Thus, as long as 'the epistemic needs of scientific communities tend to be similar across time and across scientific perspectives, these standards too tend to stay the same across time and across scientific perspectives. Simplicity was a value for Newton no less than for Lavoisier. As such, these standards are constitutive of scientific inquiry across historically and intellectually situated scientific perspectives. This is important because these standards allow scientific communities to assess the ongoing performance of scientific knowledge claims across time. Scientific knowledge claims concerning specific laws of nature that continue to fare well on the scores of simplicity, strength and their balance over time are justifiably retained in subsequent scientific perspectives; they are discarded, otherwise. Yet there is a sense in which these standards also vary (not among individuals, as Kuhn had it) but among scientific perspectives. Thus, while deeply constitutive of scientific inquiry and fairly stable across theory change, standards of performance adequacy are also subject to interpretive shifts across scientific perspectives' (Massimi 2018f, pp. 153–154).

[19] 'Mass conservation was an axiom in the best system at Lavoisier's time (where strength was a large-meshed net that allowed for caloric to feature among non-nomic facts). And it continues to be an axiom in our best system now, despite our standards of simplicity and strength having been drastically recalibrated to exclude the deduction of caloric-related phenomena. Momentum conservation was a law in the Cartesian Best System, and it continues to be a law in our current Best System, where strength has been redefined to encompass not just elastic collisions in classical mechanics, but also deep inelastic scattering, jet fragmentation, and hadronization in high-energy physics' (Massimi 2018f, p. 155).

see Massimi 2017d). All is needed to make the point is a mundane observation: if laws are somehow indexed to a scientific perspective (with SSB understood themselves as standards subject to historical variation), we do not have any tether to anchor the physical conceivability exercise to reality. Aristotle could conceive of heavy bodies falling down to their natural place as much as Galileo could conceive of heavy bodies *not* falling down to their natural place. All that was needed for this transition was a change in the known laws of nature (from Aristotelian physics to Galileo's own law for free fall).

But this objection trades on an ambiguity. Defending a perspectival version of the Best System Account of laws is not buying into any claim that lawhood may come and go as our scientific perspectives come and go. Otherwise *npBSA* would indeed look like a re-enactment of the 'lunacy of the ratbag idealist'.[20] But neither Lewis's *BSA* nor my *npBSA* can justifiably stand accused of this. Being a law is contingent upon our Best System and its perspectival standards of simplicity, strength, and balance. Yet the specific (causal or non-causal) *lawlike dependencies* among relevant features of specific phenomena that act as target systems for perspectival modelling are *not* contingent.

One should not confuse lawhood with lawlikeness. Lawhood is what makes something a law of nature and it is contingent on whichever Best System particular epistemic communities work with at any particular time. But the lawlikeness that one sees in phenomena (or, better, among specific features thereof) is grounded in nature and is not on wheels. To disambiguate the two, consider the following example.

The lawhood of the Pauli principle has to do with there being a Best System of knowledge (quantum mechanics) in which it features as an axiom (and the spin–statistics connection as a theorem). Consider now a range of different phenomena: for example, how electrons' distributions in orbitals (call it x) explain the periodicity of chemical elements (call it y); or how nucleons' distributions in shells/orbitals (x) explain magic numbers and the stability of certain nuclides (y). The lawlikeness between features x and y in these two phenomena is not

---

[20] 'The worst problem about the best-system analysis is that when we ask where the standards of simplicity and strength and balance come from, the answer may seem to be that they come from us. Now, some ratbag idealist might say that if we don't like the misfortunes that the laws of nature visit upon us, we can change the laws—in fact, we can make them always have been different—just by changing the way we think! (Talk about the power of positive thinking.) It would be very bad if my analysis endorsed such lunacy. . . . The real answer lies elsewhere: if nature is kind to us, the problem needn't arise. I suppose our standards of simplicity and strength and balance are only partly a matter of psychology. It's not because of how we happen to think that a linear function is simpler than a quartic or a step function. . . . Maybe some of the exchange rates between aspects of simplicity, etc., are a psychological matter, but not just anything goes. If nature is kind, the best system will be robustly best—so far ahead of its rivals that it will come out first under any standards of simplicity and strength and balance. We have no guarantee that nature is kind in this way, but no evidence that it isn't. It's a reasonable hope' (Lewis 1994, p. 479).

contingent on any perspectival Best System Account of laws. The periodicity of chemical elements or the stability of certain nuclides observed in the Earth's crust and meteorites exists in nature and would exist even if we did not have a Best System such as quantum mechanics with the Pauli principle in it.

Lawlikeness is something that a perspectival realist shares with the empiricist and the realist. Regardless of which particular metaphysical view one endorses about 'the laws of nature', that natural phenomena display lawlike dependencies among particular features is a fact that experimentalists, engineers, chemists, ecologists, psychologists, and so on, know. Divergences may arise between the *bona fide* Humean empiricist who might regard these lawlike dependencies as merely supervenient on non-nomic facts in a Humean mosaic; the non-Humean empiricist who might be willing to include causal lawlike dependencies (I stand with them); and the dispositional essentialist who read these lawlike dependencies as the expression of nature's essential properties and their dispositions.

But the fact remains that there are these lawlike dependencies. How we turn some (but not others in other domains) into axioms or theorems of a perspectival series of Best Systems over time does not change this fact. The known laws of nature that play a role in physically conceiving scenarios are always a subset of the larger set of (mind-independent and perspective-independent) lawlike dependencies existing among relevant features of phenomena. In other words, lawhood (as a function of the perspectival series of Best Systems) is always a subset of the aforementioned primitive lawlikeness of phenomena—just as physically conceivable paths that track modal knowledge are a subset of a larger and time-sensitive space of branching inferential paths that epistemic communities learn to navigate across scientific perspectives (recall Figure 5.2).

As we gain new evidence about phenomena, we adjust the standards of simplicity, strength, and balance of our Best System accordingly. The changes can be gradual and difficult to detect, as with the move from Aristotle's principles about free fall to the medieval impetus theory via Arabic science, and from there to Galileo's law of free fall (see Massimi 2010 for detail). Zooming out, one gets the feeling of major revolutions in the passage from, say, Aristotelian physics to Newtonian mechanics. But that passage was made possible by important steps taken by Abu'l-Barakāt al-Baghdādī, Buridan, Oresme, and Galileo, among many others (see Massimi 2016). This is what my definition of *npBSA* is designed to capture: that is, the portion of lawlike dependencies in phenomena that retain over time the status of axioms or theorems *across the perspectival series* of Best Systems—satisfying ever-evolving SSB_{sp1}, SSB_{sp2}, SSB_{sp3}, and so forth.

Thus, it would be a category mistake to conflate the flexibility of perspectival standards of simplicity, strength, and balance with the lawlikeness to be found in natural phenomena. Climate scientists studying global warming are latching onto real (perspective-independent) lawlike dependencies among

relevant features of selected climate signals—be they at play in the equilibrium of the geothermal temperature gradient for borehole data or ratios of
isotopes in palaeoproxies, and so on. That the Stefan–Boltzmann law drives
the analogical reasoning that treats Earth as if it was a spherical black body
radiating energy, and that the Stefan–Boltzmann law is itself a theorem in
our current Best System of laws but was not a theorem at the time of, say,
Galileo or Aristotle does not change the fact that this and associated other
lawlike dependencies hold among the relevant features of selected climate
signals under investigation. Nor does it affect the causal lawlike dependency
between anthropogenic greenhouse gases and the effects associated with it—
from ice melting to ocean heat uptake—namely the stable event called 'climate sensitivity'.

This is a more general point that I want to stress also in response to the second
worry that my account neglects areas of scientific inquiry that do not trade in
laws of nature. Again the key issue is not to conflate lawhood with lawlikeness.
Recall Frith and Morton's developmental contingency modelling for dyslexia in
Chapter 4.c. We saw how a plurality of causal models are available to correlate
features across different domains: behavioural, cognitive, and biological, with
the environment affecting all three. This modelling highlights how a range of
prima facie similar behavioural symptoms can hide very diverse pathways to the
phenomenon of 'difficulties with reading'.

Identifying the possible pathways within the developmental framework is
key given comorbidity of dyslexia with other problems such as slow temporal
processing and the attention deficit. In none of the causal models, and none
of the literature on dyslexia I have read, I encountered reference to any 'law of
nature'.

Yet developmental psychologists and cognitive neuroscientists do use reliable and robust procedures for identifying what—in my philosophical lingo—
I call causal lawlike dependencies among relevant features. Let us go back to
Figure 4.c.1 and Figure 4.c.2 in Chapter 4.c and their different causal models for
difficulties with reading at the behavioural level. In both cases, the models point
to a neurobiological difference—cerebellar abnormality in Figure 4.c.1 and temporal lobe and frontal cortex abnormality in Figure 4.c.2. These differences may
lead to a number of problems in cognitive processes.

In Figure 4.c.1 these are timing/sequence deficit which in turn causes phonological deficit and motor control deficit. In Figure 4.c.2 there are two separate
cognitive deficits: attention and phonological, where the latter is not downstream from any further cognitive deficit in this model. All these cognitive processes in turn causally affect a range of behavioural tasks. Some are common in
the two figures: from difficulties with reading to difficulties with naming speed.
Others are specific to each causal model: difficulties with motor development

and with balance in Figure 4.c.1; difficulties with planning and school achievement in Figure 4.c.2.

By looking at the differences in the performance of specific behavioural tasks, psychologists can establish how the deficits vary from one individual to another, and discern the relevant pathway for these tasks. Educationalists can use these models to implement more tailored teaching strategies better suited to the individual needs of each child. Neurobiologists can work to better understand the differences among these possible neurobiological bases for similar symptoms.

The graphs in these causal models are designed to highlight the relevant causal lawlike dependencies among features across a number of domains—neurobiological, cognitive, and behavioural. There might not be a 'law of neurobiology'—analogous to say the Stefan–Boltzmann law or the spin–statistics theorem—that tells neuroscientists why a particular feature of the neurobiology causes particular cognitive deficits; or why the latter in turn cause the behavioural phenomena that one might observe, and so forth.

But it would be hasty to conclude on this basis that therefore there is no lawlikeness in phenomena such as difficulties with reading; or, better, between, say, phonological deficit in the cognitive domain and difficulties with reading and writing. There are lawlike dependencies (probabilistic ones) between specific cognitive deficits and performance of associated behavioural tasks. Psychologists routinely use these to make diagnoses for SLD, and could not do so without them.

In general, the known laws of nature that matter for physical conceivability and perspectival modelling need not be enshrined in any Best System. Some scientific fields thrive with perspectival series of Best Systems. Others are more parsimonious or altogether deflationary about it. Nonetheless, lawlike dependencies among relevant features of phenomena of interest are still present. Woodward's (2003, 2007) causal interventionism comes to mind here to better grasp the nature of these lawlike dependencies in causal models for dyslexia and other special sciences, for example.

I will return to these causal (and non-causal) lawlike dependencies when I spell out the realism that I see emerging from these modelling practices. But before then, recall that the main philosophical idea I said I was going to unpack in this chapter was the following:

(I)   Perspectival models model possibilities by *acting as inferential blueprints* to support a particular kind of conditionals, namely *indicative conditionals with suppositional antecedents*.

I have now explained how perspectival models act as inferential blueprints. In the final section I turn to indicative conditionals and the modal nature of the inferences at play.

5.7. Indicative conditionals and subjunctive conditionals: a division of modal labour

Let us take stock. I see perspectival models acting as inferential blueprints that can deliver modal knowledge by inviting us to physically conceive certain scenarios. But the way I see this taking place is not in terms of models supporting counterfactual conditionals, as sometimes one finds in the literature (e.g. Rice 2015, 2018). It is time for me to explain this point. Counterfactual reasoning invites us to ask *what-if-things-had-been-different* kind of questions, as when one says, 'I would have gone to the cinema, if I had not had to work last weekend'. The *if*-clause refers to a state of affair that is known to be not realised (e.g. I did have to work after all last weekend), and the main clause invites us to think what *would have happened if things had been different* (I would have gone to the cinema).

There might be modelling situations where, no doubt, this exercise of asking *what-if-things-had-been-different* kind of questions applies. But, I contend, this is not what goes on when we reason with perspectival models. To see why, consider the Coupled Model Intercomparison Project (CMIP5) climate models of Chapter 4.b, where Representative Concentration Pathways (RCPs) invite us to imagine certain greenhouse gas (GHG) concentration scenarios so that one can ask questions about global warming along the lines of, e.g. 'Let us conceive that GHG concentration peaks by mid-century and declines before 2100 (as with RCP2.6). What might global warming look like in 2100?'. In this example, one is not asking a *what-if-things-had-been-different* kind of question where the antecedent includes a state of affairs that is known not to have occurred and the main clause asks how things could have gone differently.

Instead, one is asking to suppose or imagine (*physically conceive*, in my lingo) a certain scenario (e.g. concerning GHG concentrations in the year 2050 and onwards) and then reason about how global warming would be affected under this scenario. The year 2050 has not arrived as I write this book. What GHG concentrations are going to be like in 2050 is something we can only make reasonable assumptions about as of today. Thus the reasoning here at play is distinctively different from counterfactual reasoning. Reasoning with perspectival models involves using the models and the particular scenarios they invite us to imagine or physically conceive to make inferences about phenomena of interest. The inferences in question cover predictions, long-term projections, diagnoses, or calculations of specific parameters of interest, among others. They take the form of indicative conditionals with suppositional antecedents.

Think again of the kind of inferences scientists make using the perspectival models I described in the case studies in Chapters 4.a, 4.b, and 4.c. For example, consider the shell model of the nucleus. The goal in this case is not to make long-term projections (as with climate models), but to explain nuclear stability among other phenomena. The inferences can be expressed with the following indicative conditionals, among others:

(E.1) If the nucleus is physically conceived$_{LE}$ as per shell model, then natural radioactive chains will end with lead ($Z = 82$).

(E.2) If the nucleus is physically conceived$_{LE}$ as per shell model, particular nuclides with magic numbers will show stability.

With developmental contingency models for dyslexia the inferences are for diagnostic purposes. Once again we can express them with indicative conditionals as already indicated in Ch 4.c:

(E.3) If language development is physically conceived$_{LB}$ as per DCM in Figure 4.c.1,[21] there will be comorbidity between difficulties with reading and difficulties with balance.

(E.4) If language development is physically conceived$_{LB}$ as per DCM in Figure 4.c.2, there will be comorbidity between difficulties with reading and difficulties with planning.

And so on. The whole point of DCM is to allow psychologists, neurobiologists, and educationalists to make relevant and appropriate inferences, as indicated in Chapter 4.c. Physical conceivability enters the *If* clause in the above indicative conditionals. The main clause of the indicative conditional typically has a present or future tense, that is, 'is', 'will be'. However, as already pointed out in Chapter 4.c, it is not a necessary truth that if someone has a phonological deficit, they will also necessarily display difficulties with reading. Depending on a number of environmental factors—from access to good schooling to the transparent nature of the language in question (e.g. Italian vs English)—the child might develop good compensatory strategies and not display difficulties with reading by, say, the age of 13. Therefore, as already mentioned in Chapter 4.c, the main clauses of the indicative conditionals here at play are best read as hiding a modal verb:

(E.3) If language development is physically conceived$_{LB}$ as per DCM in Figure 4c.1, there *may* be comorbidity between difficulties with reading and difficulties with balance.

---

[21] One can take the DCM in Figure 4.c.1 as expressing a physically conceivable scenario with well-defined pathways linking behavioural phenomena with their respective cognitive and neurobiological bases. In my language, this would be an example of physical conceivability$_{LB}$ *within* the (loosely defined) boundaries of lawlike dependencies in this case.

(E.4) If language development is physically conceived$_{LB}$ as per DCM in Figure 4.
c.2, there *may* be comorbidity between difficulties with reading and difficulties
with planning.

Let us take a closer look at how the inferential reasoning supported by perspectival models is expressed by indicative conditionals with suppositional
antecedents like these.

Philosophers of science have long been fascinated by subjunctive and counterfactual conditionals, especially when it comes to modality and the modal
reasoning surrounding laws of nature. Indicative conditionals are usually not
regarded as involving objective modality (see Williamson 2016, p. 465), by contrast with subjunctive conditionals. And we often use subjunctive conditionals
where we could have got by with indicative conditionals, as when one says,

(E.1*) If the nucleus were physically conceived$_{LE}$ as per shell model, then natural radioactive chains would end with lead ($Z = 82$)

where the subjunctive in (E.1*) does not add any extra modal dimension to
what is expressed by the indicative conditional (E.1). The fact of the matter is
that natural radioactive chains can be explained on conceiving the nucleus along
the lines of the shell model. We do not really need the subjunctive mode to remark that if the nucleus is physically conceived$_{LE}$ in a certain way, then natural
radioactive chains end with lead; or that if language development is physically
conceived$_{LB}$ as per DCM in Figure 4.c.2, there is comorbidity between difficulties
with reading and with planning. Using the subjunctive mode in these contexts is
a loose way of speaking that does not add any genuine extra modal information.

However, I do take the indicative conditionals like the ones just mentioned
to be expressing *implicit* modal information. Following Angelika Kratzer (2012,
Ch. 4), I see them as 'bare indicative conditionals' with an implicit modal verb
such as e.g. 'may' or 'might'.[22] They are epistemic modals, as logicians and
philosophers of language call them. Kratzer sees the *if*-clause in indicative
conditionals as a restrictor for the scope of the modal verb in the main clause.
From a heuristic (rather than semantic) point of view, one can see the *if*-clause as
a supposition under the suppositional view of indicative conditionals.[23] On the
latter view, what I call physical conceivability can be regarded as a way of making

[22] Kratzer (2012, Ch. 4, pp. 101–104) refers to the epistemic modal MUST in her example of how
bare conditionals have unpronounced modal operators. In what follows I am interested in the epistemic modals 'may', 'might', 'can', and 'could' instead. I see these epistemic modals at work in the
main clauses of indicative conditionals supported by perspectival models.

[23] See Edgington (2020) for details; see also Williamson (2020) for an extensive analysis of the
semantics and heuristics of conditionals with a suppositional antecedent. I am grateful to Timothy
Williamson for helpful comments.

a supposition and exploring the consequent: let us suppose/conceive that p, and ask ourselves whether q will be/might be the case.

The physically conceiving I see at play in perspectival models can be viewed as an invitation to *suppose* a given scenario and ask us to make a judgment about the consequent (main clause of the conditional) under the assumption that the supposition is satisfied. After all, recall that perspectival models are exploratory and the inferential reasoning associated with them can be regarded as largely suppositional. In the example (E.2), nuclear physicists around 1930s-1950s were supposing/conceiving of a given scenario (shell model) to explore the possible stability of certain nuclides with particular magic numbers.

Here we need to keep separate two issues that easily become entangled. Most of the literature on indicative conditionals has centred on the psychology of conditionals: what happens to our system of beliefs when we evaluate conditionals by imagining/supposing an antecedent and adjust our background beliefs and knowledge in the light of the supposition to judge whether the consequent is likely. To do that, one usually performs a version of the Ramsey test.[24] The literature in logic, psychology, and semantics surrounding the Ramsey test need not detain us, though. For my goal is not to contribute to any formal framework for the semantics or probabilistic logic of indicative conditionals.

My focus is on the model-based inferential reasoning that scientists make with perspectival models, and I see indicative conditionals as playing centre stage in it. I am interested in how we reason with perspectival models to get to know the world; and what we can legitimately conclude *about reality* by reasoning this way. Make no mistake, though. I do not see the burden of realism as resting on these indicative conditionals. For they are common and widespread in colloquial language as well as in mathematics, not just in scientific modelling. They do not bear any hefty metaphysical import up their sleeves. They only express common and general ways of reasoning involving some kind of imagining, supposing, or what I call physical conceivability. Thus, the burden of realism ought to be placed somewhere else, more precisely on the particular ontology of nature that I articulate in Part II of the book, starting from Chapter 6. But in the meantime, what else can be said about these indicative conditionals at play in reasoning with perspectival models?

To start with, they enter into enthymematic arguments where some premises are not explicitly stated. Think of the number of assumptions that have to feature as premises for scientists to be able to use any of the aforementioned indicative conditionals to draw conclusions. Developmental psychologists use (E.4) as

---

[24] Frank Ramsey (1929, p. 247) presented the following scenario: 'If two people are arguing "If p, will q?" and are in doubt as to p, they are adding p hypothetically to their stock of knowledge, and arguing on that basis about q; . . . they are fixing their degrees of belief in q given p'.

short for 'If one conceives of language development along DCM in Figure 4.c.2 *plus* auxiliary assumptions a, b, c (e.g. contributing environmental factors such as ineffective schooling, and no transparent natural language), then there *may* be comorbidity between difficulties with reading and with planning'.

Second, suppositional antecedents can deliver true consequents via enthymematic argument even when the additional hidden premise is taken from theories that later turn out to be false, or inconsistent with current science. This is an important aspect of what I call the *conditionals-supporting inferences* delivered by perspectival models. Indicative conditionals express the way in which epistemic agents *think and talk* in inferential reasoning supported by perspectival models. They provide the ropes that, intertwined, make the chain of inferential reasoning robust and resilient over time and across scientific perspectives even when additional premises include theories that later turned out to be false. I shall give an example of this in Chapter 10, when it comes to inferences about the nature of the electric charge and, indirectly, the natural kind electron around 1897–1906.

Taking a long view, the task ahead for perspectival realism is to elaborate the inferential view that explain how and why scientists did come to agree that a certain historically identified grouping of phenomena is a natural kind (what I call an *evolving kind*) in spite of disagreements about how to think of some of these phenomena and notwithstanding long abandoned theoretical /modelling premises. I will argue that these inferences often take the form of indicative conditionals: they invite epistemic communities to walk in what I'd like to call the 'inferential garden of forking paths', in a game of giving and asking for reasons (to echo Brandom 1998) as to why particular groupings of phenomena seem to go hand-in-hand. We will examine specific examples of this inferentialist view of natural kinds in Chapters 8–10.

Third, there is a famous problem about the truth conditions of indicative conditionals with a suppositional antecedent. As David Lewis (1976) noted: thinking that q is probable on the supposition that p is not equivalent to believing that q is probably true. Lewis's objection was levelled against interpretations of the Ramsey test in terms of probabilistic semantics for indicative conditionals.[25] In what follows, I read 'if p, q' as 'may (if p, q)' where, following Timothy Williamson (2020, p. 136), one can take p as a supposition that restricts on heuristic (rather than semantic) grounds the epistemic modal in the main clause. For example, in physically conceiving the nucleus as per shell model, we

---

[25] See Stalnaker (1968) for a defence of truth conditions for conditionals in terms of semantics of possible worlds. Gärdenfors (1986) has offered an alternative interpretation of the Ramsey test as a test of acceptance, and on his view a conditional whose antecedent is a supposition is construed as having not truth conditions but conditions of acceptance based on what he calls a 'belief system'. For an overview of this literature (which falls outside my remit), see Arlo-Costa (2019).

automatically exclude scenarios where the nucleus is conceived of as per liquid drop model when assessing whether 'may (if p, q)'—that is, e.g., whether particular nuclides with magic numbers may show stability as in (E.2).

Although the covert modal in the main clause is epistemic, as explained, the main clause expresses sometimes causal, other times explanatory, or, at other times, objective relations: for example, *that* greenhouse gases cause global warming; *that* magic numbers explain nuclear stability; *that* there is comorbidity between difficulties with reading and with balancing. In other words, q in 'may (if p, q)' can be understood as a proposition that picks out sometimes causal, other times explanatory, or, at other times, objective relations. And these relations are in turn underwritten by (perspective-independent) lawlike dependencies, as discussed.

The covert epistemic modal in the main clause does not therefore express the sheer belief or thought of any particular epistemic agent: it captures instead possibilities concerning causal/explanatory/objective relations, respectively, within the limits afforded by perspectivism. For no epistemic community can reclaim a God's eye view to talk about what is causally, explanatorily, or objectively possible *simpliciter* or in absolute terms. There is nonetheless a lot that situated epistemic communities can think and say at any historical junction about what is causally possible, explanatorily possible, or objectively possible within well-defined domains of inquiry.

Consider the following example. The sentence 'Atomic number 82 is a magic number' is either true or false, regardless of which scientific perspective and perspectival model one operates with. Consider now the covert epistemic modal in it

(i) 'Atomic number 82 *may* be a magic number'.

One can easily imagine Elsasser and Guggenheimer (see Chapter 4.a) back in 1933 uttering (i) and assessing (i) as false on the supposition that the nucleus was conceived as per their shell model and on the basis on the available evidence at the time (recall that they did not have evidence for high Z being magic numbers). Yet in 1948, Goeppert Mayer assessed (i) as true in the light of new available evidence and on the supposition of her improved shell model, including a strong spin–orbit coupling. Elsasser and Guggenheimer were not wrong in assessing (i) as false back in 1933. Given their evidence and the early shell model, they could not countenance that 82 may be a magic number. But after 1948, it would be odd to insist on assessing (i) as false.

This is an illustration of what I mentioned above as a time-sensitive epistemic accessibility relation between what is physically conceivable at any point in time $t_e, \ldots, t_s$ and what is (explanatorily in this example) possible (since magic numbers Z explain nuclear stability). It is this kind of time-sensitive branching modal knowledge (recall Figure 5.2)—underpinned by lawlike dependencies in nature—that perspectival models deliver over time and across scientific perspectives. Although the modals involved in these indicative conditionals are epistemic in that they reflect the particular state of knowledge and perspectival models available to any epistemic community at any one time to *think* and *talk* about what is possible, whether there are indeed causal, explanatory, or objective relations as *actual* relations and—a fortiori—as *possible relations* (in a non-epistemic sense of possibility) depends on stable events in nature.

As I explain in Chapter 6, I understand the stability of events in terms of their lawlikeness as a primitive relation—this is the *realist* tether in my story. *That* there are actual causal relations between greenhouse gases and global warming is a fact about nature captured by 'climate sensitivity'. The multifactorial phenomenon 'global warming' is nothing but climate sensitivity that is reliably identified and reidentified through a plurality of inferences concerning various climate signals qua 'stable events' (as described in Chapter 4.b). *That* there are actual explanatory relations between magic numbers and nuclear stability is also a fact about nature captured by a number of stable events concerning natural radioactive chains and abundances of particular nuclides, as discussed in Chapter 4.a. In sum, as I have already warned my reader, indicative conditionals and their covert epistemic modals do not bear any hefty metaphysical import up their sleeves for my perspectival realism. Lawlike dependencies do.

Lawlike dependencies are perspective-independent. They do not change with the scientific model or the scientific perspective. They are not affected by changing scientific language or concepts. And they do not have to be enshrined in the Laws of Nature in a Best System (although often they are in the physical sciences; less so in other areas). That there are such lawlike dependencies is a fact about nature, not about us, or our scientific perspectives. Iron filings would still be attracted by the lodestone whether or not Faraday had spotted the phenomenon, and introduced 'lines of force' and a law of electromagnetic induction to describe it. Lawlike dependencies play a key role in cashing out realism in Part II of this book. But for now let me return one more time to indicative conditionals and highlight a further (and related) aspect concerning their truth conditions and assertability conditions.

Angelika Kratzer has argued that truth conditions and assertability conditions may diverge when modal claims depend on underspecified contextual parameters:

> In statements with unmodified epistemic modals, truth-conditions and assertability conditions can come apart. It may happen that, for one and the same modal statement, truth-conditions are inherently vague, while assertability

conditions are relatively sharp. For assertability conditions, speaker's evidence is what counts, but that's not necessarily so for truth-conditions. . . . It is a general property of circumstances of evaluation that they can remain un- or underspecified. We saw that even if the truth-conditions for a conditional assertion can remain underdetermined, the corresponding assertability conditions do not have to be. The presumption of cooperativeness allows us to rely on the assertability conditions, rather than the truth-conditions, to convey the information we want to convey and obtain the information we are seeking. (Kratzer 2012, pp. 101 and 104)

Like Kratzer, I see truth conditions and assertability conditions coming apart whenever relying on the evidence available to an epistemic community at any particular historically and culturally situated scientific perspective. Even if in the original context of use truth conditions for such conditionals are not well defined, their assertability conditions can often be sharp. Consider again the following indicative conditional:

(E.2) If the nucleus is physically conceived$_{LE}$ as per shell model, particular nuclides with magic numbers will show stability.

The assertability conditions for (E.2) were well defined around 1930–1950 in light of the evidence available about isotopic abundances. However, the truth conditions for this conditional were not sharply defined back in 1933 in the absence of a more complete theoretical understanding of the nuclear structure, which came only later with the 1948 shell model and the unified model of Bohr, Mottelson, and Rainwater (Chapter 4.a).

Asking for truth conditions for a conditional like (E.2) is therefore asking for a grand vista on knowledge claims and how they get retained or withdrawn over a long period and across many scientific perspectives. Truth conditions for knowledge claims are open-ended and ongoing, not a done deal business to be delegated and ratified by any specific scientific perspective.

Assertability conditions are, by contrast, much easier to identify at any particular time and within any particular perspective. They are within epistemic communities' perspectival grasp in a way that truth conditions are not. This does not mean there are *no* truth conditions for these claims. Of course, there are. Yet they should not be understood in terms of perspective-indexed truthmakers as if scientific perspectives could provide truthmakers for knowledge claims. Nor should truth conditions be regarded as relativized to scientific perspectives as if the propositional content expressed by a given claim remained the same but took different truth values in different perspectives.

There are perspective-independent facts about nature that ultimately act as truth conditions in adjudicating scientific knowledge claims. But I do not identify such perspective-independent facts with 'truthmakers' as in Armstrong's[26] classic truthmaker theory, whereby truthmakers are states of affairs that provide an *ontological ground* for truth. I endorse instead a standard correspondence theory of truth tempered by cross-perspectival assessment. As I argued elsewhere (Massimi 2018e, p. 354), I see truth conditions (understood as rules for determining the truth values based on features of the context of use) to depend on a scientific perspective in which the claims are uttered and made. Yet, crucially, I impose the restriction that such claims must be assessable from the point of view of other (subsequent or synchronous) scientific perspectives. Each scientific perspective acts both as a context of use and a context of assessment. In the latter role, each perspective offers a standpoint for evaluating the ongoing performance adequacy (to echo again Jay Rosenberg's terminology, 2002, Ch. 4) of claims of knowledge originally put forward in other (preceding or synchronous) scientific perspectives.

Consider once again examples from the history of science. There we have innumerable examples of knowledge claims. Some were retained from one scientific perspective to the following ones. Others were withdrawn because they failed to satisfy the standards of performance adequacy set by their own original context of use, when assessed from the point of view of other (historically subsequent) perspectives.[27] This Janus face of each scientific perspective—acting both as a context of use and as a context of assessment—is relevant to understanding how the conditionals-supporting inferences licensed by perspectival models can be *truth-conducive over time*.

My view chimes, then, with Kratzer's when she remarks that truth conditions may often be vague in the original context of use and may diverge from assertability conditions. It is only through what she calls a 'presumption of co-operativeness' among epistemic agents that one can rely on the assertability

[26] According to Armstrong (1997), a truth such as <*a* is *F*> should be understood as having as its truthmaker the state of affairs that [*a* is *F*], whereby such a state of affairs, in Armstrong's original formulation, was not understood as the mereological sum of individual *a* and universal property *F*-ness (for a different view which takes *a* and *F*-ness as overlapping in the instantiated state of affairs [*a* is *F*], see Armstrong 2004).

[27] 'For meeting standards of performance-adequacy fixed by the original context of use is, of course, not sufficient to establish the truth of any scientific knowledge claim. If scientific perspectives were allowed to sanction the truth of their own knowledge claims by their very own lights and standards, perspectival truth, would be bankrupt. Every scientific perspective could legitimize its own knowledge claims in the name of its own (genuine or presumed) standards of performance-adequacy. Ancient Greek crystalline spheres could be said to satisfy explanatory standards for the generation and corruption of entities in the sub-Lunar sphere; as much as phlogiston could be said to satisfy standards of predictive power for combustion and calcination. Contextual truth-conditions (qua standards of performance-adequacy) must be bridled to avoid perspectival truth, to accrue too easily' (Massimi 2018e, p. 357).

conditions within one's own scientific perspective (rather than grand-vista truth conditions spanning several perspectives) to engage in the inferential game.

What about cases where we do not have clear-cut cross-perspectival assessments? Suppose there is no easy way to retain knowledge claims couched in one model within a particular scientific perspective into those of another.[28] This might well be the case in several examples from the history of science. We might just have to live with the fact that there are situations where no cross-perspectival assessment for knowledge claims takes place maybe because on a long vista some of the forking branches (recall Fig. 5.2) turn out to be dead branches (more on this in Ch. 10). However, the main take-home message of this entire discussion about indicative conditionals should be clear. Nowhere have I suggested that knowledge claims should 'converge across perspectives'. Perspectival realism is not a brand of convergent realism because *there is nothing to converge to*. All we are left with are chains of indicative conditionals through which our inferential reasoning with perspectival models routinely takes place.

This inferential reasoning guides epistemic communities to navigate the garden of modal knowledge claims, based on their scientific perspective, its experimental tools for gathering evidence and data, and its modelling practices. Conditionals-supporting inferences are therefore *truth-conducive* to the extent that the paths taken in the inferential game of asking why particular groupings of phenomena go hand-in-hand *track*—across time and across scientific perspectives—*real lawlike dependencies* in the phenomena. Truth is *not* the yearned-for trophy at the end of scientific inquiry. Nor is it the overarching aim of cross-perspectival assessment.

Yet there is still a sense in which 'knowledge claims can be said to be true across scientific perspectives'. This is what elsewhere I call perspectival truth$_4$: 'combining a contextualist, yet still bona fide realist account of truth within a perspective' (Massimi 2018e, p. 357). Tracking real lawlike dependencies among features of phenomena is what secures that. This tracking finds in turn its semantic expression in subjunctive conditionals of the form 'were x the case, y would be the case'.

Here we see an interesting division of modal labour between subjunctive conditionals and indicative conditionals. As already mentioned, subjunctive conditionals convey the real (non-epistemic) possibility of some y occurring, were the antecedent x to hold. I see subjunctive conditionals semantically at

[28] Mazviita Chirimuuta (2019, p. 143) gives one such example from modelling the brain by comparing the *intentional perspective* with the *dynamical perspective* in neuroscience: '[E]ach neuron represents or codes for some state of affairs in the extra-cranial world. . . . An alternative is to model the dynamical evolution of the neural system but to seek a relatively simple set of equations governing it. This is the *dynamical perspective*, and it often (but not always) comes with the denial that neurons code for or represent anything'.

work in tracking lawlike dependencies as we move across perspectival models and scientific perspectives. It is the subjunctive mode that speaks to the resilience of the relevant lawlike dependencies in nature.

By contrast, the indicative conditional mode speaks to *our epistemic attitudes* when we judge whether something is likely to occur (as a causal, explanatory or objective relation) in the physically conceivable scenario described by the model. This is the realm of perspectival models acting as inferential blueprints. But it is ultimately the *lawlike dependencies* in the relevant phenomena that underpin the *truth-conducive* nature of conditionals-supporting inferences as we walk in the inferential garden of forking paths. One may think of lawlike dependencies and associated subjunctive conditionals as the signposts guiding epistemic communities down some paths rather than others at each forking junction.

Lawlike dependencies among relevant features and truth-conducive conditionals-supporting inferences will be crucial to support the view of natural kinds (Natural Kinds with a Human Face) that aligns with the brand of realism I wish to defend. I argue that our realist commitments typically emerge from a number of perspectival models (and the broader practice of perspectival modelling) for modally robust phenomena and their associated conditionals-supporting inferences. This is how epistemic communities learn over time to cooperate, to group modally robust phenomena into kinds, and to discern empty kinds from evolving kinds. To go back to the architectural metaphor of the blueprints, lawlike dependencies are relational instructions (e.g. roof height 5 metres from the ground; chimney elevation 1.5 metres from roof) that allow the modelling exercise and associated inferences over time to open up 'windows on reality'. But these are matters for Part II of the book.

PART II
THE WORLD AS WE PERSPECTIVALLY MODEL IT

6

From data to phenomena

6.1. Three philosophical traditions

We have come a long way from Chapter 1. In Part I, I described how our encounter with the natural world is always through scientific perspectives.
Scientific perspectives—as I understand them—are the historically and culturally situated scientific practices of real scientific communities at any given historical time. An integral part of these scientific practices are the experimental,
theoretical, and technological resources available to any scientific community at
any time to *reliably* make those scientific knowledge claims; and second-order
(methodological-epistemic) principles that can *justify* the *reliability* of the scientific knowledge claims advanced.

Whilst specific to any given scientific perspective, these modelling practices
also tell a story about how different epistemic communities come to make
inferences about the phenomena of interest and how therefore they open up a
'window on reality'. But what is to be said about these phenomena? What does
the natural world look like when encountered through perspectival modelling?
This is the task ahead in Part II, 'The World as We Perspectivally Model It'. In the
next six chapters, I spell out the details of the realist story in perspectival realism.

Perspectival realism is not a top-down kind of realism: from the best theories in mature science to states of affairs in the world. It works bottom-
up: from data to phenomena, and from phenomena to natural kinds. This
is the kind of realism that can work with perspectival pluralism. Realism is
arrived at over time from a kaleidoscope of historically and culturally situated
scientific perspectives.

In the previous chapters, I took a closer look at how a plurality of lines of inquiry, experimental evidence, and datasets feed into perspectival models so
that these models can act as inferential blueprints for various communities to
compare their findings, make more reliable inferences, and ultimately advance
knowledge claims about specific phenomena of interest. Perspectival realism, as
I see it, is *realism about these modally robust phenomena*.

Specifically, this chapter investigates how to go from data to phenomena. The
passage from phenomena to natural kinds will be the focus of Chapters 7 through
10. Are phenomena a plausible candidate at all for realism? Or do they belong to
a kind of phenomenalism that is no match for realism about science? I approach

Perspectival Realism. Michela Massimi, Oxford University Press. © Oxford University Press 2022.
DOI: 10.1093/oso/9780197555620.003.0009

these questions starting from three grand philosophical traditions: empiricism, realism, and constructivism. Highlighting the key insights in each is helpful for charting the way forward for perspectival realism.

The empiricist tradition includes Locke's defence of nominal essences, Bacon's method of observing and enumerating phenomena, Newtonian experimentalism in optics and chemistry, and, much later, Mach's criticism of absolute space. Empiricism finds its apt twenty-first-century incarnation in the Quinean 'tribunal of experience'. From observation statements (the 'protocol sentences' of the Vienna Circle) to the observable phenomena of Bas van Fraassen, it emphasizes the *empirical roots* of scientific knowledge.

On this view, knowledge is never knowledge of essences (or essential properties), but only of empirical occurrences. Whether the natural world comes pre-packaged with essential properties—be they categorical properties or dispositional ones—and modal capacities is a metaphysical claim that to the empiricist is not amenable to scientific answer. Scientific knowledge is knowledge that *originates* from empirical occurrences; that is *about* empirical occurrences; and it ultimately *responds* to empirical occurrences only.

Behind the second grand tradition—realism—there are well-founded concerns about the limits of the empiricist tradition to probe the epistemic machinery of science. Scientists do not just register empirical occurrences, cluster data, and use them to measure something or arrive at some general conclusion. Science is also about theorizing; coming up with hypotheses about unobservable entities that might be responsible for the observable phenomena; making inferences, finding out laws of nature, and offering scientific explanations. The ontological landscape of science is swarming not only with sparks in spark chambers but also with charged particles *causing* the sparks; not only with phenotypic traits but also with the DNA and RNA sequences that *encode* them. Behind each observable phenomenon—the realist argues—there are often unobservable entities, and these entities belong to kinds (different kinds of charged particles, different kinds of nuclei, different kinds of DNA sequences for zebras, lemons, hellebores, etc.).

The realist tradition has long argued for the *truth* of our best scientific theories in mature science as a way of securing the validity of our inferences and explanations. If we do not want our scientific inferences and explanations to run idly, realists maintain, entities and kinds must populate the ontological landscape of science. Science is about discovering new *entities*. For example, J.J. Thomson discovered the electron, Lavoisier discovered oxygen, and scientists at CERN discovered the Higgs boson. Our epistemic practices would be fruitless unless the world comes pre-packaged with a well-defined realm of entities and natural kinds characterized by properties that can ground the modal nature of our scientific knowledge.

Scientific knowledge, in this realist tradition, is indeed knowledge of the *modal* nature of entities and natural kinds: of what-would-occur-if-*x*-occurred, how-would-entity-*y*-behave-were-*y*-in-condition-*z*, and so on. It is hard for modal claims to gain hold in a world of bare empirical occurrences. Hence realists appeal to metaphysically richer scenarios where entities come fully equipped with categorical or dispositional properties, capacities, or potencies that embed as much as possible the required modal tools.

Yet these seem in turn inadequate to capture an important insight that has long been emphasized by a third tradition, that of constructivism (and, to some degree, the empiricist tradition, especially van Fraassen's constructive empiricism): namely, human beings have no privileged epistemic access to a world of essences, dispositions, potencies, unobservable entities, and so forth. Our scientific knowledge is knowledge of particular epistemic communities at particular times and places, and not knowledge of ever-lasting metaphysical entities in some Platonic realm.

This third tradition comes in many flavours and is harder to sum up. The variety that is most relevant here to my discussion includes Quine's ontological relativity, Goodman's worldmaking, and Putnam's internal realism. Quine's (1968) version of ontological relativity takes its cue from the anthropologist trying to understand what the native term '*gavagai*' stands for, without a translation manual. Any ostensive act associated with the utterance '*gavagai*' could be equally understood as denoting undetached rabbit parts or a multitude of temporal rabbit stages. This indeterminacy of reference is not the outcome of an impoverished evidential basis for translation, Quine argued. And it cannot be bypassed by pointing at more examples. It is instead the inevitable expression of how ontology and principles of identity and individuation (at play in nouns like 'rabbit') may work differently across different epistemic communities. Where one rabbit begins and another rabbit ends is pretty much relative to how different languages and cultures carve up the world.

Goodman (1978) went further, in claiming that we do make facts but not like, say, a baker makes bread, or a sculptor makes a statue. In Goodman's view, we make facts any time we construct what he called a 'version' of the world (via art, music, poetry, or science). A more prosaic, but no less effective, way of constructing a version of the world is by clustering objects according to a particular shape and giving them a name, as we do, say, with constellations. As Putnam (1996, p. 181) expresses it, 'Nowadays, there is a Big Dipper up there in the sky, and we, so to speak, "put" a Big Dipper up there in the sky by constructing that version'.

This Quine–Goodman–Putnam tradition relates to another strand that begins with Thomas Kuhn. No one more than Kuhn has stressed how the historian of science is like Quine's anthropologist in having to understand the lexicon of a past epistemic community. By contrast with Quine, however, Kuhn (2000, p. 49) concluded that radical interpretation (instead of radical translation) was at play in these situations. Past scientific lexicons are not translatable: different lexicons give access to different worlds that never entirely overlap. Quine's radical translator becomes, in Kuhn's hands, a language learner (see Kuhn 1990).

Accordingly, we did not discover that water is H_2O when Lavoisier discovered oxygen. Nor has the term 'water' been referring all along to some essential property of consisting of H_2O (via some kind of Putnamian causal baptism impermeable to conceptual changes) because, Kuhn argued, before 1750 the term referred only to *liquid* water and the distinction between liquids, solids, and gases became physical rather than chemical only after the Chemical Revolution. Under this tradition, scientific knowledge is, then, knowledge that *pertains to us* qua communities of historically and culturally situated epistemic agents. It is knowledge that originates from and reflects our finite linguistic and conceptual resources.

Among feminist philosophers of science, this led to a reassessment of truth, objectivity, evidence, and progress in science. The important work of Haraway (1988), Harding (1991), Kourany (2002), Longino (1990, 2002), Solomon (2001), and Wylie (1997, 2003) made it clear that taking the human standpoint seriously requires a profound re-thinking of realism and pluralism, the mechanisms through which scientific theories get confirmed, and our scientific beliefs are justified.

In what follows, I use the insights emerging from these three main traditions to explore the ontology of nature that a perspectival realist can endorse. The ontology that best fits perspectival realism has to take into account that scientific knowledge has *empirical roots*, a *modal nature*, and is *historically and culturally situated*. Perspectival realism thus shares with van Fraassen's the view that we *construct* models that reflect a human and perspectival point of view (see van Fraassen 2008). From data models to theoretical models, our scientific knowledge is empirical all the way up: it originates from empirical occurrences and has to respond ultimately to empirical occurrences.

But perspectival realism does not take scientific knowledge to be *just* about empirical occurrences. Perspectival realism is a form of *realism* in endorsing the *modal nature* of scientific knowledge. By contrast with other forms of realism, however, perspectival realism makes no grand claims about modality and its physical seat in powers, potencies, capacities, categorical properties, or dispositions. This is a thin rather than a thick form of realism.

It shares with Philip Kitcher's (1992) 'real realism' the Galilean strategy of starting with familiar situations to establish the *reliability* of our methods. It shares some of the motivations for structural realism (see French 2014; Ladyman and Ross 2007) in rethinking realism along lines that are more congenial to conceptual change and the problem of referential continuity over time. It is sympathetic to Stathis Psillos's (2012) causal descriptivism as a semantic strategy for reconciling Kuhn on conceptual change and Putnam's causal baptism.

And of course perspectival realism is a realist form of *perspectivism* after all, sharing with the third tradition the insight that scientific knowledge pertains to situated epistemic communities. This does not, however, translate into ontological relativity, worldmaking, conceptual relativism, or 'living in a new world' scenarios. Instead, its full implications will become evident in Chapter 7 when I discuss the perspectival realist approach to natural kinds—what I call *Natural Kinds with a Human Face* (NKHF).

My realist commitment is to the data, the phenomena, and the natural kinds. Where I differ from standard realists is in the perspectival story I tell about how one forms such realist commitments, all the way up from data to phenomena to kinds. Such a realist commitment originates *from within scientific history*. It is situated and perspectival every inch of the way, without being any less real.

6.2. Data and phenomena

A perspectival realist shares with the empiricist the intuition that scientific knowledge is not knowledge of metaphysical essences. Our scientific knowledge is empirical from beginning to end. Yet this does not imply forsaking all the good that comes from a realist approach. Modality is one such good. Insisting that scientific knowledge is empirical goes no way towards answering questions about what *could, would,* or *should* happen in nature. Scientific models capture what *may* be the case about some phenomena of intertest. Scientific theories explain and predict what *would* happen to it. These claims have a *modal* flavour that demands a richer account than the empiricist claim that phenomena *P were, are,* and *will be* as model *M*/theory *T* says they are.

Dispositions, categorical properties, potencies, and universals have all been deployed to various degrees by scientific realists. Empiricists in turn have put in place an impressive machinery to respond to the realists' concerns: David Lewis's (1973) possible worlds semantics is a case in point. However, the choice between endorsing the *empirical roots* of scientific knowledge *or* making sense of its *modal* nature is in fact a false dichotomy.[1] In the following sections, I point to this conclusion via a discussion of what I call *the evidential inference problem*: when

---

[1] In this respect, I join various philosophers who in different ways have resisted this dichotomy being imposed on us: from French's ontic structuralism (2014), to Ladyman and Ross's naturalized metaphysics (2007), and Bueno and Shalkowski's modal epistemology (2014), to mention just three examples.

do empirical occurrences *provide evidence for* phenomena? I begin with some comments on the terms 'empirical occurrences' and 'phenomena'.

In what follows, I shall use the term 'data' as a proxy for 'empirical occurrences'. The latter might suggest that our experiential encounter with nature is via things that *occur* rather than things *epistemic communities* select, model, and use for inferences. Scientists collect datasets and use them as *evidence* for phenomena. Only recently have philosophers of science started paying attention to the nature of 'data'. In her influential *Data-Centric Biology: A Philosophical Study*, Sabina Leonelli (2016, pp. 77–78; see also Leonelli 2019) has laid out a relational account of data, whereby 'data are objects that (1) are treated as potential evidence for one or more claims about phenomena, and (2) are formatted and handled in ways that enable its circulation among individuals or groups for the purpose of analysis'. Under this view, (a) data are portable and their reliability depends on social activities; (b) data are material artefacts whose scientific significance changes when the media change (e.g. photographic plate vs computer-aided reconstruction, etc.); and (c) data are non-local and man-made without, however, compromising their ability to serve as reliable evidence for claims about phenomena (pp. 88ff.).

Leonelli's notion of 'data journeys' latches onto and departs from another influential account originally due to James Bogen and James Woodward (1988). They drew a clear distinction between data and phenomena. The former, they argued, provide evidence for the existence of phenomena. From bubble chamber photographs to records of reaction times in psychological experiments, data are idiosyncratic to specific experimental contexts in a way that phenomena are not. Bogen and Woodward's notion of phenomena as stable and repeatable characteristics emerging out of different data marked the beginning of a new trend, in which phenomena are regarded as robust entities that scientific theories have to explain and predict. This worked against a 'thinner' notion of phenomena familiar from the empiricist tradition, most notably that of Pierre Duhem (1908/ 1969), who saw phenomena as appearances: for example, the appearance of retrograde motion that became the object of study of Ptolemaic astronomy and a century-long tradition of what Duhem called 'saving the phenomena'.

From a metaphysical point of view, Bogen and Woodward share realist intuitions about phenomena: they exist 'out there' in nature, not as mere appearances but as stable and repeatable features emerging across a variety of experimental contexts and data. As Woodward (1989, p. 438) nicely puts it: 'Detecting a phenomenon is like looking for a needle in a haystack or . . . like fiddling with a malfunctioning radio until one's favourite station finally comes through clearly'.

Despite the variety of the *causal factors* involved in data production, Woodward stresses the importance of controlling and screening off influences (e.g. potential confounding factors, background noise) that may undermine the reliability of phenomena detection.[2] Thus, the epistemological framework is reliabilism while the metaphysical framework is close to that of Ian Hacking's experimental realism. Hacking too (1983, p. 221) has defended a realist view of phenomena on experimental grounds: 'A phenomenon is commonly an event or process of a certain type that occurs regularly under definite circumstances'. Phenomena can be created in a lab thanks to scientists' ability to intervene in causal properties of entities, as when weak neutral currents are created with the electron gun PEGGY II, which exploits the causal properties of circularly polarized light on a gallium arsenide target to produce electrons.

In what follows, I articulate a view of phenomena that is kindred in spirit to Hacking's. Like him, I see phenomena as being the outcome of an inference. Where I depart from Hacking's experimental realism is in the emphasis I place on perspectival modelling rather than experiments in delivering phenomena, and on the lawlikeness of events rather than causal properties of entities in offering the realist tether. Perspectival realism is not realism about causal properties to intervene in but realism about modally robust phenomena inferred via perspectival data-to-phenomena inferences. I side, then, with Leonelli in taking data as the starting point of a thoroughgoingly empirical view of scientific knowledge. I endorse Bogen and Woodward's stance in taking phenomena (rather than data per se) as the ontological unit for the perspectival realist's commitment.

Yet there is more to be said about phenomena, and in particular about data-to-phenomena inferences, than reliabilism might offer. I see data-to-phenomena inferences as perspectival in ways that are sympathetic to Leonelli's relational account of data. After all, my working definition of scientific perspective includes:

(i) the body of *scientific knowledge claims* advanced; (ii) the experimental, theoretical, and technological resources available to *reliably* make those scientific knowledge claims; and (iii) second-order (methodological-epistemic) principles that can *justify* the *reliability* of the scientific knowledge claims so advanced.

Data-to-phenomena inferences are perspectival in that it is not enough to have experimental, theoretical, and technological resources available to *reliably* make claims of scientific knowledge. Second-order principles that can *justify* the (*reliably formed*) methods for those scientific knowledge claims are also required. That is what a scientific perspective provides over and above reliabilism.

[2] Here I draw on Massimi (2011a). Reprinted by permission from Springer, *Synthese* 182, pp. 101–116 'From Data to Phenomena: a Kantian Stance', Michela Massimi, Copyright Springer Science + Business Media B.V. 2009. For a more recent discussion by Woodward on causation, see Woodward (2021).

In the following sections, I give epistemic communities their due. I do not treat phenomena as ready-made, nor data as causally downstream from them.[3] My account packs as much modal force into phenomena as is required for explanation and prediction in science. The final outcome is an ontology that gives both empiricism and modality their due. The argument for it takes the name of *evidential inference problem*.

6.3. The evidential inference problem

Data are routinely taken as providing *evidence for* phenomena. To echo Woodward (1998, p. S166), dataset D_j is evidence for a phenomenon P_i if the following indicative conditional holds:

(6.1) If D_j is found, then P_i is real.

The indicative conditional (6.1) is different from the ones I mentioned in Chapter 5 as it does not have a suppositional antecedent. The antecedent expresses a factual claim about dataset D_j being experimentally found, or harvested through various technological means. However, as with the indicative conditionals of Chapter 5, here too the main clause can be regarded as being implicitly modalized (à la Kratzer). In this case, though, the relevant implicit epistemic modal is a 'must' rather than a 'may', 'might', 'can', or 'could'. The main clause of (6.1) is almost an injunction for epistemic communities to conclude that the phenomenon P_i must be real whenever confronted with the evidence of dataset D_j:

(6.1*) If D_j is found, P_i must be real.

Under what conditions do datasets *provide evidence for* particular phenomena, the sort of reliable evidence that constitutes knowledge that the phenomenon is indeed real? How to understand exactly (6.1*) and the relation between the dataset and the phenomenon? Practitioners in any given scientific field of inquiry might adhere to instructions of the form

(6.1 **) 'If D_j, then conclude that P_i is real!'

But what kind of instruction is this?

---

[3] My view of phenomena finds a natural companion in Mieke Boon's (2020) account of phenomena, which, like mine, gives emphasis to the construction of scientific models and what she calls the 'disciplinary perspective' (in Kuhn's sense).

Does it enjoin us to take P_i as real on *pragmatic* grounds, as when I may believe that P_i is the case purely because it 'makes sense' or accommodates the available evidence given the experimental resources I have? Or is the instruction meant to invite me to take P_i as real on *epistemic* grounds, as when I want to find out whether the phenomenon P_i is indeed likely to be real given dataset D_j, regardless of whether it makes sense given the particular experimental or technological resources involved in harvesting D_j? Clearly for realism about phenomena to be defensible, one needs to understand (6.1**) as an epistemic imperative.

But then the question becomes: on what *distinctively epistemic grounds* do communities come to know that some phenomena are indeed real, given that the evidence of particular datasets is harvested through inevitably perspectival scientific practices? How to go beyond the specific experimental and technological resources available to conclude that some phenomena are indeed real? And how to distinguish real phenomena from non-real ones in the light of limited experimental evidence? For example, the retrograde motion of planets 'made sense' of the evidence available to ancient Greek astronomers. But it would be hasty to conclude that their datasets about positions of planets vis-à-vis the fixed stars counted as *evidence for* the reality of retrograde motion on distinctly epistemic (rather than pragmatic) grounds. Examples like this are familiar. Understanding what goes wrong with them proves more difficult.

One common intuition defended by Gideon Rosen (2007), for example, is that imperatives of this nature are distinctively epistemic if they are associated with reliability claims of the following form:

(6.1.R) If D_j, then reliably P_i.

(6.1.R) is not a shorthand for the bare indicative conditional (6.1), or its covert modalized version (6.1*). (6.1.R) goes a step further in trying to pin down what the evidential relation between D_j and P_i is, upon which the epistemic imperative (6.1**) is based.

(6.1.R) is a reliability claim that conveys Woodward's insight that 'the detection and measurement procedure should be such that different sorts of data D_1 ... D_m are produced in such a way that investigators can use such data to *reliably track* exactly which of the competing claims $P_1 ... P_n$ is true' (Woodward 1998, p. S166, emphases added). One can imagine this as some kind of matching game with different datasets providing reliable pointers for different phenomena (e.g. D_1 for P_1, D_2 for P_2, etc.). How should the reliability claim (6.1.R) be understood then? When does a dataset (or sets) offer reliable evidence for a phenomenon?

I argue that the most promising way of understanding how such evidential inferences work is by taking phenomena to be modally robust. Building on

Massimi (2011a),[4] I suggest that we take phenomena to be modally robust because they pack as much modal information as is needed to allow scientists to make inferences from the data to warranted conclusions about what may be the case. Phenomena are the outcome of perspectival modelling that helps us to physically conceive something so as to carve out the space of what is possible in nature.

On my view, the modal robustness of phenomena is therefore not supervenient on some Humean mosaic of non-modal matters of fact about sparse natural properties. Nor is it an epiphenomenon of dispositions, categorical properties, potencies, or essences—Gilbert Ryle's 'hidden goings on'—that populate metaphysics books. Modal robustness is built from the ground up in the notion of phenomena. How does this bear on what I am going to call the *evidential inference problem*? My solution requires re-thinking the notion of phenomena and appreciating the role of scientific perspectives and perspectival modelling in data-to-phenomena inferences. I shall ultimately defend a phenomena-first approach to ontology.

The relation between data and phenomena is far from straightforward. Data are not empirical proxies for phenomena. They do not come with labels and instructions for what they stand for. Data are pretty much where one wants to find them—in a swarm of bees, a flock of birds, or ripples in a lake.

Whether or not epistemic agents *see* some things as data depends on whether they see them as *providing evidence* for some phenomena. So circularity affects the evidential relation in (6.1.R). We cannot conclude that P_i is the case just on the basis of collecting data D_j, if what makes D_j relevant data depends on it being able to offer *evidence* for P_i. How, then, should we understand the reliability claim?

One might think data provide evidence for phenomena by *representing* them. No circularity arises if one takes data as representational proxies for phenomena in (6.1.R). But this buys into the *representing-as-mapping* assumption (see also Leonelli 2016 for arguments against the representational role of data). Matching data with phenomena is often presented as if it were a one-to-one mapping game. Imagine one of those activity books for children where there are pictures of habitats and moveable stickers of animals that one has to match: camels in the desert, penguins in Antarctica, and crocodiles in the jungle. Trying to infer

---

[4] In Massimi (2011a, emphases in original), I put forward what I called a 'Kantian stance' on phenomena as a '*conceptually determined appearance*', namely an appearance that has been brought under the categories of the understanding. . . . Phenomena are not ready-made in nature, instead we have somehow to *make* them' (p. 109). In the rest of this chapter, I stand by my original characterization of data-to-phenomena inferences, which is still Kantian in taking phenomena as the (modally robust) objects of our scientific knowledge (albeit stripped of the Kantian-sounding label '*conceptually determined appearance*'—for there is no need to hark back to outmoded Kantian language). Nor is my account meant to be in any way patterned upon Kant's own view (which I discuss in Massimi 2017c), but it is only loosely inspired by it. And I expand on this original intuition by clarifying the perspectival nature of the data–phenomena inferences and the role I see modality playing in phenomena.

matching connections between data and phenomena along similar lines is mis-guided. For there is no representation-as-mapping going on between data and the phenomena they offer evidence for.

In Chapters 4 and 5, we saw how perspectival varieties of model pluralism are representational insofar as they are first and foremost *exploratory*. For example, it would be odd to say that Marie Clay's data about children's improved reading performance under suitable educational interventions provides evidence for the phenomenon of difficulties with reading by 'representing' it. If anything, one should say that such data (within the broader educationalist perspective) allowed Marie Clay to *explore* the phenomenon of difficulties with reading by examining the range of effective educational interventions. 'Representing' is not the right philosophical term here.

This is a more general point about the evidential relation between data and phenomena. For example, Faraday's data about iron filings do not *represent* (in the sense of being similar to or resembling or mapping onto) the phenomenon of magnetism they provide evidence for. Data about fMRI images of the brain cortex provide evidence for functional connectivity, without yet representing in the sense of necessarily mapping onto 'actual physical connections in the under-lying neuronal substrate' (Schwarz et al. 2008, p. 914).

Evidential inferences are long, complex inferential chains. And the eviden-tial burden is not carried out single-handedly by *representation*. Evidential inferences are instead guided by reasonable expectations informed by past knowledge of what might or might not be the case. Thus, to return to our main question: when is the reliability claim (6.1.R)—'If D_j, then reliably P_i'—valid for the imperative (6.1**) 'If D_j, then conclude that P_i is real!' to hold on distinctively *epistemic grounds*?

6.4. The evidential inference problem dissolved

Here is a formal rendition of my expanded argument for a solution to the ev-idential inference problem. The solution, I contend, lies in a phenomena-first ontology, which is at a distance from both stringent empiricist and metaphysical realist readings of phenomena. The argument goes as follows:

[1] 'If D_j is found, P_i must be real' (indicative conditional 6.1* with epi-stemic modal)
[2] 'If D_j, then conclude that P_i is real!' (imperative 6.1**)
[3] For the imperative in [2] to hold on epistemic (rather than sheer pragmatic) grounds, namely for us to *know* that P_i is real (rather than P_i 'makes sense' or accommodates the available evidence), something ought to be said about the

dataset D_j in the antecedent and how it reliably provides evidence for the phenomenon P_i.

[4] If there is a way of establishing that D_j does reliably provide evidence for P_i (6.1.R reliability claim), then the imperative in [2] holds on epistemic grounds.

First lemma: There is no way of establishing that D_j reliably provides evidence for P_i if P_i is understood along stringently empiricist lines as an augmented set of empirical occurrences, on pain of inductive circularity (see Section 6.5).

Second lemma: There is no way of establishing that D_j reliably provides evidence for P_i if P_i is understood along metaphysical realist lines as teeming with causal properties (or, equivalently, what one might want to call causal powers, essences—the 'hidden goings on' in Ryle's terminology)[5] conferring dispositions, on pain of epistemic bootstrapping (see Section 6.6).

[5] Epistemic communities typically do reliably conclude that P_i when D_j is found.

[6] To conclude on epistemic grounds that P_i is real when D_j is found can therefore be based neither on augmented sets of empirical occurrences of the same type (*first lemma*) nor on phenomena-teeming-with-causal-properties-etc. (*second lemma*).

[7] To conclude on epistemic grounds that P_i is real when D_j is found must therefore have to do instead with how epistemic communities occupying a plurality of scientific perspectives have the resources to both *reliably* advance and *justify* the *reliably formed* methods behind the knowledge claims about P_i.

[8] It is the ability of epistemic communities occupying a plurality of scientific perspectives to feed data into inferences licensed by perspectival models (where applicable) that justifiably underpins the reliability claim (6.1.R)—see Section 6.7.4.

---

[5] This qualification matters to my argument. What follows is *not* an argument against metaphysically more modest views of causal properties (say, David Lewis's view on sparse natural properties some of which might be of causal nature). As I shall further explain in Chapter 9, the overall realist view I am defending might be compatible with a modest realist reading of properties along Lewisian lines (much as this is not my own view given the phenomena-first ontology I advocate in what follows). This is instead an argument against metaphysically more substantive views where causal properties are identified with the dispositions they confer to manifest certain behaviours in the presence of the right stimuli (the so-called 'dispositional identity thesis', or DIT). As mentioned in Chapter 3, DIT says that what makes a property the property that it is—in other words what constitutes its 'essence'—is the set of dispositions associated with it. Different authors use different terminologies. Thus what I henceforth call 'causal-properties-with-dispositions' are sometimes referred to as causal powers; other times, they are referred to as dispositional essences; and at yet other times, the properties are taken as genuinely dispositional without being further grounded. For my purpose here, I shall not make a distinction among these different metaphysical approaches because the argument is meant to be a general argument against metaphysical realist views of a dispositional flavour, broadly speaking.

[9] The link between (6.1.R) and (6.1**) is no surprise and no miracle either, once phenomena are understood as stable events that are modally robust across a variety of perspectival data-to-phenomena inferences (see Sections 6.7.2 and 6.7.3).

[10] For the imperative in [2] to hold on epistemic (rather than sheer pragmatic) grounds, the data-to-phenomena inferences captured by the indicative conditional (6.1*) have to be perspectival (as per [8]).

This perspectival view of how data reliably provide evidence for phenomena is philosophically at a distance from the strict empiricist one, although it shares with it the key commitment to the empirical roots of knowledge. It agrees with Bogen and Woodward's realist characterizations of phenomena as appearing through various models and different kinds of data. But it departs from metaphysical-realist readings of phenomena as proxies for 'hidden goings on'. It builds on Leonelli's relational analysis of data and Hacking's view of phenomena. But by contrast with these accounts, it places epistemic communities and their scientific perspectives centre-stage in inferring phenomena from data. In the rest of this chapter, I tease out and substantiate some of these premises. In Section 6.5, I explain the first lemma in [4] by showing how a stringent empiricist view of phenomena would not do. In Section 6.6, I attend to the second lemma in [4]. And from Section 6.7 onwards, I articulate the positive perspectival realist view of a phenomena-first ontology.

6.5. No stringently empirical phenomena

In the empiricist tradition, phenomena have been taken as synonymous with appearances, or things as they appear to us and our senses. In Pierre Duhem's (1908/1969) beautiful reconstruction from Ptolemy to Proclus, ancient Greek astronomers saw their task as that of 'saving the appearances', well aware that very different hypotheses could do so.[6] The same epistemic attitude is still evident in Osiander's Preface to Copernicus's *De Revolutionibus*, where the heliocentric view is presented as another hypothesis that can save the appearances. In Comte's

---

[6] 'Very different hypotheses may yield identical conclusions, one saving the appearances as well as the other. Nor should we be surprised that astronomy has this character: it shows us that man's knowledge is limited and relative, that human science cannot vie with divine science.... In more than one respect, Proclus' doctrine can be likened to positivism. In the study of nature *it separates, as does positivism, the objects accessible to human knowledge from those that are essentially unknowable to man.* But the line of demarcation is not the same for Proclus as it is for John Stuart Mill.... By extending to all bodies what Proclus had reserved for the stars, by declaring that only the phenomenal effects of any material are accessible to human knowledge whereas the inner nature of this material eludes our understanding, modern positivism came into being' (Duhem 1908/1969, pp. 21–22, emphasis added; see Massimi 2010 for a discussion).

Cours de philosophie positive (1830–42/1853), the observable phenomena become the basis for an order of things and a natural classification around which
five main sciences revolve: astronomy, physics, chemistry, physiology, and what
Comte calls 'social physics'.[7] In Ernst Mach's hands, phenomena as objects of the
senses became the fundamental building-blocks of physics.[8]

More recently, van Fraassen's *The Scientific Image* breathed new life into this
venerable empiricist tradition. On his view, observable phenomena are phenomena *observable-to-us* qua human beings whose sense perception acts as
a kind of measuring device. Van Fraassen's constructive empiricist view has
shaped the debate over the past forty years. He has since clarified and drawn a
distinction between phenomena and appearances: 'Phenomena are observable,
but their appearance, that is to say, *what they look like in given measurement or
observation set-ups,* is to be distinguished from them as much as any person's
appearance is to be distinguished from that person' (van Fraassen 2008, p. 284,
emphasis in original).

Appearances are simply the contents of measurement outcomes, not to be
confused in turn with sense perceptions or subjectively experienced impressions
(p. 276). Reality 'consists of smelly, colourful, noisy (observable) phenomena',
but 'appearances are the way phenomena "look like" in a given measurement set-
up, and hence from a particular vantage point' (Massimi 2009, p. 326), as in art,
to take the metaphor of perspective that van Fraassen masterfully spearheaded
in his 2008 book. Thus, on this view as I earlier reconstructed it (Massimi 2009,
p. 326), we ' "save phenomena" by embedding perspectival appearances (as
given by a certain instrument, measurement set-up, or frequencies in a data
model) into another abstract structure, the surface model, which "smooths" and

[7] 'All observable phenomena may be included within a very few natural categories, so arranged as
that the study of each category may be grounded on the principal laws of the preceding, and serve as
the basis of the next ensuing. This order is determined by the degree of simplicity, or, what comes to
the same thing, of generality of their phenomena. Hence results their successive dependence, and the
greater and lesser facility for being studied. It is clear, a priori, that the most simple phenomena must
be the most general: for whatever is observed in the greatest number of cases is of course the most
disengaged from the incidents of particular cases' (Comte 1830–42/1853, p. 25).

[8] '[T]here is every reason for distinguishing sharply between our theoretical conceptions of phenomena and that which we observe. The former must be regarded merely as auxiliary instruments
which have been created for a definite purpose and which possess permanent value only with respect to that purpose.... [A]ll theoretical conceptions of things—caloric, electricity, light-waves,
molecules, atoms and energy—must be regarded as mere helps or expedients to facilitate our consideration of things.... For example, when I ascertain the fact that an electric current having the strength
of 1 Ampère develops 10(1/2) [sic] cubic centimetres of oxyhydrogen gas at 0 degree Celsius, ... I am
readily disposed to attribute to the objects defined a reality wholly independent of my sensations. But
I am obliged, in order to arrive at what I have determined, to conduct the current through a circular
wire having a definite measured radius.... I maintain that every physical concept is nothing but a
certain definite connection of the sensory *elements,* ... and that every physical fact rests therefore on
such a connexion. These elements ... are the simplest building-stones of the physical world that we
have yet been able to reach' (Mach 1897/1914, pp. 186–192, emphasis in original).

"idealizes" the measurement outcomes, and eventually embed the surface model into theoretical models'.

One common thread in this empiricist tradition is that the term 'phenomenon' is a proxy for empirical occurrences of the same type. From Proclus's stars, to Mach's oxyhydrogen gas and van Fraassen's Moons of Jupiter, the observed/observable phenomena stand for a bundle of empirical occurrences.

What of the evidential inference problem if phenomena P_i are understood along broadly empiricist lines? 'Of course data provide evidence for phenomena', someone might reply. Consider Mach's example in footnote 8: 'Just run the next electrolytical experiment to confirm that an electric current of 1 Ampère produces oxyhydrogen gas in exactly the quantity that Mach describes'. The data on oxygen and hydrogen bubbles provide evidence for the phenomenon of electrolysis: 'If you *see* bubbles in water coming out at the two ends of the electric wire, you must reliably infer that P_i'.

Yet when are there *enough data* to conclude that P_i is real? At what point of the data analysis should the evidential line for inferring phenomena be drawn? The problem of induction creeps into any stringently empiricist picture of data-to-phenomena inferences. How many bubbles-in-liquid would be enough to reliably establish on *epistemic grounds* that there is a genuine phenomenon here?

This is a problem of how to *reliably* draw inferences from a limited sample of occurrences (the data) to an augmented set of empirical occurrences *of the same type* (i.e. the phenomena understood along stringent empiricist lines).[9] And it is hard to resist the conclusion that the reliability claim (6.1.R) inevitably involves an inductive leap.

Likewise, how can phenomena as an augmented set of empirical occurrences *of the same type* be in turn expected to be projectible and to license conclusions about future empirical occurrences? An inductive circularity would creep into the reliability claim (6.1.R) and the way it is used to infer from past token datasets to future token datasets of the same type:

$$\left[D^n_{t_1}, D^n_{t_2}, D^n_{t_3} \right] \rightarrow \left\{ \mathbf{P} = \left[D^n_{t_1}, D^n_{t_2}, D^n_{t_3} \dots D^n_{t_k} \right] \right\} \rightarrow D^n_{t_{15}}$$

where the subscripts indicate the times of harvesting the datasets and the superscript the type of dataset. No amount of data *in and of themselves* can reliably support the conclusion that P_i is real, if P_i is tantamount to an augmented empirical collection of datasets of the same type. There is no way of establishing that D

---

[9] This is a classic argument against strict empiricism and one that, for example, Armstrong (1985) has levelled against empiricist readings of laws of nature and why they cannot deliver nomic necessity.

reliably provides evidence for P_i if P_i is understood as an augmented set of empirical occurrences on pain of inductive circularity (*first lemma* in my argument from Section 6.4). Therefore, there is no distinctively epistemic grounds for the imperative (6.1**) to hold under a stringent empiricist reading of phenomena. Yet there is so much that the empiricist tradition has historically contributed to our understanding of how data and phenomena are related. I owe to the empiricist tradition of van Fraassen the insight that questions about scientific knowledge should always start from its empirical roots, looking at how data enter into data models and working all the way up into how data models in turn fit into theoretical models. Understanding science through the lenses of model construction is the greatest debt that perspectival realism owes to the empiricist tradition. But it is not the only one.

The attention, care, and detail to perspectival representations that van Fraassen (2008) has brought to general attention is and remains a cornerstone for perspectival realism. There is more that unites my perspectival realism to van Fraassen's constructive empiricism: a natural reluctance to populate reality with what Ryle called the 'hidden goings on'. It is to the latter that I turn next.

6.6. No ready-made phenomena teeming with causal properties either

How about a more realist-sounding view of phenomena? The main novelty of Bogen and Woodward's notion of phenomena lies precisely in their use of reliabilism (see Goldman 1986) to support a (broadly) realist metaphysics. The distinctive feature of reliabilism is to provide justification for beliefs about phenomena in a way that is independent of our knowing the causal mechanisms behind them. Phenomena exist in the world and causally produce data in a reliable way regardless. Knowing the causal mechanism is neither necessary nor sufficient to infer phenomena from data.[10]

My view shares many features with Bogen and Woodward's account and with Teller's friendly amendments to it (Teller 2010). I take phenomena (as opposed to scientific theories and theoretical posits) as the starting point for the type of realism I want to defend. However, I differ from Bogen and Woodward in one important respect. I locate the role of reliabilism within the wider scope

---

[10] Woodward (1998, p. S176) stresses this point: 'The idea that one can often empirically establish that (4) a detection process is reliable without (5) deriving its reliability from some general theory of how that process works and/or why it is reliable is supported by a number of episodes in the history of science. . . . Galileo advanced a number of empirical arguments showing that his telescope was a reliable instrument in various astronomical applications even though he lacked a correct optical theory that could be used to explain how that instrument worked or why it was reliable'.

of what I call a scientific perspective, since I defined a scientific perspective from the outset as including '(ii) the experimental, theoretical, and technological resources available to *reliably* make those scientific knowledge claims' but also '(iii) second-order (methodological-epistemic) principles that can *justify* the *reliability* of the scientific knowledge claims so advanced'. Let me explain why.

Consider the following counterexample. An experimenter may come to believe that a phenomenon P_k is real by a reliable process, that is, a process that generates true beliefs about P_k from data with high frequency, although the means by which the experimenter undergoes this process might be unreliable. For instance, the experimenter may have learned a reliable data analysis process from a colleague whose statistical knowledge is unreliable. Although the belief-forming process is reliable, we would not say that the experimenter *knows* that *phenomenon* P_k is real. For the experimenter cannot *justify* why the belief-forming process is indeed reliable.

As a response, one might appeal to second-order processes. If there were a way of keeping in check the unreliable source of belief-forming by appeal to second-order factors, over time unreliable first-order processes would be discarded.[11] However, there is a problem with delegating the judgement of reliability for first-order processes to second-order processes. The problem, which affects reliabilism in general, is bootstrapping.[12]

As Jonathan Vogel (2000) has pointed out, reliabilism cannot sanction its own legitimacy. Suppose someone believes what a spectrometer says about the spectrum of a chemical substance, without having justification for believing that the spectrometer is reliable. Suppose that the spectrometer happens to function very well. Then, an experimenter looks at it and forms the belief 'In this case, the spectrometer reads "X" for substance a, and X', where X is the proposition that substance a has a certain spectrum. Since the experimenter's perceptual process of reading the spectrometer is presumably reliable (they do not suffer from hallucinations), given the assumption that the spectrometer is functioning correctly, one can say that the experimenter is justified in believing that the spectrum of substance a is indeed X. Therefore, the experimenter can deduce that 'On this occasion, the spectrometer is reading accurately'.

Suppose the experimenter repeats this many times, without ever checking whether the spectrometer is indeed reliable (without ever checking whether it is properly wired, or it is functioning as it should, etc.). By induction, the

---

[11] For example, see Goldman (1986, p. 115) on this point: 'A second-order process might be considered meta-reliable if, among the methods (or first-order processes) it outputs, the ratio of those that are reliable meets some specified level, presumably greater than .50'.

[12] In what follows, I loosely draw on Massimi (2011a). Reprinted by permission from Springer, *Synthese* 182, pp. 101–116 'From Data to Phenomena: A Kantian Stance', Michela Massimi, Copyright Springer Science + Business Media B.V. 2009.

experimenter infers that 'The spectrometer is in general reliable', and hence uses it to measure the spectrum of other unknown substances. The experimenter would fall prey to bootstrapping, according to Vogel (2000). The moral of the story is that reliability cannot be detached from *causal knowledge*.

Unless one knows already how the phenomena that one is looking for should look, how can one appraise whether data provide *reliable* evidence for them? In a way, this is a re-elaboration of what Harry Collins (1985/1992) has called the *experimenter's regress*. In order to prove that an experimental process furnishes reliable evidence for a given phenomenon, it has to show that it identifies the phenomenon correctly. But to identify the phenomenon correctly, one has to rely on the experimental process.

There seems to be a more robust realist strategy where the reliability claim (6.1.R) does not fall prey to a circular bootstrapping because it is understood in terms of the holding of dispositional properties. For example, one may take the view that measurement devices are successfully deployed to detect dispositions, and indirectly the causal properties of the phenomena at stake. After all, causal properties are identified with the dispositions they confer on objects, according to the dispositional identity thesis (DIT) mentioned in footnote 5.

On this metaphysically thick view of phenomena qua bearers of causal properties, one may argue that the coincidence of measurement outcomes in different experimental contexts is a clear indication that the causal property at issue is real and that the phenomena are just *dispositional manifestations* of it. Even if our causal knowledge is incomplete, one is justified in endorsing the imperative 'If D_j, then conclude that P_i is real!' in (6.1**) because of the way one forms reliable beliefs about underlying causal properties conferring particular dispositions on phenomena. The reality of the phenomenon is downstream from the reality of the underlying metaphysical substratum of causal properties and their dispositional manifestations in the right environment.

This dispositional realist's take about phenomena-qua-bearers-of-causal-properties relies on an inference to the best explanation. Phenomena-qua-bearers-of-causal-properties are the best explanation for the success of our scientific instruments in delivering reliable beliefs. Consider the following example. A dispositional realist would say that electrons have the causal property electric charge. On this view, the imperative (6.1**) is likely to hold on epistemic grounds as long as the belief that P_i is real is reliably generated via a suitable detection procedure. Believing that electrons have negative electric charge is the best explanation for the success of our detection procedure in producing reliable beliefs.

That seems to fit J. J. Thomson's discovery of the electron via experiments on cathode rays in 1897. Thomson's experimental set-up revealed that rays were bent in the presence of an electric or magnetic field. He was able to measure the

charge-to-mass ratio of what at the time he called 'corpuscles' via a series of la-
borious inferences starting from the data concerning the deflection of the green
fluorescence inside the exhausted glass tube.

From the details of this episode, to which I return in Chapter 10, on a dispositional realist account of phenomena the experimenter may conclude that there
are phenomena in nature (e.g. electric currents I) having certain causal properties (e.g. electric charge Q) that manifest themselves via particular dispositions
(*Dis*). One such disposition is the ability of electric charge of being deflected in
presence of an electric or magnetic field. For example, one can imagine an experimenter uttering the following:

(A)   'On this occasion, the ammeter measures electric current I in ampere as
       yyy and since $I = Q/t$, then $Q = xxx$ coulomb'. ·

On a dispositional realist view, one would be justified in believing in the reality of
the phenomenon electric current I as the best explanation for the success of the
relevant scientific instrument (e.g. an ammeter) in detecting/reading I. However,
I contend, such a procedure risks being subject to a further particular version of
epistemic bootstrapping, as follows. Let us use 'K' to stand for 'Experimenters
know that . . .', following Vogel (2000, and 2008, p. 519, on which I loosely
draw here):

(1) $K(P_i)$ (where P_i: the phenomenon of electric current I carries electric
       charge Q) RELIABLE PROCESS.
(2) K (If D_j, then reliably P_i) RELIABILITY CLAIM (6.1.R) FROM DATA TO
       PHENOMENA.
(3) K (D_j: the ammeter reads '$I = yyy$ amperes' at time t_l) PERCEPTION
(4) K (D_k: electrical discharge in a glass tube deflected by a magnetic field)
       PERCEPTION.
(5) K (D_j & D_k: ammeter reads '$I = yyy$ amperes' at time t_l & electrical
       discharge in a glass tube deflected by a magnetic field) LOGICAL
       INFERENCE FROM (3) AND (4).
(6) K (ammeter reads accurately at t_l) LOGICAL INFERENCE FROM (3)
       AND (5) (under the assumptions that the ammeter functions well *and* the
       experimenter's perceptual system is not deceptive).
(7)   REPEAT THE OPERATION SEVERAL TIMES.
(8) K (ammeter is reliable) by INDUCTION.

But, of course, the experimenters cannot claim to *know* that the ammeter is indeed reliable on this basis. Instead, they would have to independently test the
ammeter, check that it was correctly wired, have a correct understanding of how

an ammeter works, and a model to reliably construct such a measuring device before they could conclude that the ammeter reliably tracks the phenomenon electric current-with-causal-property-electric-charge.

This has consequences for the envisaged metaphysically thick version of phenomena-with-causal-properties-conferring-dispositions. The view is affected by what Vogel calls 'rollback': if the experimenters do *not* after all know that the ammeter is reliable [not-(8)], they cannot claim either to *reliably* know $K(P_i)$ on this basis [not-(1)]. Therefore there is no way of establishing that D_j reliably provides evidence for P_i if P_i is understood along metaphysical realist lines as teeming with causal-properties-conferring-dispositions on pain of epistemic bootstrapping: *second lemma* in [4] of the argument for the evidential inference problem dissolved (Section 6.4). In other words, there is no distinctively epistemic ground for the imperative (6.1**) to hold under a dispositional realist reading of phenomena.

To conclude, to *know* that there are phenomena-with-causal-properties-conferring-dispositions, more is required than the kind of process reliabilism at work in this scenario. I stress that this is an epistemological argument against reliabilism as a privileged epistemic stance to secure a metaphysical realist view of phenomena (of a dispositional flavour). It functions also as the opening wedge for a perspectivist rejoinder about how to understand data-to-phenomena inferences, moving beyond the dichotomy of stringent empiricism and metaphysical realism.

6.7. Rethinking phenomena

In the past two sections, I made the point that in order to conclude on *epistemic grounds* that phenomenon P_i is real on the basis of dataset D_j, the phenomenon in question can be regarded neither as an augmented set of empirical occurrences of the same type nor as a phenomenon teeming-with-causal-properties-conferring-dispositions (*first* and *second lemmas*, respectively, in premise [4] of the evidential inference problem dissolved presented in Section 6.4). It is now time to spell out my positive view behind steps [7]–[10] of the argument.

I build on earlier work (Massimi 2007, 2008, 2011a) and put forward an alternative way of thinking about phenomena that takes them in their own right. It places phenomena centre-stage when it comes to ontological commitments for the perspectival realist. Robust views of phenomena (such as the aforementioned ones by Bogen and Woodward and by Hacking) share a kindred realist approach. However, the view presented here has a distinctive Kantian flavour.

I am not of course harking back to any Kantian or quasi-Kantian system of categories and principles of understanding. The Kantian aspect of my view of phenomena is instead to be located in their modal robustness. I ultimately argue that to conclude on epistemic grounds that P_i is real on the basis of D_j, one should look at how epistemic communities occupying a plurality of scientific perspectives have the resources to tease out inferences from datasets to stable events across a variety of perspectival data-to-phenomena inferences. To articulate this I first briefly present (Section 6.7.1) some of the recent literature where the need for a different conception of phenomena has clearly emerged.

6.7.1. The 'quiet revolution'

That phenomena are more substantive than sheer appearances has commonly been recognized. Often the rationale for a different approach to phenomena has come from the literature in the philosophy of physics. Interpretations of quantum mechanics have proved a fertile ground in this respect. Laura Ruetsche (2011, p. 353, emphasis in original), for example, has advocated what she calls a 'pragmatized' notion of physical possibility whereby 'sets of possible worlds are indexed (or indexed *as well*) to circumstances of application within the "anchor" world'. That is how she describes the modal commitment that she sees as associated with scientific theories:

> When physical possibility is pragmatized, there is a single way the world is, but (as gleaned by physics) there is not a single set of ways it might be. Instead, there are many sets of ways it might be. Taking these sets seriously is part and parcel of commitment to a physical theory: to call something an 'electron' is to take on commitments regarding how it would behave in a variety of circumstances, and to offer explanations, evaluations, and further theory constructions in ways constrained by those commitments. . . . I am contending that 'the' modal dimension of a theory is really manifold: at least some successful scientific theories, theories of QM_∞, are best understood not as determining a single set of possible worlds . . . but as encompassing many such sets. . . . [T]his is no accident. . . . [I]t is a central theoretical virtue . . . to foster manifold spaces of physical possibility, spaces that are naturalized and pragmatized. Again, this is not inconsistent. It is not saying theories contradict themselves with respect to how the world is. It is saying good theories are good theories because of a certain inbuilt flexibility about how the world might be. (Ruetsche 2011, pp. 353–354)

Lydia Patton (2015, p. 3444) has aptly called this modal feature of scientific theories 'modal resourcefulness'. It is the ability of a theory to function as a guide to varying modal contexts 'without requiring a unifying physical interpretation of the theory as a depiction of reality'. Patton has further argued for a trade-off between the modal resourcefulness of a scientific theory and the causal explanations/descriptions associated with it, and for re-thinking the exploratory nature of science not according to the Quinean 'web of knowledge', but according to the new metaphor of mining:

> Explanations constructed in particular material contexts, using specific phys-
> ical interpretations of a theory, are mined until they no longer pan out, but
> mining may go on beyond the point at which a unifying, coherent web of phys-
> ical interpretation can be woven. . . . Ruetsche emphasises the conventional el-
> ement of what we call an 'electron', for example. But it is true that, once we have
> decided to call something an electron, and have chosen a specific description
> of physical interpretation, those choices constrain the inferential power of a
> given background theory of the electron. That choice is like deciding to mine a
> particular seam. The metal we are after is the entity of interest; the reinforcing
> beams and structure are the laws assumed to be in place. . . . [I]f we find no gold
> in a particular place, we will close the shaft, even if it is structurally sound: the
> laws may apply, but there is no more of what we are seeking to be found there.
> But if the shaft collapses, we know that the laws do not hold. . . . The best mines
> are those that allow us to draw inferences about the possible constructions of
> other mines. (Patton 2015, p. 3458)

Both Ruetsche and Patton locate the modal manifold of physical possibilities and the variety of inferences it licenses within theory-building. My analysis is con-genial with theirs, but shifts the focus away from scientific theories to the phe-nomena themselves. I ultimately locate modality not in scientific theories but in the phenomena themselves.

In Part I, I described how phenomena are often inferred through a plurality of perspectival models. Physically conceiving something about something else is central to perspectival modelling as a guide to modal knowledge This exercise is constrained by laws of nature or, better, lawlike dependencies (the mine shafts in Patton's analogy). Perspectival models as exploratory models span the space of what is possible not because the world can inconsistently be in all these different ways, but because behind each facet of this modal mani-fold lies a potential 'inferential seam' that can be fruitfully explored. What is (ontologically) required is a new notion of phenomena to go with it, with mo-dality built into it.

Re-conceptualizing phenomena as the seat of modal inferences is a novel sug-gestion. Philosophers of physics have variously urged rethinking fundamental ontology along inferentialist and perspectival lines. Richard Healey (2017, p. 203, emphasis in original) has defended what he calls 'pragmatic inferentialism' as the view that 'concepts of classical physics, of the rest of science, and of daily life *all* get their content from how they help determine the inferential role of statements in which they figure'. By the same token, he argues, when it comes to quantum theory, quantum states and observables-as-operators do not acquire their content by representing

> purported elements of physical reality denoted by corresponding magnitude terms. .... The function of a quantum state (represented by a wave function, vector, or density operator) is not to represent properties of a physical system to which it is assigned, nor anyone's knowledge of its properties. What gives a quantum state its content is not what it represents but how it is used—as an *informational bridge*. (p. 206, emphasis in original)

And when it comes to the foundations of quantum mechanics, no one more than Carlo Rovelli has contributed to a fully relational and perspectival quantum mechanics over the past twenty-five years since the seminal work in Rovelli (1996). Relational quantum mechanics shifts attention away from the language of substance and properties towards the relations and interactions between physical systems. All properties are only relational properties. Accordingly, there cannot be a single description of the external world, but only perspectival descriptions concerning the relational properties of any given physical system. Rovelli (2021) returns and expands on this theme by connecting it with general reflections on the epistemology of science:

> Science, we may say, is only an extension of the way in which we see: we seek out discrepancies between what we expect and what we gather from the world. We have visions of the world and, if they do not work, we change them. The whole of human knowledge is constructed in this way. (p. 163)

And if we switch from quantum mechanics to time itself, with its statistical-mechanical underpinning, the link with physical reality is once again perspectival.[13] The asymmetry of time—typically explained by the assumption of a past low entropy—becomes a '*perspectival* phenomenon, like the rotation of the sky' as Rovelli has persuasively argued for (Rovelli 2017, p. 286; see also Rovelli 2018).

Jenann Ismael shares a similar view of causal phenomena:

> The fact that even seemingly fundamental concepts are shaped by contingencies about our circumstances in the world is not surprising. Our

---

[13] Dieks (2019) offers another interesting perspectivalist take on quantum mechanics. And so does Steven French (2002, 2020) in relation to the phenomenological literature.

concepts are, after all, *our* concepts. When we model the world we make all kinds of distinctions that are invidious from a cosmic perspective but that have practical or epistemic significance for us. But it can be surprising how deep that parochialism runs. To discover that the direction of causation is frame-dependent in this sense is to discover that the idea that earlier events bring about later ones is a matter of point of view, an artefact of the epistemic lenses through which we view them, not intrinsic to the field of events but imposed by distinctions that we make because they have practical and epistemic importance to us. That is a quite astounding surprise to pre-theoretic assumptions about the world. (Ismael 2016, p. 264, emphasis in original)

Ismael distinguishes between what she calls the 'perspectivalist insight' and a generalization of it that might go under the name of causal perspectivalism, to echo Huw Price (2007). It is an invitation to 'put ourselves in the picture' and think about why creatures like ourselves should represent the world in causal terms, for example.

Price's causal perspectivalism is indeed another prime example of a different way of thinking about realism. Without denying the reality of causal asymmetry (or of time asymmetry), a causal perspectivalist would argue that this phenomenon is to be explained solely with reference to our situated nature. I share the fundamental insight of a fully naturalistic type of perspectivism which Price compares to Kant's own Copernican revolution:

We ask ourselves 'What kind of reality would look like *this*, from the particular standpoint we humans happen to occupy?'. . . . I do not want to eliminate causation altogether from science, but merely to put it in its proper place, as a category that we bring to the world—a projection of the deliberative standpoint. Causal reasoning needn't be bad science, in my view. . . . Some perspectives simply cannot be transcended. By offering a modest, pragmatic, agent-centric view of causation, perspectivalism thus . . . foments revolution, but a quiet revolution, in the spirit of Kant's Copernican revolution, that avoids the mysteries of 'monarchist' metaphysics without the anarchic nihilism of causal eliminativism. It dethrones causation, certainly, but saves it, for all ordinary purposes, by revealing its human face. (Price 2007, pp. 290–291, emphasis in original)

In the rest of this chapter, I want to do to modality—a long-standing stronghold of the realist tradition—what Ismael and Price have done to causation: avoid the mystery of the 'hidden goings on' without the modal nihilism

Figure 2.1 Diego Velázquez *Las Meninas*. © Photographic Archive Museo Nacional del Prado

Figure 2.2 Diego Velázquez *Las Meninas*. Mirror detail. © Photographic Archive Museo Nacional del Prado

Figure 2.3 Jan van Eyck. *Portrait of Giovanni (?) Arnolfini and his Wife*. 1434 © The National Gallery, London.

Figure 2.4 Jan van Eyck. A detail from *Portrait of Giovanni (?) Arnolfini and his Wife*. 1434 © The National Gallery, London.

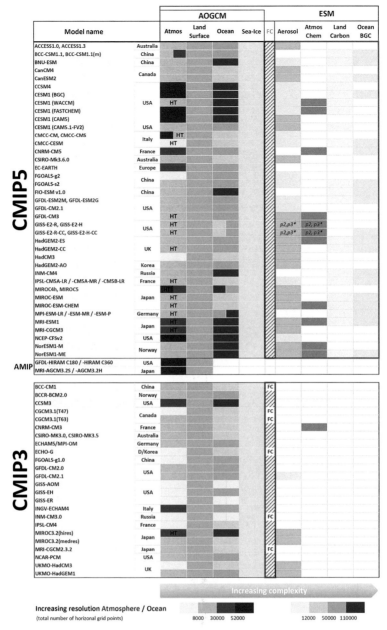

Figure 4.b.1 Table with the Coupled Model Intercomparison Project Phase 5 and Project Phase 3, highlighting the main models involved in the assessment of each of the main components at play in the Atmosphere-Ocean General Circulation Models (AOGCMs) and Earth System Models (ESMs). Copyright: IPCC AR5-WG1, p. 747, Table 9.1 from Chapter 9, Flato, G., J. Marotzke, B. Abiodun, P. Braconnot, S.C. Chou, W. Collins, P. Cox, F. Driouech, S. Emori, V. Eyring, C. Forest, P. Gleckler, E. Guilyardi, C. Jakob, V. Kattsov, C. Reason and M. Rummukainen, 2013: Evaluation of Climate Models. In: *Climate Change 2013: The Physical Science Basis. Contribution of Working Group I to the Fifth Assessment Report of the Intergovernmental Panel on Climate Change* [Stocker, T.F., D. Qin, G.-K. Plattner, M. Tignor, S.K. Allen, J. Boschung, A. Nauels, Y. Xia, V. Bex and P.M. Midgley (eds)]. Cambridge University Press, Cambridge, United Kingdom and New York, NY, USA. Reproduced with permission.

We acknowledge the World Climate Research Programme's Working Group on Coupled Modelling, which is responsible for CMIP, and we thank the climate modelling groups (listed above) for producing and making available their model output. For CMIP the U.S. Department of Energy's Program for Climate Model Diagnosis and Intercomparison provides coordinating support and led development of software infrastructure in partnership with the Global Organization for Earth System Science Portals.

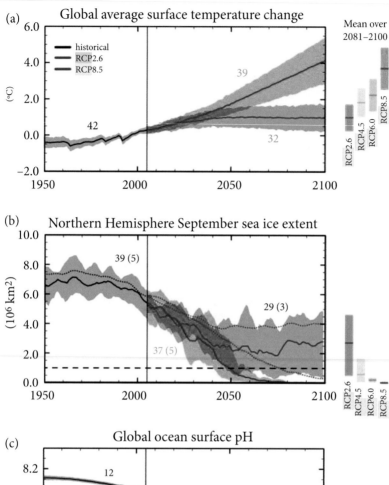

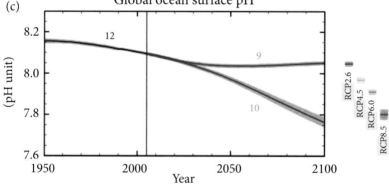

Figure 4.b.2 Copyright IPCC AR5-WG1, p. 21, Figure SPM.7 from IPCC, 2013: Summary for Policymakers. In: *Climate Change 2013: The Physical Science Basis. Contribution of Working Group I to the Fifth Assessment Report of the Intergovernmental Panel on Climate Change* [Stocker, T.F., D. Qin, G.-K. Plattner, M. Tignor, S.K. Allen, J. Boschung, A. Nauels, Y. Xia, V. Bex and P.M. Midgley (eds)]. Cambridge University Press, Cambridge, United Kingdom and New York, NY, USA. Reproduced with permission.

We acknowledge the World Climate Research Programme's Working Group on Coupled Modelling, which is responsible for CMIP, and we thank the climate modelling groups (listed above) for producing and making available their model output. For CMIP the U.S. Department of Energy's Program for Climate Model Diagnosis and Intercomparison provides coordinating support and led development of software infrastructure in partnership with the Global Organization for Earth System Science Portals.

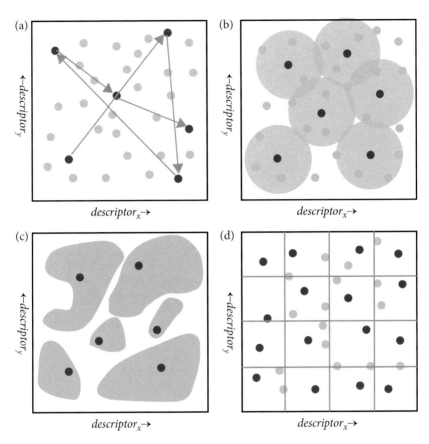

Figure 7.1 Four different computational methods for clustering molecules in chemoinformatics: (a) dissimilarity-based compound selection (DBCS); (b) sphere exclusion; (c) clustering; and (d) cell-based selection. N. Brown (2016) *In Silico Medicinal Chemistry. Computational Methods to Support Drug Design,* RSC Theoretical and Computational Chemistry Series N. 8., p. 125. Reproduced by permission of The Royal Society of Chemistry (https://pubs.rsc.org/en/content/ebook/978-1-78262-163-8).

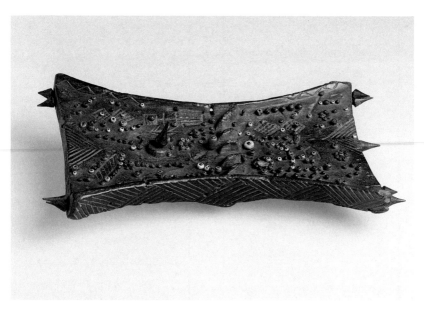

Figure 11.1 Luba. *Lukasa Memory Board*, late nineteenth or early twentieth century. Wood, metal, beads, 10 × 5 3/4 × 2 1/4 in. (25.4 × 14.6 × 5.7 cm). Brooklyn Museum, Gift of Marcia and John Friede, 76.20.4. Reproduced by permission of Brooklyn Museum.

of stringent empiricism.[14] Dethroning modality, for sure, from its hidden realm of causal properties and dispositions, but ultimately saving it by revealing its human face.

6.7.2. A phenomena-first approach to ontology: stable events and their lawlikeness

What are phenomena, then, under the view I am proposing? Here is a definition:

> Phenomena are stable events indexed to a particular domain of inquiry, and modally robust across a variety of perspectival data-to-phenomena inferences.

Let me unpack this. First, phenomena are events: they are not things, entities, structures, facts, or states of affairs. Important recent philosophical work has clarified the ontology of events, and how they differ from other more familiar metaphysical notions (see Faye et al. 2001; Shipley and Zacks 2008).[15] Events *happen* or *occur* over a period of time (short or long).

Sometimes events are part of a causal chain. Sometimes they occur spontaneously. At other times events are performed by us (or other non-human animals). By contrast, things, entities, or structures enjoy an ontological status whose identity is not affected by temporal, causal, or performative considerations. Phenomena qua events include things that happen to us (e.g. a shower of cosmic rays reaching my body), alongside things that *human beings* (or other non-human animals) make happen (e.g. anthropogenic CO_2 emissions).

Phenomena as events include, for example, the beta decay of a radioactive atom; the stability of some nuclides; the hadronization of quarks; the engineering of a new synthetic molecule; the carcinogenesis caused by gene mutations; the global warming of planet Earth; phonological awareness; pollination of flowers; echolocation in belugas; the GDP of a nation, among myriad other examples.

Our epistemic encounter with some of these phenomena is through perspectival models as discussed in Chapters 4.a, 4.b, and 4.c. Others (e.g. the pollination of flowers or echolocation in belugas) are less dependent on models but still

---

[14] I take Wolfgang Spohn's (2018) view on modality to be similar to the direction I am heading towards, with some important differences. For he takes his view as a 'statement of Humean projectivism' as well as a counter-programme to David Lewis's Humean supervenience. My view takes its lead from a Kantian insight about modality as a secondary quality. Accordingly, I take phenomena rather than states of affairs (and their objects, properties, and relation) as the minimal ontological unit.

[15] For a helpful summary, see Casati and Varzi (2020).

dependent on situated practices (e.g. ecology and electrophysiology) for their identification and analysis.

Phenomena qua events should not be conflated with facts or states of affairs. Phenomena are not atemporal facts in some B-series block universe à la McTaggart (1908), as when one says that Poland was invaded in 1939 or that Charlemagne was crowned in 800 CE, whereby in a block-universe view these facts were true *then* as they remain true *now* and will continue to remain such in the year 2100. Phenomena qua events are not states of affairs either, if one understands states of affairs à la Armstrong (1993) as particulars instantiating universal properties. Phenomena qua events have instead a processual, dynamic nature. They happen over time and they stretch along McTaggart's tensed A-series (if one really wants to use the reference to the A- and B-series for purely explanatory and clarificatory purposes here—to be clear, my account does not depend or rely on it in any way). Their happening has a processual nature, rather than being the mere instantiation or exemplification of some abstract universal properties in some particulars. So far this definition of phenomena qua events should not be surprising for it resonates with Hacking's aforementioned definition and chimes with David Nicholson and John Dupré's (2018) view of processes in philosophy of biology.

What is distinctive about my definition of phenomena qua events is their being indexed to a particular domain; the emphasis on stability; and their modal robustness. In Part I, I showed why perspectival pluralism is needed in areas of inquiry where one is modelling target systems *across domains*. Here they are again with a new additional example. Mutation in the APC gene take places at the genetic level,[16] but colorectal carcinogenesis happens at the level of tissues and organs. Phonological decoding occurs at the cognitive level; activation of the Broca area at the neurobiological level. Large-scale ocean heat uptake occurs in the hydrosphere; glaciers melting in the cryosphere.

Hydrosphere, cognitive level, genetic level, and so forth, function as *domains of inquiry* that allow epistemic communities to sieve, so to speak, the stable events that get identified as the relevant phenomena. Phenomena are those stable events that can be recognized in a swarm of data and across different data-to-phenomena inferences. The process of identification and re-identification of stable events requires a distinctive domain. Let me illustrate this with an example.

First, clearly not every event is a candidate for a phenomenon. The event of my son's visit to the zoo is not a candidate for a phenomenon. That is not because the event itself does not generalize to some kind of 'event-type' of which yesterday's

---

[16] I am here using the term 'level' in a rather loose way as synonymous with 'domain', given the controversy surrounding the very notion of levels in biology (see Eronen 2013; Potochnik and McGill 2012; and Potochnik 2021).

visit to the zoo was an 'event-token', so to speak. For it clearly does. One can easily think of 'visiting the zoo' as an abstract event-type of which my son's visit to Edinburgh's zoo yesterday was a token. Thus, the reason as to why this event is not a candidate for a phenomenon has to be looked for elsewhere.

Being indexed to a particular domain is key to identifying events that are candidates for phenomena and sieving them aside from those that are not. Although it might be indexed to a particular spatio-temporal location (say, Edinburgh's zoo yesterday), the event of my son's visit to the zoo does not have a proper domain of inquiry. As such, it is not a candidate for a phenomenon.

While domains of inquiry give us a sieve for identifying *bona fide* event candidates for phenomena, the stability itself of the event has nothing to do with the identification and re-identification in a given domain. Stability is *not* interchangeable with repeatability. Indeed there are one-off stable events which are nonetheless *bona fide* phenomena even if they happen only once: for example, the radioactive decay of an atom, or once-in-a-lifetime palaeoclimatic event resulting in a particular coral fossil record.

The stability of an event that is a *bona fide* candidate for a phenomenon has to do with what in Chapter 5 I called lawlikeness. An event is stable if there is a lawlike dependency among relevant features of it. One can think of this stability along Woodward's interventionist lines. For example, the stretching of an elastic spring is a stable event because there is a lawlike dependency (enshrined in a Best System to which Hooke's law belongs, in this case) between applied force and elastic displacement. And, crucially, such lawlike dependency does not change by changing metals, the humidity of the environment, and so on. Likewise, the melting of glaciers is a stable event because there are lawlike dependencies between incoming and outgoing radiative energy responsible for it. Echolocation in, say, belugas is a stable event because there are again lawlike dependencies in physiology between emitting high-frequency sounds and return echoes used to determine the distance from an object. It is lawlikeness that makes these events stable and therefore candidates for phenomena.

As explained in Chapter 5, lawlikeness is different from lawhood: not all lawlike dependencies in nature have a counterpart in a Best System of laws of nature (and perspectival series thereof). There are, for example, clear lawlike dependencies at play in the pollination of flowers, even if there might not be anything equivalent to Newton's law of gravity or Stefan–Boltzmann's law to codify them. In brief, I see lawlikeness not as a property that presupposes some kind of modal underpinning (along the lines of dispositional realists, who would read lawlikeness off dispositions or causal powers). Lawlikeness is a primitive relation of stable events in nature. And it is also the realist tether in my story. The natural world comes pre-packaged with stable events—stable because lawlike. Melting of glaciers, phonological awareness, flower pollination, and peaks at 125 GeV are all stable events because they are lawlike regardless of whether or not there might be names for them or even a Best System of laws in given fields.[17]

The lawlikeness at play in stable events is semantically captured by subjunctive conditionals as described in Chapter 5, Section 5.7. In the hands of epistemic communities, subjunctive conditionals are the semantic tools to guide and direct inferential reasoning about what might be the case in any given domain of inquiry. They are the signposts guiding communities over time through the inferential garden of forking paths. Lawlikeness as a primitive relation grounds, then, a first-tier modality at play in how an elastic spring *would* stretch *if* a force were applied to it, or how a glacier *would* melt *if* the balance between incoming and outgoing radiative energy were affected by GHGs. How to go from stable events to phenomena? And how to distinguish this first-tier modality, rooted in lawlikeness as a primitive relation among features of stable events, on the one hand, from a second-tier modality, or what I call the *modal robustness* of phenomena, on the other? These are the topics of the next section.

6.7.3. Empowered phenomena: modal robustness as a secondary quality

Let us now consider the second central feature of my definition: modal robustness. Data do not necessarily lead to phenomena. Consider, in biomedical research, the data overload of 60,000 papers on one protein, p53, a tumour suppressor in carcinogenesis (see Hager and Gu 2014). Identifying the variety of possible protein kinases[18] remains a major challenge in developing new drugs against cancer. One of the problems of inferring phenomena from data in this case is that there are over 500 known human kinases and tens of thousands of possible protein targets: 'Experiments require months to establish a single novel kinase–protein relationship, and then years to fully elucidate the relationship's biological impact' (Spangler et al. 2014, p. 1879).

Or consider the vast amount of data (1 petabyte) collected in a particle detector like the CMS at CERN just in 2012 and released as public open data in 2017.[19] The

---

[17] In this respect, my view echoes Brading's 'law-constitutive' approach in her analysis of Newtonian phenomena (see Brading 2012) and also Friedman's reading of Newton's methodology and its relevance to the realism debate (Friedman 2020). See Schliesser (2021) for an insightful discussion of these themes within Newton's metaphysics.

[18] The protein p53 has been known to play an important role in suppressing tumours by enhancing the natural defences of human cells (including destroying cancerous cells before they spread). But the efficacy of the mechanism through which p53 acts as a gate-keeper for carcinogenesis is hugely sensitive to possible modification of this protein—called phosphorylation—in which a phosphate molecule bonds to the protein molecule, impeding its gate-keeper activity. Phosphorylation is caused by enzymes called 'protein kinases' and they are of particular interest for the pharmaceutical industry, which aims to develop drugs to control and tackle mechanisms behind carcinogenesis.

[19] See https://home.cern/news/news/experiments/cms-releases-more-one-petabyte-open-data.

Higgs boson was one of the phenomena that this batch of data provided evidence for. Yet there is a big gap between the data produced and the number of new phenomena they *could* potentially provide evidence for: from supersymmetrical particles to particles of more exotic kinds that particle physicists have been actively seeking for decades.

How, then, do epistemic communities conclude that a phenomenon P_i is real, given certain data? And why do some of these data-to-phenomena inferences prove to be a goldmine, whereas others eventually become the equivalent of abandoned mine shafts? Three other elements of my definition are relevant to answering these questions: *modal robustness*, the *inferential* nature of phenomena, and the *perspectivity* of the data-to-phenomena inferences.

Modal robustness first. In how many different ways *can* a phenomenon manifest itself? A phenomenon, I contend, is a stable event that has the modal resources to *happen* in many different possible ways. Stability goes hand-in-hand with modal robustness. Indeed the two come together in a two-tier view. A first-tier modality is to be identified with the lawlikeness of the event, where, as already mentioned, I understand lawlikeness as a primitive relation among features of the event. Lawlikeness secures the stability of the event indexed to a domain of inquiry (i.e. its being stable in virtue of lawlike dependencies holding among features of it).

Phenomena are stable events that have an additional element: what I call modal robustness understood as a second-tier epistemic form of modality. Modal robustness expresses the many ways in which epistemic communities *infer* the relevant phenomenon by connecting often diverse datasets to the stable event.

The modal robustness of phenomena, as I see it, is neither an epiphenomenon of underlying causal properties nor the manifestation of dispositional essences. It is not a side-effect of surplus theoretical structure either. I'd like to think of modal robustness not as an intrinsic property of phenomena but as a secondary quality arising from the following triadic relation among:

(I)   the stability of the relevant event;

(II)   the data that provide evidence for it;

(III) situated epistemic communities able to tease out the network of perspectival inferences from the data to the stable event.

The lawlikeness of the event is expressed by subjunctive conditionals. As discussed in Chapter 5, Section 5.7, perspective-independent lawlike dependencies underwrite sometimes causal, other times (non-causally) explanatory, or, at other times, objective relations holding in the event. The stability of the event speaks therefore to there being causal, explanatory, or objective relations as *actual* relations in nature and—a fortiori—as *possible relations* (in a non-epistemic sense of possibility).

By contrast, the epistemic modality at play in modal robustness is captured by indicative conditionals. The division of modal labour that I described in Chapter 5, Section 5.7, finds its natural explanation in this two-tier view that I see as built from the ground up into the notion of phenomena. A perspectival realist is a realist about *empowered phenomena*.

More to the point, it is only when phenomena are understood along these lines that the evidential inference problem gets dissolved. The link between the epistemic imperative

(6.1**) If D_j, then conclude that P_i is real!

and the reliability claim

(6.1.R) If D_j, then reliably P_i

is no surprise once phenomena are understood along these lines (premise [9] in the argument in Section 6.4). For the burden of the reliable inference is neither outsourced to 'hidden goings on' (on pain of epistemic bootstrapping), nor is it delegated to an augmented set of empirical occurrences of the same type (at the cost of inductive circularity). The burden lies squarely with epistemic communities and their ability to work together over time and devise inferential blueprints that enable them to reliably infer that P_i must be real if D_j is found. Perspectival modelling across a plurality of scientific perspectives is what allows epistemic communities over time to reliably encounter nature as teeming with modally robust phenomena. Let us look at some other examples in addition to those discussed in Ch. 4.a, 4.b and 4.c.

The decay of the Higgs boson is a stable event indexed at 125 GeV. Stability takes here the form of a clearly identifiable peak underwritten by laws in physics that explain how smashing protons at high energy results in a certain number and type of particles being detected. The phenomenon is modally robust in that it *can happen* in more than one way: via four-lepton decay, two-photon decay, bottom quarks, two leptons and a photon (the so-called Dalitz decay).[20] At the ATLAS experiment, the Higgs boson is searched for in 'channels', defined by the Higgs boson's decay to other particles such as two photons, two Z-bosons (which further decay to four leptons), two b-quarks, or a muon pair and a photon, and so on. Of these, two photons and four leptons are the channels that contributed to the original discovery in 2012.

To claim that a new phenomenon has been identified, particle physicists typically look for an excess of events in these final-state particles. The distributions for the Higgs boson decay to four leptons (m_{4l}), and to two photons ($m_{\gamma\gamma}$), from

---

[20] See https://home.cern/news/news/physics/atlas-finds-evidence-rare-higgs-boson-decay.

the public ATLAS analyses, with the full data collected in 2011–2012, are shown in Figure 6.1.

In these plots, the black points that are labelled 'data' are frequency counts actually observed in the ATLAS experiment. So the y-axis is a count of events seen in the experiment for each mass value in GeV shown on the x-axis. The red and purple histograms in Figure 6.1(1) labelled 'background' show the frequency of data that would have been expected to be produced from other known physical processes. A difference can be clearly seen between the 'data' and the expected background. It is this difference that establishes that a previously unknown particle has been produced.

Compare the four-lepton decay (1) with the two-photon one in (2). In the latter case, the background is determined by performing a statistical fit to the experimentally observed distribution. In this channel, there are large numbers of observed events and the statistical errors are small. Furthermore, the background in (2) has a simple shape where the numbers of events steadily decrease with increasing mass. And here too the Higgs signal 'bump' at 125 GeV stably appears. Methodologically, these are two distinct pathways to the Higgs boson (see Massimi and Bhimji 2015 for a discussion).

Traditionally, one would say that the peak at 125 GeV is the signal for the particle. Or if you like, the peak is the phenomenon that makes its debut in data models like those in Figure 6.1 (1) and (2), and the Higgs particle is the 'hidden going on' behind the peak. But I have been resisting this language, which is reminiscent of a 'two-worlds' view: the world of the phenomena vs the world of the

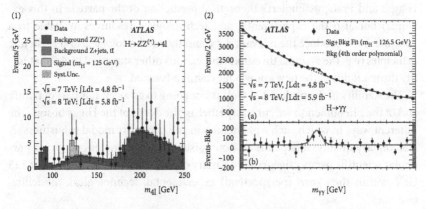

Figure 6.1 The distribution of the invariant mass of the Higgs decay to (1) four leptons (m_{4l}), and (2) two photons ($m_{\gamma\gamma}$), respectively, in the ATLAS detector (ATLAS Collaboration 2012, Fig. 2 and Fig. 4 © Copyright 2012 CERN, Published by Elsevier B.V. Open Access under CC-BY-NC-ND license (https://www.sciencedirect.com /science/article/pii/S037026931200857X?via%3Dihub) (For interpretation of the references to colours in this figure, the reader is kindly referred to the Open Access eBook version).

real entities behind the phenomena. Elsewhere (Massimi 2007), looking at a different episode in particle physics (the discovery of the *J/psi* particle), I urged going beyond this dichotomy. What I called there 'unobservable phenomena' (in the context of the observable/unobservable discussion) I here prefer to call modally robust phenomena: phenomena that *can happen* in different ways. The peak at 125 GeV is the event whose stability in all decay channels is underwritten by the lawlikeness of the underlying resonance mechanisms (i.e. the peak is the effect of the energy of incoming particles coinciding with the mass of a reso-nant particle—the Higgs boson in this case). The decay of the Higgs boson is the modally robust phenomenon that can take place in a number of different decay channels. The Higgs boson in turn is a kind of particle: it is a natural kind, not a phenomenon. And discussions of natural kinds will have to wait until Chapter 7.

This modally robust phenomenon is *inferred* from data via sophisticated practices that involve theoretical modelling, computer simulations, and sta-tistical analyses (see Morrison 2015 for an excellent in-depth analysis of these details). What makes these particular data-to-phenomena inferences perspec-tival is not just the underlying theory (the Standard Model in particle physics) but also the two main experimental set-ups at CERN, where the phenomenon was independently identified. The ATLAS and CMS experiments are inde-pendent experiments being carried out by two distinct communities of particle physicists, both using data from proton–proton energy collisions at the LHC. The experimental design of the machines is also different.

Our epistemic access to new kinds of particles is conditional on a scientific perspective that includes not just a body of knowledge claims (including Peter Higgs's and François Englert's theoretical prediction of the particle in this ex-ample) but also the experimental and technological resources available to a community to produce the relevant data, and the methodological and epistemic principles (e.g. the 5-sigma threshold and various other statistical tools) that *jus-tify* the *reliability* of the new knowledge claims advanced.

The stability of the peak at 125 GeV is not going to go away. It is part of nature and of the phenomenon we currently label as 'the decay of the Higgs boson'. The different ways in which such a phenomenon *can* occur (its modal robustness) is a secondary quality: it depends on how epistemic communities occupying par-ticular scientific perspectives relate a variety of datasets to the stable event at 125 GeV within their own (perspectival) experimental, technological, modelling practices.

The latter *are* subject to change over time. We might by 2200 have found new physics Beyond the Standard Model; we might develop new generations of par-ticle accelerators that do not rely on proton–proton collisions; we might also develop new statistical techniques and maybe impose more stringent or less stringent constraints than 5 sigma. For the particle physics community in the year 2200 to be able to hold on to the phenomenon of the decay of the Higgs boson, it will have to be the case that datasets in the form of proton–proton collisions (or a new equivalent), analysed through triggers and statistically checked against the 5-sigma threshold (or a new equivalent), still allow them to make this perspectival inference to the stable event in question.

Consider other examples of modal robustness. Carcinogenesis is a modally robust phenomenon that *can occur* in many different ways across a variety of cells and tissues under a great variety of circumstances. Mutations in particular genes (like the APC gene mutations) are the stable events, their stability under-written by lawlike dependencies such as that between the random changes in amino acids and the resulting change in the functionality of a protein. The presence of a stable event (e.g. APC gene mutations) does not automatically translate into a phenomenon (e.g. carcinogenesis), unless there are localized, highly context-sensitive causal mechanisms (or pathways) that facilitate the transition (see Plutynski 2018 for an extensive analysis). Teasing out the inferential network about the highly localized causal mechanisms in each case is the job of geneticists, pathologists, cellular biologists, histologists, and so forth.

Or, to give another example, the transfer of pollen to the ovules is a stable event across a variety of botanical species. Pollination is a modally robust phenomenon that could occur in a variety of different ways: including hummingbirds hovering, honeybees dancing, bumblebees sensing static electricity, and midges pollinating cacao tree flowers while laying eggs in rotting cocoa husks. And here too teasing out the highly contextual inferences from the data to the stable event in each case is the job of beekeepers interested in the flowering peak season no less than the phytogeneticists interested in plants' DNA, of the ecologists who care about the ecological niche for the plant no less than the botanists who classify the plants.

This is what makes phenomena an interesting ontological category. The modal exercise central to the scientists' ability to predict, explain, retrodict, and infer is located right at the level of the phenomena. In the next section, I take a closer look at how scientific perspectives do the heavy lifting in this plurality of data-to-phenomena inferences behind modal robustness.

6.7.4. Perspectival data-to-phenomena inferences

I started this chapter with the promise of explaining what a perspectival realist is a *realist about*. I presented a phenomena-first approach to ontology going beyond stringent empiricism and versions of metaphysical realism. I gave a definition of phenomena in terms of stable events and modal robustness. I hope I have clarified why the definition goes beyond just 'stable events' or even just data.

Stable events do not have modal robustness. Phenomena do. That is why phenomena are dynamic, under the view I am presenting. Parsing or trying to insulate the stable event from the space of modal knowledge claims in which it is located would deprive it of its modal robustness.

I have also insisted that while I see stability in terms of lawlikeness as a primitive relation, I see modal robustness as a secondary quality. That the Higgs boson *could* decay in two photons as opposed to four leptons, or two bottom quarks, is the outcome of how (I) that *stable event* (a peak at 125 GeV) at a particular domain is (II) evinced from proton–proton collisions data (III) via a network of perspectival inferences by historically and culturally situated *epistemic communities* (the CMS and ATLAS groups at CERN).

Similarly, that a particular kinase *could* phosphorylate lots of possible proteins is the outcome of how (I) a *stable event in a particular domain* (e.g. a structural loop in the kinase regulatory region)[21] is evinced from (II) data[22] via (III) a network of perspectival inferences by the relevant epistemic communities of molecular biologists, organic chemists, geneticists, and pharmacologists.

To say that modality is a secondary quality is to stress how the modal features that are so crucial to scientific discourse about phenomena depend both on the *stability of the event* and on *epistemic communities occupying one or more scientific perspectives* who are able to observe, detect, identify and re-identify the stable event *from one or more datasets* through often different inferential routes. The good news is that modality is no longer mysterious, or parasitic upon hidden dispositional properties or causal powers. Explaining how the Higgs boson *can* decay in four leptons; a kinase *can* phosphorylate a protein; a neuron axon *can* fire—none of these need call on 'hidden goings on'. There is no ultimate metaphysical foundation to modal robustness under the ontology I propose.

Which is not to say that the decay of the Higgs boson, or phosphorylation, or pollination, or any other example are any less real as phenomena than under a metaphysically rich account. On the contrary, whether or not there *are* stable events is a fact about nature independent of us, of our scientific history, and the resources of particular situated epistemic communities. However, discourse about how particular phenomena *could, would,* or *should* happen is a function of how epistemic communities relate data to the stable events to provide *evidence*

---

[21] For example, the A-loop in kinases displays a remarkable number of possible behaviours that are of particular interest to pharmaceutical companies trying to produce anti-cancer drugs designed to interfere with and block the underlying mechanisms responsible for cancer. The active conformation of the A-loop can in turn be affected by a variety of mechanisms involving binding other proteins (cyclin, for example, responsible for cellular mitosis—see Kobe and Kemp 2010).

[22] The data in this example come from human cyclin being expressed by organisms such as the *Escherichia coli*, purified, and crystallized to obtain diffraction data, which are then subject to analysis and simulation reconstruction (see Jeffrey et al. 1995).

for phenomena. They must be able to identify and re-identify the relevant stable events across a plurality of datasets, via a sophisticated network of inferences. This is how the evidential inference problem outlined in Section 6.4, steps [8]–[10], gets dissolved.

But stable events would be exactly the same with or without the particular scientific perspectives that human beings have historically developed. *This* is realism. At the same time, what makes a stable event a 'phenomenon' *does* depend on us as epistemic agents. Teasing out the space of inferences does depend on the scientific perspective in use and associated perspectival models as inferential blueprints (wherever applicable), and technological and experimental resources. Indeed, it is a prerogative of historically and culturally situated epistemic communities. Windows on reality open up by walking along the inferential garden of forking paths, where at every turn and junction lawlike dependencies point communities towards the directions to choose, which path to explore, and which one to leave behind.

Two objections persist. First, many philosophers will no doubt feel uneasy about the move I am proposing: namely, relocating modality from categorical properties, dispositions, capacities, or potencies to a secondary quality of phenomena that includes reference to epistemic communities. Am I not resurrecting yet another version of the 'lunacy of the ratbag idealist' (Lewis 1994, p. 479), a remnant of an outmoded Kantian take about modality? Am I not reintroducing subjectivity into nature? Second, what is distinctively perspectival about this inferentialist view? Couldn't we just say that there are data-to-phenomena inferences that epistemic communities perform without any need to appeal to perspective?

In response to the first objection, I draw attention to the epistemic good that comes from introducing epistemic communities into discussions about modally robust phenomena. A phenomena-first ontology does not buy into any two-worlds view: the world as it appears and the world as is; or the phenomena vs the noumena. It re-orients instead discussions about realism around epistemic communities, rather than around how best theories in mature science map onto given states of affairs from a presumably neutral standpoint.

As for the worry about subjectivity in science, I return to my own motivations. My pre-philosophical realist leanings originated from worries of a concerned citizen in a society where trust in science was being eroded under the pressure of powerful lobbies. As I write, the COVID-19 pandemic is raging and some people are burning 5G mobile phone masts they believe are somehow related to the spread of the virus.

In a society that begins to resemble an Orwellian dystopia, one had better think again about the role of epistemic communities in scientific knowledge. Trust in the expertise of relevant epistemic communities—and their ability to work together on the inferential practices that deliver those phenomena—is the best way to defend realism.

My argument places epistemic communities centre-stage. Relocating modal robustness from the metaphysical 'hidden goings on' to phenomena inferred by epistemic communities from data is not reintroducing subjectivity into science. It is a way of empowering epistemic communities to speak to power, and reclaim as their own expertise, reliable methods, and epistemic know-how—a great and wonderfully diverse variety of them—that has been hard won over centuries.

Now, what is perspectival about all this? Could we tell a similar inferentialist story eschewing mention of scientific perspective altogether? The simple answer is 'no'. The phenomena-first ontology is empiricist and perspectival from beginning to end. Inferences do not take place in a vacuum. *Whose* inferences? *By whose* lights? *With whose* experimental, modelling, technological resources?

Data-to-phenomena inferences are perspectival all the way, in the social and collaborative process I see as so central to science, and to realism. They are rooted in well-defined scientific practices that have the resources to *reliably advance* knowledge claims and to *justify* their *reliability*. This is what scientific perspectives do.

Data-to-phenomena inferences are therefore perspectival because they are the outcome of a ubiquitous scientific epistemology: what in Part I, Chapters 4 and 5 I described as perspectival modelling, where models themselves act as inferential blueprints. Far from being shadows on a wall, mere things-as-they-appear-to-us, phenomena are an ontological category in their own right and the seat of a rich and intricate network of both subjunctive and indicative conditionals-supporting inferences because they are always embedded within (not just one but) a plurality of scientific perspectives. It is in this sense that phenomena are the outcome of *perspectival* data-to-phenomena inferences.

7

Natural Kinds with a Human Face

7.1. From phenomena to kinds: intersecting perspectives

Stanford Industrial Park, adjoining Stanford University, was booming in the early 1960s. Tenants included the Syntex Corporation, which relocated there, and the Syntex Institute for Molecular Biology, where Carl Djerassi in the Chemistry Department became friend with geneticist and early computer enthusiast Joshua Lederberg. In 1963, the Center for Advanced Studies in Behavioral Sciences at Stanford organized a meeting on computer models of thought. Among the participants was the AI pioneer Ed Feigenbaum, who had studied with Herbert Simon at the Carnegie Institute of Technology and who had just co-edited with Julian Feldman an influential book entitled *Computers and Thought*.

This marked the start of a fruitful collaboration between Feigenbaum, Djerassi, and Lederberg that led to the so-called DENDRAL project (Lederberg 1964, 1965). DENDRAL was one of the first applications of computer science to organic chemistry. It became a model for thinking about how heuristic programming can be applied to an empirical science. DENDRAL opened the door to automated discovery, raising the question: can computers help us discover new natural kinds?

DENDRAL (see Lindsay et al. 1980) was an algorithm designed to exhaustively enumerate topologically possible arrangements of atoms in line with general rules of chemical valence. Mass spectroscopists seek to infer molecular structure from molecular mass. Physical and chemical properties of any compound are not just determined by which atoms it contains. Structure is equally important. The water molecule, for example, consists of one atom of oxygen and two atoms of hydrogen. The six valence electrons of the oxygen atom combine with the single electrons of each hydrogen atom not in some random way, but in two covalent H–O bonds and four 'lone pairs' with a specific H-O-H angle. Bonds and lone pairs form a tetrahedron around the nucleus of the oxygen, with bond lengths and angles constraining the atomic arrangements. This molecular arrangement explains some key macroscopic properties of water.

In more complex compounds, many isomers[1] may be consistent with a given mass number. The problem is particularly acute in organic chemistry, where

---

[1] Two isomers consist of the same number and type of atoms but have different physical and chemical properties because of their molecular structure.

Perspectival Realism. Michela Massimi, Oxford University Press. © Oxford University Press 2022.
DOI: 10.1093/oso/9780197555620.003.0010

quadrivalent carbon can bind covalently with several other atoms at once. Mass spectroscopy provides an indirect source of information about molecular structure and in so doing it facilitates the identification of unknown compounds in environmental pollutants for example, or forensic samples. High-speed electrons bombard target molecules, breaking them up into ionized fragments, which pass through an electrostatic or magnetic field (at least in the generation of these machines at the time of DENDRAL). The particles at play in each fragment can then be detected via their mass-to-charge ratios, which appear as distinctive peaks in a mass spectrum.[2] Any molecular sample will yield ionized fragments with a particular pattern of mass-to-charge ratios, and they can be separated by passing them through electrostatic and magnetic fields.

This is an exceedingly subtle, bottom-up inferential exercise that goes from data (i.e. peaks in a mass spectrum) to phenomena, and from phenomena to natural kinds. The data are those provided by mass spectroscopy. The phenomena are the ionized fragments. The kinds are the different chemical compounds (sometimes isomers) that need to be inferred from the fragments.

Nothing about this is straightforward. Not all chemical bonds cleave in the same way (single bonds are easier to break than double bonds, for example). Several molecules get fragmented at the same time. The ionized molecular fragments have to be accelerated and deflected by either an electrostatic or a strong magnetic field (in mass spectrometers by deflection) so that they can be sorted into different mass-to-charge ratios. A detector then registers the abundance of ions of different mass-to-charge ratios and plots the mass spectrum of the compound accordingly.

Complex molecules produce mass spectra with many peaks. But which particular combination of atoms, in what arrangement, is responsible for the observed peaks? For example, a certain mass-to-charge ratio can be used to infer that the compound must have two atoms of carbon, six of hydrogen, and one of oxygen. But the chemical compound in question can be either dimethyl ether (CH_3-O-CH_3), or ethanol (CH_3-CH_2-OH), which have very different physical and chemical properties.

DENDRAL was designed to facilitate such inferences in more complex cases, using graphs to represent what in a ball-and-stick model would be the atoms and bonds of the chemical compounds.[3] It was developed to explain molecular

---

[2] Many thanks to Julia Bursten, Marcel Jaspars, and Jon Turney for helpful comments. See Bursten (2020b) for a helpful discussion on the topic.

[3] Lederberg (1987, p. 7) recalled how 'various arithmetic tricks were devised that took account of valence rules, plausibility of composition, the negative and positive packing fractions of O and N, and the abnormal proportional discrepancy of H, to keep the search down to a manageable scope. For paper and pencil work (in 1964) this was embodied in a handbook of some 50 pages. . . . Even that small book was later . . . obsoleted by an algorithm that depended on a one-page table with just 72 non-zero entries, and a few arithmetic steps easily done on a 4-function hand calculator'.

structure by packing in expert knowledge about organic chemistry and auto-
mating the process of making inferences from the available empirical data about
mass spectra to the phenomena (ionized fragments), and from there to the relevant chemical compound (e.g. an isomer).

This heuristic programming allowed to explore the space of possible molecular arrangements within well-defined chemical rules. The programming was
interactive, and allowed scientists to revise the rules at any stage or add further
constraints at any round of hypothesis generation. In the words of Lederberg:

> DENDRAL-64 is a set of reports to NASA . . . that outlines an approach to
> formal representation of chemical graph structures, and a generator of all pos-
> sible ones. . . . The DENDRAL generator was then designed so that only one ca-
> nonical form of a possible automorphic proliferation is issued, greatly pruning
> the space of candidate graphs. . . . DENDRAL is remarkably neatly structured
> (as implied by its name) as a generator of trees of candidate structures . . . .
> These can easily number in the billions or more, in practical cases: the efficiency
> of the program depends on the pruning of impossible or implausible cases, as
> early as possible. . . . To give a . . . example, if N (nitrogen) is absent, we don't
> generate molecules that may contain N, then retrospectively eliminate each of
> those twigs. (Lederberg 1987, pp. 9 and 12)

This is another illustration of a perspectival model, in this case an algorithm-
aided representation of the chemical space of possible compounds. DENDRAL
took on the task of physically conceiving molecular structures compatible with
empirically given mass spectra within broad lawlike constraints. Being able to
explore the space of possible molecular structures for the same group of atoms
and delivering modal knowledge about which chemical isomer might be at stake
was delegated to DENDRAL as a way of facilitating inferences from data to phenomena and from phenomena to kinds.

But there is another reason why the story of DENDRAL matters for the
inferentialist ontology that I see as central to perspectival realism: it was an interdisciplinary research programme. As Lederberg recalls:

> I had no idea how one would go about translating these structural concepts into
> a computer program. . . . It was fortunate indeed that Ed Feigenbaum came to
> Stanford just at this time. . . . Stanford University, in the 1960s, was a fortunate
> place to be for the pursuit of scientific innovation, and equally for a highly in-
> terdisciplinary program. Computer science, medical science, chemistry were
> all in a surge of rapid expansion and new opportunity. . . . Lindley Darden's
> discussion of the 'history of science as compiled hindsight' [Darden 1987] elo-
> quently captures my own perspectives. My interest in AI has little to do with my

background as a biologist, a great deal with curiosity about complex systems that follow rules of their own, and which have great potentialities in preserving the fruits of human labor, of sharing hardwon tradition with the entire community. In that sense, the knowledge-based-system on the computer is above all a remarkable social device, the ultimate form of publication. (Lederberg 1987, pp. 9, 13, and 15)

The interdisciplinarity behind DENDRAL reinforces a key aspect of my discussion. Modelling what is possible is indeed a social and cooperative inferential exercise. DENDRAL offered an inferential blueprint for chemists, mass spectroscopists, and computer scientists to engage with one another and bring their respective expertise to bear on the task of facilitating phenomena-to-kinds inferences. The inferential ability to go from mass spectra to the chemical compounds that might be at play is not the prerogative of the chemist, or the mass spectroscopist, or anyone else. It is a collective endeavour. DENDRAL enabled different epistemic communities to work together, with their respective perspectival practices.

The chemical kinds that we know are not the output of some primordial Putnamian baptism of archetype samples, whose microstructural essential properties were discovered later on. Rather, they are the long-term open-ended inferential outcomes of intersecting scientific perspectives.

How, then, did perspectives intersect here? DENDRAL opened the path to chemoinformatics. Like any interdisciplinary programme, chemoinformatics is not the mereological sum of chemistry plus informatics. It has a disciplinary outlook of its own, with distinctive methodologies, epistemic approaches, and remit that partially overlap with those of both chemistry and informatics. Projects such as Dial-a-Molecule[4] and the AI^{3}SD Network (Artificial Intelligence, Augmented Intelligence for Automated Investigation for Scientific Discovery)[5] in the UK are testament to the medical and industrial interests associated with such exploratory searches. Chemoinformatics—with its wider experimental practice and methodology—is another example of perspectival modelling. Let us see why by going back to one of my examples in Chapter 6: phosphorylation as an example of a modally robust phenomenon.

Phosphorylation is a common modification of proteins, which is often behind a variety of carcinogenic mechanisms. Enzymes called 'protein kinases' carry phosphate molecules and target a number of proteins. One of the challenges in developing anti-cancer drugs is that there are hundreds of known human kinases, and tens of thousands of possible target proteins. Years of biochemical

---

[4] See https://generic.wordpress.soton.ac.uk/dial-a-molecule/the-network/.
[5] See http://www.ai3sd.org/.

experiments are required to fully elucidate the mechanisms in each case. Thus, in current cancer research, AI-led efforts help identify the greatest possible number of kinase–protein relations so that new potential pharmaceuticals can be produced to target a variety of carcinogenic mechanisms. This is a subtle exercise that requires assessing the toxicity of possible new drugs, and cost-effective methods for synthesizing them. Chemoinformatics comes into these assessments, as Figure 7.1 illustrates.

For example, the dissimilarity-based compound selection (DBCS) method in Figure 7.1.a selects compounds that satisfy the criterion of minimum average distance from every other point in the dataset at stake to identify eligible subsets of molecules over the whole space of possible combinations. By contrast, the sphere exclusion method in Figure 7.1.b treats the diverse points (molecules that are sufficiently dissimilar in structure) as centroids of the compounds. This method scales very well and is very cost-effective, but tends to penalize diversity in the compound synthesis.

The cell-based selection method in Figure 7.1.d in turn tends to sample the whole space of molecular structures rather than selecting specific portions on the basis of diversity considerations or scaling factors. It selects synthesizable compounds whose molecules are more evenly distributed at the cost of losing computational efficiency when the number of molecules goes up. Finally, the clustering method in Figure 7.1.c partitions molecules into groups on the basis of similarity considerations.

These different methods (and associated algorithms) furnish another example of perspectival pluralism. The methods are part of the experimental, technological (AI-led), and theoretical resources available within chemoinformatics as a scientific perspective in its own right for making reliable knowledge claims. Methodological and epistemic principles are in place to justify the reliability of the knowledge claims, including considerations about the toxicity of the compound and cost-effective production. Few possible chemical combinations delivered by any of these methods translate into new drugs. Scientists use different algorithms to physically conceive several scenarios about how new molecules can be arranged, so as to deliver knowledge about which new synthetic compound is possible. The general lesson is summarized by Nathan Brown (2016, pp. 130–131):

> As with many challenges in the field, there is no single, right answer for every occasion. . . . Diverse sets could be the ones that cover the extremities of the space or that distribute evenly over the entirety of the space. As with all modelling methods, it is important to understand the application prior to selecting the algorithms and other methods since the potential application will affect these decisions. If possible, multiple methods should be used, and importantly,

224 THE WORLD AS WE PERSPECTIVALLY MODEL IT

the results visualised to identify whether 'natural' clusters are being identified. While it would be nice to have a generally applicable clustering or diversity selection for all applications, this is wishful thinking and it is still necessary ... to fully consider the range of approaches and desired outputs.

All this has far-reaching implications for how to think about natural kinds. My discussion so far has concentrated on phenomena. But natural kinds have been the traditional battleground of debates on realism in science. Are kinds natural so long as they are found in nature? Do kinds harvested through algorithms such as DENDRAL count as 'natural kinds'? What makes a kind *natural* anyway? Is it some microstructural essential property? Or is it because it belongs to some useful taxonomy endorsed by some scientific community for some specific purpose?

The story of DENDRAL, and its legacy for chemoinformatics, challenges some deep-seated philosophical intuitions about natural kinds that cut across the realism/anti-realism divide. Traditionally, there is a division between taking natural kinds as natural divisions carved in nature (realism about kinds), or as conventional labels attached to a bunch of things that someone somewhere has deemed as sufficiently similar (anti-realism/conventionalism about kinds). There are of course many more nuanced views. Some realists endorse natural divisions in nature but refuse to associate them with essential properties. Some anti-realists are realists about individual things and entities but conventionalists about taxonomic classifications, for example.

One trend in philosophy of biology has challenged realist orthodoxy about natural kinds. Probing taxonomic classifications has made clear the inadequacy of thinking about natural kinds as defined by a set of essential microstructural properties.[6] This has brought a revival of nominalist approaches to natural kinds following a tradition going back to John Locke and re-energized by Hacking (1991, 1999, 2007a). And it has also ushered in different varieties of realism about kinds—see Dupré's promiscuous realism (1981, 1993) and Boyd's homeostatic property cluster kinds (1991, 1999a, 1999b) where realism about individuals or about properties is combined with a good dose of pragmatism about classifications.

Where does perspectival realism sit in this vast and nuanced landscape? What is the relation between the modally robust phenomena introduced in Chapter 6 and the more familiar notion of natural kinds? How is the phenomenon of the

---

[6] It goes beyond my scope and goal to discuss this voluminous literature here. The interested reader is referred to Beebee and Sabbarton-Leary (2010), Bursten (2014, 2018), Dupré (1981), Ereshefsky (2004, 2018), Ereshefsky and Reydon (2015), Kendig (2016a), LaPorte (2004, Ch. 3); MacLeod and Reydon (2013), O' Connor (2019), and Slater (2015); Khalidi (2013) and Magnus (2012) both offer overviews of the debate.

decay of the Higgs boson related to the natural kind the *Higgs boson*? How is the phenomenon of nuclear stability related to kinds of nuclei, from iron to lead? Or that of pollination to kinds of flowering plants? Or of phosphorylation to kinds of proteins?

Let us take microstructural essentialism and conventionalism as two extremes of a nuanced continuum of philosophical views. In what follows, it is not my intention to review this vast literature or chart this whole territory, but only to mark the salient differences of the view I defend from these two classical, long-standing, opposite views. If the perspectival realist concedes that there are natural kinds carving nature at its joints in virtue of some kind-defining microstructural essential properties, the phenomena-first approach would prove ultimately parasitic upon more traditional realist views.

Yet siding with the conventionalist in denying kind-defining properties and in thinking of natural kinds as convenient labels attached to a set of phenomena would open a wedge between perspectival pluralism and the promise of realism.

I am going to deal with this conundrum by introducing a different way of thinking about natural kinds, which is novel although it draws on the insights of other philosophical views. For example, I share with Hacking the sentiment that this is the twilight of the debate on kinds, and that the complexity of the challenges posed by human and social kinds might well be insurmountable. I also share the common wisdom that there cannot be a single metaphysical account for the bewildering variety of kinds to be found in the natural world. I aim for a sophisticated type of realism that I see available to the perspectivalist, who is not fazed by the prospect of taking phenomena as an ontological starting point. It will not be a one-size-fits-all approach to what is a natural kind. It is not a universal metaphysical account offering necessary and sufficient conditions for kind membership. The local moves I have made in Part I will find their counterparts in local moves in this second part of the book. I draw attention to a range of epistemological practices that can help us re-jig the way *we think and talk* about natural kinds so that realism about kinds is downstream from these epistemological considerations, rather than from a metaphysics-first approach.

In my discussion so far I have eschewed any talk of properties. I have resisted the temptation to think that perspectivism is just the claim that different properties are ascribed to the same target system when seen from different points of view, or from different models. My discussion on perspectival models as inferential blueprints and on perspectival data-to-phenomena inferences has taken us far away from the traditional metaphysical starting point of these discussions: that there is a world of properties (be they dispositional or categorical) as a given and that kinds can be seen either as carving them at their joints or as clustering them in some convenient way. The ontology I have defended is *inferentialist* all the way up, and places centre-stage *epistemic communities* with their situated scientific perspectives. Where to go from here?

I take my phenomena-first approach as the springboard for a thoroughgoingly inferentialist and perspectival view of natural kinds. Let us return one more time to chemoinformatics. The success of practices that involve dialling molecules and designing new synthetic compounds for pharmaceutical purposes is not decided by whether or not they unveil some hidden chemical substances in nature. It is measured instead by how these practices allow scientists to *make inferences* from the available data to phenomena and finally to kinds.

There is no single correct way of making such inferences. Some computational methods are more efficient and cost-effective than others. Some are more representative. Others strive for diversity (looking for molecules that are at the extremities of the chemical space). Purpose ultimately guides the choice of the algorithm and method amid the combinatorial explosion of possible drug-like molecular objects: which synthesizable compound can have particular pharmaceutical applications? Which is not toxic? Which one can be produced in large quantities and cost-effective way?

The view of natural kinds that I am about to lay out places centre-stage intersecting scientific perspectives in opening for us a 'window on reality'. In this respect, I join a recent trend that has emphasized the role of epistemological rather than metaphysical considerations when thinking about natural kinds. I see Kendig's discussion on 'kinding', Bursten's methodological role for kinds, Chang's epistemic iteration in chemical natural kinds, and Knuuttila's approach to synthetic kinds as kindred approaches to mine. What is novel here is that kinds are the outcome of ever-expanding collections of modally robust phenomena that epistemic communities encounter over time via perspectival data-to-phenomena inferences. In other words, the view I will articulate from here to Chapter 10 stresses the key role that intersecting scientific perspectives play behind our *talk* and *thought* about natural kinds.

My goal in this chapter is to offer examples from scientific practices (past and present) as a prelude to my *inferentialist* view of natural kinds which echoing Putnam (1990) I am going to label as 'Natural Kinds with a Human Face' (NKHF), and it goes roughly as follows:

(NKHF)

Natural kinds are (i) historically identified and open-ended groupings of modally robust phenomena, (ii) each displaying lawlike dependencies among relevant features, (iii) that enable truth-conducive conditionals-supporting inferences over time.

In the next section, I review a set of functions typically associated with natural kinds in the philosophical literature. I will call them (A) naturalism, (B) unanimity, (C) projectibility, and (D) nomological resilience. Depending on which function takes precedence, different philosophical views of natural kinds emerge. I then consider four counterexamples, one for each of these views. The counterexamples concern what I am going to call *engineered* kinds, *evolving* kinds, *empty* kinds, and *in-the-making* kinds. The view of NKHF is designed to shed light on them all.

7.2. What are natural kinds for?

Scientific realism traditionally begins with homely metaphysical considerations. There is an external world, independently of us human beings, and this world comes pre-packaged with natural kinds: water, gold, hydrogen, but also lemons, zebras, hellebores, and snowdrops, among a myriad of other examples.

For scientific realists, natural kinds mirror divisions in the natural world that do not depend on our language, the evolution of our conceptual resources, or which taxonomic classification happens to be in place. Natural kinds are what there is. A realist about science seeks strategies that guarantee that—to the best of our knowledge—we accurately describe these natural kinds. Thus, to be a scientific realist about the electron theory is to believe that what the theory says about the electron accurately describes a group of entities that we have reasons for thinking form a natural kind.

But what is a natural kind? Defining kind membership by a list of necessary and sufficient properties has proved fraught with difficulties. Granted a rough-and-ready definition of 'mammal' as a 'lactating animal, with fur, and not laying eggs', the discovery in New South Wales of the platypus at the end of the eighteenth century troubled zoologists. The platypus has fur and mammary glands but lays eggs.[7] Does it still count as a mammal? Or is it closer to a duck?

How about isotopic varieties of water? Do they still count as water despite very different chemical properties? Deuterium oxide, for example, also known as 'heavy water' is toxic and is used in the production of the hydrogen bomb (see LaPorte 2004, Ch. 4). And what to make of the decision of the International Astronomical Union in 2006 to downgrade Pluto to the rank of a 'dwarf planet'[8] when the definition of what counts as a 'planet' changed? Note also the wildly contingent nature of higher taxa where gulls and terns form sub-families,

---

[7] I discuss this example in Massimi (2012c). See also Moyal (2004).

[8] In 2006, the International Astronomical Union re-classified Pluto as a 'dwarf planet' because its gravitational mass is not large enough to clear debris on its pathway.

kingbirds and cuckoos correspond to genera, while owls and pigeons make up whole orders, as Dupré (1981) pointed out.

In spite of these difficulties, natural kinds have traditionally been regarded by philosophers as delivering on four main functions in science.

A. *Naturalism.* Beyond the Platonic metaphor of 'carving nature's joints', natural kinds identify 'functionally relevant groupings in nature' (Quine 1969). Quine believed that searching for logical principles of similarity for kinds was a doomed enterprise. Kinds nonetheless are 'part of our animal birthright' (p. 123), of our subjective and survival-adaptive spacing of qualities into classes or groupings. From sorting wild berries and mushrooms into edible/non-edible to sorting elementary particles into hadrons and leptons, the story of natural kinds is the story of how 'our innate subjective spacing of qualities accords so well with the functionally relevant groupings in nature' (p. 126).[9]

According to Quine, we learned to hone functional groupings on the basis of their ongoing inductive success or failure in serving practical and epistemic needs—from the classification of chemical elements in terms of atomic number to the classification of animals in clades. Quine presented natural kinds as the survival-adaptive outcome of how human beings have successfully learned to navigate the world around them.

B. *Unanimity.* Zoologists might disagree about specific morphological features of a given platypus specimen, and oceanographers about the percentage of deuterium oxide present in the Pacific Ocean or Atlantic Ocean. Astronomers might debate whether Pluto indeed counts as a planet. But as long as *mammal*, *water*, and *planet* form natural kinds, there is something everyone can agree on. Natural kinds are designed to identify features common to a class of entities.

Unanimity figures in traditional realist accounts such as the Kripke–Putnam account of natural kinds (see Kripke 1980, p. 124). Kripke argued that if we were to discover tomorrow that the mineral found in the mountains of America, South Africa, and Russia does not in fact have atomic number 79 but is instead fool's gold (iron pyrite), it would be wrong to insist that it would still be gold.

---

[9] I have addressed Quine's naturalism as part of my earlier defence of a type of naturalized Kantianism about kinds (Massimi 2014). In that article, I was interested in responding to a series of classical objections against Kantianism about kinds and widespread conflation of Kantianism with constructivism. In this chapter, I do not engage with this topic as such, but I build and expand on some of these original ideas.

He argued that properties essential to kind membership are not observable (e.g. the yellow colour of gold) but microstructural properties featuring in theoretical identity statements such as 'water is H_2O' or 'gold is the element with atomic number 79'.

In Putnam's (1975) Twin Earth scenario, the stuff that fills oceans and lakes shares the superficial properties of water on Planet Earth but its chemical composition is XYZ rather than H_2O. Putnam argued that we would not count it as water because having the microstructural property of being H_2O is necessary for something to be water. Kripke and Putnam oppose naïve descriptivism, the view that the reference of natural kind terms is fixed by a description of the set of properties representative of the meaning of the natural kind term in question.[10]

For we do not gain a priori knowledge of natural kinds by grasping the meanings of natural kind terms. If anything, we have a posteriori knowledge of natural kinds. We empirically discovered that water is H_2O even if we take it now to be necessary for something to be water that it has to share the same microstructural kind relation to a sample in a presumed original causal baptism, according to Putnam's causal theory of reference.

Natural kinds in this Kripke–Putnam realist tradition are defined by microstructural essential properties, which presumably offer a common platform for unanimous judgements, shorn of all the historical accidents and contingencies of how any particular epistemic community might (or might not) have come to know that water is H_2O, or gold is the element with atomic number 79.

C. *Projectibility.* This has traditionally been a defining feature of the realism debate about natural kinds. The idea originates from Nelson Goodman's new riddle of induction (Goodman 1947) and the risk it poses to the success of our inductive inferences. Any number n of positive instances for a generalization such as 'All emeralds are green' up to a specified time t_1 inductively support a pair of conclusions: namely, that 'All emeralds are green' and 'All emeralds are grue' after time t_1, where something is 'grue' if it is either examined before time t_1 and found to be green or it is examined after time t_1 and found to be blue.

'Gruified' inferences were at the heart of a problem Goodman saw for any theory of natural kinds: how to demarcate between projectible predicates like 'green' and non-projectible ones like 'grue'? Goodman himself did not go much further than explaining the difference in terms of what he called 'entrenchment'.

---

[10] For a recent defence of a cluster version of descriptivism, see Häggqvist and Wikforss (2018); and for a defence of Putnam's semantic argument, see Hoefer and Martí (2019). See Beebee and Sabbarton-Leary (2010) for a collection of essays on the topic.

Some predicates (like green) are more entrenched in our languages than others (like grue). But for others, projectibility became an ongoing concern.

A theory of natural kinds has to make sense of the success of our inductive inferences. Projectible natural kinds can avert the risk of Goodman's grue scenarios. As Richard Boyd (1991, p. 131) stressed, 'Even for the purposes of guessing we need categories of substance whose boundaries are not (or not just) "the workmanship of the understanding"'. To deliver on projectibility, Boyd (1990, 1992, 1999a, 1999b) proposed the homeostatic property cluster kinds (HPCK) account. Natural kinds, on this view, are imperfect and fuzzy clusters of co-occurrent properties supported by a homeostatic mechanism rather than a clear-cut set of properties acting as necessary and sufficient conditions for kind membership.

Some philosophers of biology (Brigandt and Griffiths 2007; Currie 2014; Ereshefsky 2012) have explained projectibility by appealing to the notion of homologues: traits or features observed across species and traceable to a common ancestor (e.g. 'human arms, bat wings, and whale fins are homologues, they are the same character—the mammalian forelimb'; Ereshefsky 2012, p. 383). In the philosophy of the social sciences too, the notion of 'cultural homologue' has proved a helpful tool to explain the projectibility of social kinds.[11]

D. *Nomological resilience.* Another traditional function of natural kinds is strictly related to projectibility: natural kinds are taken as supporting laws of nature, a feature I call *nomological resilience*. After all, how can natural kinds license successful inductive inferences if not through their ability to support laws of nature? Knowing that water, for example, is a natural kind opens up the possibility of inferring that the next sample will boil at 100° and freeze at 0° Celsius. Knowing that the electron is a natural kind makes it possible to explain phenomena such as electrostatic repulsion (given Coulomb's law), electronic configurations for chemical elements (given the periodic table), and the stability of matter more generally (given Pauli's principle).

That kinds go hand-in-hand with laws of nature was famously pointed out by Ian Hacking in his influential discussion of how what he calls Mill-kinds become Peirce-kinds. Hacking (1991, p. 112) starts with Russell's view on natural kinds

---

[11] Marion Godman, for example, has defended the notion of 'cultural homologue' to describe the social kinds of anthropologists and social scientists (e.g. social democracy, among others). She defines a cultural homologue as one that 'contains *systematically arranged information or content . . .* [which] is typically a combination of factual knowledge about the world and prescriptive or practical knowledge' (Godman 2015, pp. 500–501, emphasis in original). Godman et al. (2020) have further explored the extent to which the notion of historical lineage from Ruth Millikan's (1998, 1999) influential work can be reconciled with a suitable version of essentialism about natural kinds.

as the 'class of objects all of which possess a number of properties that are not known to be logically interconnected', and notes how this definition goes back to John Stuart Mill. In *A System of Logic* (1843), Mill argued that natural kinds exist in nature and what he called 'real Kinds' were characterized by an inexhaustible number of properties that can be ascertained by observation and experiment. Charles S. Peirce later supplemented Mill's real Kinds. 'Peirce-kinds'—as Hacking calls them—refer to a class such that there is a systematized *body of laws* about things belonging to this class and 'providing explanation sketches of why things of a given kind have many of their properties' (Hacking 1991, p. 120). Natural sciences often develop Peirce-kinds from Mill-kinds, according to Hacking.

I would add that natural kinds are typically regarded as nomologically resilient. No matter how our images of some have changed over time (say, the electron from J. J. Thomson to Dirac, to QED), the nomological resilience of the class of things we call 'electrons' is important. As I have argued in Chapter 5, lawlike dependencies are key to the exercise of physical conceivability and perspectival modelling. They play an additional important role in helping epistemic communities identify relevant groupings of phenomena as belonging to a natural kind.

To recap, the four main functions of natural kinds are associated with different philosophical views about kinds. Naturalism is congenial to metaphysically deflationary accounts (à la Quine) whereby natural kinds reduce to functionally relevant groupings that latch onto natural divisions. Unanimity invites metaphysically more substantive accounts such as those that identify natural kinds with a set of microstructural essential properties. Projectibility underpins a number of realist accounts—from Boyd's HPCK to Ereshefsky's homologues in biology—with a less pronounced metaphysical slant and more attention to historical lineages. Nomological resilience, finally, can be seen to be at work across a range of views—from Putnam's metaphysically robust account[12] to Hacking's nominalism where the emphasis is more on the epistemic agent as *'homo faber'*, in Hacking's terminology: what artisans and craftspeople do, as when one makes rings with gold, necklace pendants with jade, and hydrogen bombs with deuterium oxide.

Without any presumption of offering a philosophical account of natural kinds that fits all, I do nonetheless present a philosophical stance that can explain and justify why philosophers care so deeply about naturalism, unanimity, projectibility, and nomological resilience, and why natural kinds are typically meant to deliver on these four main functions. The view I propose over the next three chapters is metaphysically deflationary like Quine's: it does not subscribe

---

[12] Hacking, for example, comments on how 'the Peirceian conception seems to rule, at present', and cites Putnam and Kripke to illustrate how Mill-kinds have been transformed into Peirce-kinds: 'There are *objective laws* obeyed by multiple sclerosis, by gold, by horses, by electricity; and what is rational to include in these classes will depend on what those laws turn out to be' (Putnam 1983, p. 71, emphasis in original; quote from Hacking 1991, p. 121).

to Kripke–Putnam essentialism, nor does it fall back into Boyd's property-realist story behind the HPCK view. It cuts across traditional philosophical dichotomies in this debate (essentialism/conventionalism; realism/nominalism). Most importantly for my story, it sheds light on how a phenomena-first ontology sits with the debate on natural kinds. In what follows, I present some examples from both scientific practice and the history of science that force us to rethink traditional philosophical stances associated with these four functions of natural kinds.

7.3. *Hachimoji* DNA and RNA: the effective naturalness of engineered kinds

Few things speak of 'natural divisions in nature' more than DNA sequencing. Since the work of Franklin, Watson, and Crick on the DNA structure and the ensuing development of molecular genetics, the nucleotides adenine A, thymine T, guanine G, and cytosine C have been regarded as the DNA natural letters in the alphabet of life. The double-helix structure of A–T and C–G pairs encode necessary information for various forms of life on planet Earth. From the taenia worm to daffodils, from kingfishers and pandas to humans, natural kinds in the life sciences have long seemed to be written out of these four simple DNA building blocks.

Putnam (1975, p. 240) claimed that for something to be a lemon (or belong to the natural kind *lemon*) it has to have the genetic code of a lemon, much as having chemical composition H_2O is necessary for something to be water. The problem, though, is that biological kinds, species, or higher taxa cannot realistically be identified just by invoking the genetic code (see Dupré 1981; Ghiselin 1974; LaPorte 2004). Higher taxa reflect contingent human decisions about classificatory boundaries—as Dupré (1981) pointed out.[13] More generally, biological species (be it *Citrus limon* or *Equus quagga* or something else) are the product of evolutionary survival-adaptive mechanisms—as evolutionary taxonomy has long studied them—that cannot be entirely reduced to a genetic code.[14]

But leaving aside the complex issue of how to think about biological kinds in the light of cladistics, evolutionary taxonomy, and so on, there is something problematic from a *metaphysical* point of view about the Putnamian microstructural essentialist story. That is, first, the presumption that natural kinds are

[13] '[B]iological theory offers no reason to expect that any such privileged relations exist, since higher taxa are assumed to be arbitrarily distinguished and do not reflect the existence of real kinds' (Dupré 1981, p. 78).

[14] It goes beyond the scope and aim of the present discussion to enter into the so-called species problem in philosophy of biology, on which a variety of philosophical views have been advanced and defended over the past two decades (see, e.g., Ereshefsky 2010; Ghiselin 1974; Griffith 1999; Hull 1978; Kitcher 1984; Millikan 1999; Okasha 2002).

metaphysically identifiable with some essential building blocks (be they chemical building blocks like atoms for chemical compounds or genetic building blocks like nucleotide pairs A–T and C–G for living organisms). And, second, the further presumption that such building blocks are somehow letters in a natural alphabet in which the book of nature is written (be it atomic numbers for chemical elements or DNA code). The reality goes beyond these narrowly conceived metaphysical boundaries.[15]

Recall here DENDRAL and how current chemoinformatics goes about exploring different methods for grouping molecules into new chemical compounds. New drugs are engineered all the time for industrial and pharmaceutical purposes. Among all the conceivable molecular combinations, only those that have passed specific tests to check for toxicity, stability, and cost-effectiveness get selected for patents and production. New chemical kinds are engineered by AI-led processes. And engineering does not just apply to chemical kinds. It increasingly applies also to kinds in the life sciences.

Consider the announcement in *Science* on 22 February 2019 of the eight-letter (*hachimoji*, in Japanese) nucleotide language for DNA and RNA (Hoshika et al. 2019). A team led by Steven Benner at the Foundation for Applied Molecular Evolution in Florida announced they had succeeded in producing synthetic DNA. In addition to the four nucleotides A–T, C–G, synthetic DNA has four additional 'letters'—purine analogues P and B and pyrimidine analogues Z and S—that would form new pairs P–Z and B–S. Previous attempts at synthesizing DNA with additional 'letters' relied on water-repelling molecules (hydrophobic nucleotide analogues). But these attempts failed because in the absence of hydrogen bonds, hydrophobic nucleotides tend to slip and distort the double-helix structure. Hydrogen bonds are required to secure the stability of additional pairs of synthetic nucleotides that can be transcribed into RNA.

Benner and his group found a synthetic DNA that not only forms stable pairs and reliably translates into RNA for protein formation, but can also potentially support molecular evolution and find applications in a variety of medical diagnostics. The discovery sends out a strong message about naturalism for natural kinds, i.e. even our most credible candidates underpinning natural divisions in nature can be synthesized and engineered. Further examples in this direction come from synthetic biology (see Kendig and Bartley 2019; Knuuttila and Loettgers 2017; O'Malley 2014), techno-science (see Russo forthcoming) and nanotechnology (see Bursten 2016, 2018).

The boundaries between the natural and the engineered are fuzzier than one might suppose. And any philosophical account of natural kinds that insists on

---

[15] I shall return to this topic in more detail in Chapter 9, where I flesh out why I do not endorse essentialism about natural kinds.

naturalism (even a thin rather than thick notion of naturalism as in Quine) has to make room for *engineered kinds*. Thus, this is the first lesson that I'd like to draw out of these examples:

Lesson no. 1: The naturalness of natural kinds is not just the product of our 'subjective spacing of qualities' (to echo Quine). It is also the result of our perspectival scientific history.

Chemoinformatics, nanotechnology, and synthetic biology are scientific perspectives on a journey to extend and redefine the boundaries of what we count as a natural kind. When a scientific perspective advances new knowledge claims about new kinds, whose reliability can be assessed and evaluated over time with cross-perspectival assessments, I see no metaphysical reason for excluding them from naturalism about kinds. Our presumed access to natural divisions in nature is no more privileged, direct, unfiltered, or unmediated than our access to kinds delivered through perspectival modelling. Which kinds count as 'natural' is ultimately a case-by-case judgement. It rests on the reliability of the historically and culturally situated practice delivering knowledge of the kinds and the methodological-epistemic principles that can justify their reliability within the perspective, with truth-conditions remaining a *cross-perspectival* affair (as per Chapter 5, Section 5.7).

7.4. Corpuscles and Faraday tubes: the unexpected unanimity of evolving kinds

The history of science challenges essentialist accounts of natural kinds. Kuhn (1990) presented what is in my view one of the most convincing historical arguments against the Kripke–Putnam semantic view. He argued that in Putnam's (1975) Twin Earth thought-experiment only a differently structured scientific lexicon could describe the behaviour of the hypothetical XYZ at all. And in that new lexicon, H_2O might no longer refer to what we now call 'water'.

Kuhn also rebutted Putnam's claim that a hypothetical *Doppelgänger* in 1750 (before Lavoisier's discovery of oxygen) would have been referring to water all along even if not knowing yet that water was H_2O. In 1750, states of aggregation (solid, liquid, and gaseous) were regarded as demarcating chemical species. Thus, *water* for an eighteenth-century natural philosopher was an elementary liquid substance, with *liquidity* being an essential property. Only after Lavoisier's Chemical Revolution did the distinction among solids, liquids, and gases become physical rather than chemical. Kuhn argued that what Kripke called 'rigid designators' (namely, names that rigidly designate the same objects) did not apply to natural kind terms such as 'water', which have in fact gone through a major conceptual and meaning change over centuries. Epistemic unanimity about natural kinds should not be expected on Kuhn's view of scientific revolutions.

I am going to use the expression *evolving kinds* to refer to natural kinds that have evolved across scientific perspectives and adapted to new scientific practices over centuries. The kinds we know and love—I contend—are all evolving kinds: they have survived endless conceptual change. This semi-Kuhnian feature is central to the view of Natural Kinds with a Human Face (NKHF). Naturalism invites us to make a presumption about functionally relevant groupings in nature. But the identification of these groupings has a history of its own, rooted in a variety of scientific perspectives and associated practices that have evolved over time and across cultures, as new technology and experiments became available, or new methodological and epistemic principles were introduced.

A perspectival realist account, then, does not delegate unanimity to rigid designators. Nor does it forsake it in the name of scientific revolutions and conceptual change. But it takes unanimity as some sort of equilibrium point in the survival and adaptation of our ever-evolving kinds across a plurality of scientific perspectives.[16] I shall discuss one of my favourite examples, the electron, in Chapter 10. J. J. Thomson, who discovered the electron, did not in fact refer to his particles as 'electrons' but as 'corpuscles'. Thomson believed that there were positive and negative electric charges whose field-theoretic behaviour was described by what he called a 'Faraday tube', working within the electromagnetic tradition.[17] Faraday tubes allowed Thomson to reconcile the discrete nature of electricity with the continuous nature of the electromagnetic field. But is our *electron*, which is now part of the current Standard Model of particle physics, the same as his? How can natural kinds realistically offer a common platform for unanimous judgements over time, if our talk and thought of them inevitably evolve over time and across scientific perspectives?

[16] I share Ruth Millikan's (1999) argument for replacing what she called 'eternal natural kinds' with 'historical natural kinds'. Millikan sees the continuity and uniformity of kinds as rooted in their historical lineage. This is particularly evident in the case of biological kinds: 'Cats must, first of all, be born of cats, mammals must have descended from a common ancestor, and so forth. Biological kinds are defined by reference to historical relations among the members, not, in the first instance, by reference to properties. Biological kinds are, as such, historical kinds. . . . [M]embers of these kinds are like one another because of certain historical relations they bear to one another (that is the essence) rather than by having an eternal essence in common' (p. 54). I agree and take on board Millikan's insight here. I would add to her observation that these historical relations are not just (or only) phylogenetic relations (cats from cats, etc.). They are not just the outcome of breeding and cladistics considerations. They are also the product of perspectival and multicultural scientific history. As I am going to argue in detail in Chapters 8 through 10, it is not just biological kinds that are historical kinds, but physical and chemical kinds too. They are all Natural Kinds with a Human Face.

[17] See Arabatzis (2006) and Falconer (1987, 2001) for excellent historical accounts of this episode. I recount it in some detail in Chapter 10.

An account of natural kinds that delivers on the promise of *epistemic* unanimity has to explain their historical evolution across a plurality of scientific perspectives. Why is it that we tend to agree in our judgements about, say, *electrons* (*water*, *gold*, *jade*, etc.), despite a variety of perspectival₁ representations across the history of science?

Recall the two notions of perspectival representation introduced in Chapter 2: i.e. a representation drawn *from a* particular point of view vs. a representation directed *towards* one or more vanishing points. I'd like to think of the unanimity in talk and thought about natural kinds as the vanishing point (if any as such exists) towards which our historically and culturally situated perspectival representations are drawn.

> *Lesson no. 2: The epistemic unanimity granted by natural kinds is not a by-product of microstructural essential properties 'from nowhere'. It is a product of perspectival scientific history, of how historically and culturally situated epistemic communities learn to engage with one another, to perspectivally model the relevant phenomena, and navigate the inferential space surrounding them.*

Hence, an explanation for the unexpected unanimity of evolving kinds has to be sought for in the epistemic grounds upon which epistemic communities come to engage with one another and agree across time on what natural kinds there are, in spite of potential disagreement in their perspectival₁ representations.

7.5. Lavoisier–Laplace's ice calorimeter: the unreasonable projectibility of empty kinds

Projectibility is a cornerstone of Boyd's HPCK account. On this view, natural kinds secure inductive inferences and explanations in science because natural kind terms refer to fairly stable clusters of co-occurrent properties in nature (which are not entirely 'the workmanship of women and men', as stressed by Boyd 2010, p. 219). Boyd's property realism about kinds is designed to offer the best explanation for their projectibility, and a safe antidote to Goodman's new riddle of induction. Yet it rides roughshod over a curious, often neglected, fact: namely, that empty kinds have often proved no less projectible than *bona fide* natural kinds.

I call *empty kinds* putative kinds whose membership eventually turns out to be an empty set. Theories of ether, caloric, or phlogiston enjoyed scientific success for relatively long periods of time (see Chang 2012a, 2012b; Ladyman 2011; Laudan 1981; Lyons 2002, 2006; Vickers 2013, 2017). Explaining the success of false theories has long been the aim of realists responding to the 'pessimistic meta-induction' from history of science (see Kitcher 1992, 1993; McLeish 2005; Psillos 1996, 1999; Stanford 2003a, 2003b). And various philosophical approaches have been developed over time to tackle this kind of prima facie counterexamples to realism about kinds. For example, Kyle Stanford and Philip Kitcher (2000) famously put forward a refined causal theory of reference to handle these counterexamples.[18] Hasok Chang (2012a, p. 247) has pointed out that there is never any stability to be expected in the act of fixing reference, and that 'the correspondence theory of reference is futile, because reference to bits of unobservable reality is just as inoperable as "Truth with a capital T"'. I shall follow Chang's advice here in not getting 'fixated' about reference-fixing for natural kind terms. The burden of perspectival realism does not rest on semantic arguments for natural kind terms. Nor does it rest on epistemic arguments for the approximate truth of best theories in mature science.

I will not speak of scientific theories, because my realism is bottom-up: from data to phenomena and from phenomena to natural kinds. Thus, instead of asking why a false theory could prove successful for a period, I am going to ask: how could empty kinds prove projectible? How could hypothetical kinds whose membership is in fact an empty set nonetheless support inductive inferences and explanations in relevant areas of inquiry?

One might bypass this question by denying that things such as caloric, ether, phlogiston, and so on, are 'kinds'. They do not exist. A fortiori they cannot be kinds. Yet these things were imagined, conceived of, supposed to exist, to have properties, and to behave in specific ways. Different models of them were built and used to gain information about different phenomena in nature. Modal knowledge was sometimes obtained using models that conceived of such things. Ditching caloric, ether, phlogiston, and so on, as falsehoods does not begin to capture the crucial role that imagined entities and putative kinds played for centuries in advancing scientific knowledge.

Should not we, then, refer to them as 'extinct' kinds?[19] Should not we be more liberal in the usage of the 'natural kind' label and accept that some of them (e.g. caloric, phlogiston, ether) did live at some point but became extinct later? Much as my take here is very much in debt to the history of science and powerful historicist criticisms of scientific realism (to use Stanford's 2015 terminology), 'extinct' would err on the side of historical generosity, in my opinion. For it would bestow the label of 'natural kind' to effectively an empty set. An empty set is

---

[18] To understand earlier uses of natural kind terms such as Priestley's 'dephlogisticated air' to refer to what we now call 'oxygen', Stanford and Kitcher argue that '*some* kind of description must play the role of samples and foils in the act of grounding reference, but whether this is a description of internal structure, causal role, causal mechanism or something else altogether will vary with the term-type and even with the term-token under consideration' (2000, p. 125, emphasis in original).

[19] I thank Julia Bursten for raising this question to me in a reading group.

always empty—yesterday as it is today. Therefore, my expression *empty kinds* comes closer to capture the good I see in some of these historicist arguments without the risk of reifying as a kind something that never was.

Let us, then, allow 'kinds' to include not just kinds known to exist, but also conceivable kinds, or hypothetical kinds, some of which will survive and become evolving kinds and some of which will eventually turn out to be empty and get discarded. How to explain the unreasonable projectibility of empty kinds for a period of time? Consider as a few examples the following indicative conditionals:

(E.1)   If caloric is a physically conceivable 'matter of fire'[20] that binds to bodies, specific heat increases with temperature.

(E.2)   If phlogiston is a physically conceivable 'combustible principle', metals turn into calxes by removing phlogiston.

(E.3)   If ether is a physically conceivable 'elastic medium' for the transmission of light, light propagates in transverse waves through it.

Empty kind terms feature in the antecedents of these conditionals. Inductive inferences and explanations could nonetheless still be given for phenomena ranging from specific heat to calcination and optical diffraction. This should not be surprising. In Chapter 5, I made the point that suppositional antecedents in such conditionals can deliver true consequents via enthymematic arguments even when additional hidden premises rely on theories that later turn out to be false (as is the case with these examples).

So the task ahead for perspectival realism is to show the inferential patterns that explain how and why different epistemic communities came to agree that a certain historically identified grouping of phenomena is (or is not) a natural kind in spite of disagreements about how to think of some of these phenomena.

The epistemic agents uttering the indicative conditionals (E.1)–(E.3) were, say, Lavoisier, Priestley, and Fresnel, respectively. Each of them was working within a well-defined scientific perspective that included experimental and technological resources to advance claims of knowledge about a number of phenomena. Those resources included Lavoisier's ice calorimeter, Priestley's nitrous air test,[21]

---

[20]   A variety of terms were used at the time, ranging from 'igneous fluid' to 'fire matter', 'heat matter', the 'principle of heat', the 'matter of fire', although in Lavoisier's *Traité élémentaire de chimie* the new official nomenclature of 'caloric' (*calorique*) was introduced. The term 'matter of fire' was a term of art in a well-defined Newtonian tradition going back to Boerhaave (1732/1735) and even Kant (1755/1986). On Lavoisier's caloric theory, see Morris (1972).

[21]   For an analysis of the role of experimentation in the Chemical Revolution and how it bears on debates about realism and perspectivism, see Jacoby (2021).

and Fresnel's optical diffraction experiments. These instrumental practices unfolded a number of inductive inferences seemingly associated with the putative kinds caloric, phlogiston, and ether, respectively. Let us zoom in on one of them: Lavoisier's ice calorimeter.

In 1783, Lavoisier and Laplace built an instrument that was designed to measure the amount of caloric bound to a body by measuring how ice melted into water. Caloric was supposed to be fixed in the external layer of ice, and in any of the subsequent more internal layers. Thus, it seemed natural to suppose that if twice the quantity of ice were melted, double the quantity of caloric would be released. This was measured through a multilayer structure that could insulate the ice in the central cavity as much as possible from the heat of the surrounding air. The stopcock that controlled the water leaking from the central cavity was separated from the run-off of the external and middle cavity and controlled by a spigot (see Heilbron 1993, pp. 101–105).

The ice calorimeter had flaws. First, not all water melted in the process. Some got retained in the porosity of the ice, and accordingly the measurements were off and difficult to reproduce. But the principle behind it was ingenious, and one that could be extended to measuring the specific heat of gases and fluid substances such as sulphuric and nitric acids (Lavoisier 1799, p. 433).[22]

The false assumption was of course that heat was a conserved quantity in these transitions of states and hence that the 'quantity of ice melted is a very exact measure of the proportional quantity of caloric employed to produce that effect, and consequently of the quantity lost by the only substance that could possibly have supplied it' (p. 423). The suspicion that caloric was in fact an empty kind was in the air already in 1798, four years after Lavoisier's death. And it was brought up in the most unexpected way in the most scientifically unassuming practice.

A former lieutenant-colonel in the American War of Independence, Benjamin Thompson was knighted, and moved to Bavaria, where he became in 1791 Count von Rumford working as grand chamberlain to the elector of Bavaria and superintendent at the military arsenal in Munich. Among his duties, he supervised the boring of cannons. In this role, he had the chance to observe 'the very considerable degree of heat which a brass gun acquires, in a short time, in being bored; . . . The more I meditated on these phenomena, the more they appear to me to be curious and interesting . . . and to enable us to form some reasonable conjecture

---

[22] '[T]he water produced by melting the ice during its cooling is collected, and carefully weighed; and this weight, divided by the volume of the body submitted to experiment, and multiplied into the degrees of temperature which it had above 32° at the commencement of the experiment, gives the proportion of what the English philosophers call specific heat' (Lavoisier 1799, p. 429). As Heilbron (1993, p. 104) notes: 'From measurements of the temperature of the gas on entry and exit, the rate of flow, and the quantity of melted ice, they could compute a value for the heat capacity of the specimen under study. In experiments performed during the winter 1783/4 but not published until 1805, they made the specific heat of oxygen to be 0.65, and the specific heat of air 0.33031(!), that of an equal weight of water'.

respecting the existence, or non-existence, of an *igneous fluid*. His answer was
that 'the heat produced could not possibly have been furnished at the expense of
the latent heat of metallic chips' and was generated instead by friction (Rumford
1798, pp. 81–83). He then proceeded to ask a series of questions:

> What is heat?—Is there any such a thing as an *igneous fluid*?—Is there anything
> that can with propriety be called *caloric*? . . . It is hardly necessary to add, that
> any thing which any *insulated* body, or system of bodies, can continue to furnish
> *without limitation*, cannot possibly be a *material substance*: and it appears to me
> to be extremely difficult, if not quite impossible, to form any distinct idea of
> any thing, capable of being excited, and communicated, in the manner the heat
> was excited and communicated in these experiments, except it be MOTION.
> (Rumford 1798, p. 98–99, emphases and capital letters in original)

There were other doubts. In England, Humphry Davy (1812), experimenting
with ice cubes that melted by friction despite the temperature being kept at
freezing point, had similar thoughts in concluding that the phenomenon of heat
(or calorific repulsion as was still called at the time) was caused by motion. It was
a good half-century before Rumford's and Davy's observations were developed
in yet another scientifically unassuming practice. James Prescott Joule came
from a wealthy family of brewers in Lancashire, who could afford among the
tutors for their children John Dalton, one of the leading chemists of the time and
a defender of the hypothesis of chemical atoms. Joule became interested in how
to improve the efficiency of the brewery, and ran a series of experiments with a
paddle-wheel machine to measure the interconvertibility of heat and work. He
used a system of strings and pulleys connected to a paddle-wheel inside an insulated copper container with different liquid substances (water, oil, and mercury).
Joule studied how much mechanical work was needed to activate the paddle-
wheel and eventually raise the temperature of the liquid in the container.

These experiments confirmed Count Rumford's insight that heat was not a
material substance being released but it was instead a kind of motion: it was indeed produced in proportion to the amount of mechanical work expended. From
these experiments, using thermometers, Joule (1850) was able to establish the
mechanical value of heat, which became known as Joule's equivalent. Rumford's,
Davy's, and Joule's observations and inferences mark the end of caloric as a putative kind. On these experimental foundations, thermodynamics and the kinetic
theory of gases were developed in the second half of the nineteenth century.

Yet the example indicates how empty kinds are more than mere idle posits in
a theory or mistaken assumptions to be eventually overthrown. They played an
important role for what Hacking called *homo faber*. Empty kinds inspired the
work of artisans, craftspeople, apothecaries, engineers, militaries, and brewers

alike. They informed the invention of machines and instruments like the ice cal-orimeter, or the paddle-wheel experiments. In spite of possible false assumptions and inaccurate measurements, these instruments advanced scientific knowledge by making possible conditionals-supporting inferences about a range of *phe-nomena*. These inferences revolved around suppositional questions such as these ones (one can think of them as a Ramsey test for the indicative conditional E.1):[23]

(a) If caloric is a physically conceivable 'matter of fire' that binds to bodies, will specific heat increase with temperature?
(b) If caloric is a physically conceivable 'matter of fire' that binds to bodies, how much ice will melt in the ice calorimeter for a substance x at temper-ature y?
(c) If caloric is a physically conceivable 'matter of fire' that binds to bodies, will heat increase indefinitely by boring a cannon?

It was through questions like these that it became possible to start measuring the specific heat of metals and gases. And it is through them (among other things) that transitions of states (from solid ice to liquid water) began to be regarded as physical rather than chemical in nature: the outcome of friction and mo-tion, rather than the release of some hidden igneous fluid. As Kuhn remarked, it was indeed only after Lavoisier that the term 'water' came to encompass not just liquid water but also ice and water vapour. Philosophical accounts of natural kinds that care about the seemingly unreasonable projectibility of empty kinds across the history of science should lay less emphasis on theories and more on the experimental practices and technological tools that were built around them.

These experimental practices are part and parcel of scientific perspectives. Lavoisier's ice calorimeter, together with the gravimetric methods of the apoth-ecaries and assayers familiar to him via his life as a tax collector (see Bensaude-Vincent 1992), were inherent in the scientific perspective he operated with. Similarly, Joule's paddle-wheel experiment was part of a scientific perspective that started with Dalton's atomism, Rumford on cannons, and Davy on ice cubes, and later intersected with Faraday's studies on the interconvertibility of elec-tricity and chemistry. These intersecting scientific perspectives offered a more general standpoint for the interconversion between mechanical work and heat as different forms of energy. In each case, the data-to-phenomena inferences were perspectival. Caloric proved to be an empty kind, by contrast with kinetic en-ergy, because it did not track groupings of modally robust phenomena over time.

[23] See Chapter 5, footnote 24. As clarified already there, my goal is not to contribute to any formal framework for the semantics or probabilistic logic of indicative conditionals, but simply to pay atten-tion to model-based inferential reasoning that scientists make with perspectival modelling. This is another case in point.

Yet the ice calorimeter was the remote ancestor of modern-day electromagnetic and hadronic 'calorimeters' used in experiments such as ATLAS at the Large Hadron Collider (LHC) that are able to detect electromagnetic and hadronic particles produced in proton–proton collisions. The persistence of the name 'calorimeter' (despite the non-existence of caloric) is testament to the seemingly unreasonable projectibility of empty kinds.

> *Lesson no. 3: The projectibility of natural kinds does not have to do with the naturalness vs concocted nature of the predicates/properties (e.g. 'green' vs 'grue') associated with natural kind terms (e.g. emeralds) in scientific statements. It has to do instead with the machines, instruments, experiments, and conditionals-supporting inferences about a range of phenomena (e.g. boring of cannons, melting of ice cubes, paddling wheels in water) that these experiments licensed (within and across scientific perspectives).*

Of course, this is only a starting point. A lot more needs be said about how these identified groupings of phenomena constitute what I call *evolving kinds*, and I attend to this task in Chapters 8 and 9.

7.6. Dark matter: the nomological resilience of a kind in-the-making

Finally, let us turn to nomological resilience. Traditionally, the ability of natural kinds to license successful inductive inferences has been linked to their ability to support laws of nature. Hacking pointed out the role of laws of nature in defining natural kinds through his distinction between Mill-kinds and Peirce-kinds, whereby Peirce-kinds sometimes develop from Mill-kinds. He identified Peirce-kinds with the Putnamian view that there are *objective laws* that gold, electricity, and so on, satisfy. Accordingly, the ability to infer from one property of the kind (say, atomic number 79 for gold) to other properties for the same kind (malleability, melting point, etc.) is grounded on a *systematized body of laws of nature*.

In this final section, I defend a role for laws of nature in natural kinds in two ways. In the absence of laws of nature, I contend that Mill-kinds turn out to be empty kinds. And with laws of nature, even in-the-making kinds enjoy the status of Peirce-kinds.

Recall that a Mill-kind (following Hacking 1991, p. 118) is a real Kind (with capital K) 'if it has a large and plausibly inexhaustible set of properties not possessed by members of K that lack [property] P'. Mill-kinds allow speakers to make inductive inferences based on the knowledge of one distinctive property P, which is the gatekeeper for innumerable others. But inductive inferences in Mill-kinds are not supported by a system of laws of nature.

Imagine you get your inferences via a sort of lottery system that randomly associates property P with a large set of other properties S (but the system might as well have associated P with a different large set of alternative properties T). Mill-kinds are genuine empiricist kinds with nominalist roots. There are no laws buttressing the connection between property P and the large (inexhaustible) set of other properties S over and above constant conjunction and co-occurrence. A problem then arises. How to tell whether the connection is purely accidental, or indicative of some genuine kindhood?

Caloric is a case in point. This empty kind was once a Mill-kind. The putative connection between the (alleged) repulsive property P of caloric (or 'calorific re-pulsion' as was known at the time) and a number of other properties S (e.g. being released in transition of states from liquid to gas, being squeezed out when ice melts) proved a versatile one. From the ice calorimeter to Carnot's steam engine, this former Mill-kind made possible far-reaching inductive inferences—until they came to a halt, and the success of the inferences turned out to be parasitic upon motions of molecules, kinetic energy, entirely different underpinnings than caloric.

This former Mill-kind did not have laws of nature backing up its inferences and it eventually faded away. 'Conservation of caloric' is not a law of nature, de-spite Sadi Carnot assuming it in the Carnot cycle: there is no such lawlike de-pendency in nature.

In such cases, Mill-kinds risk hiding empty kinds which will be revealed as time goes by. In my lingo, they do not track modally robust phenomena. And the phenomena they do seem to track do not enjoy lawlike dependencies (or the right type of lawlike dependencies). Empiricist kinds with nominalist roots can secure successful inductive inferences and explanations *only* to the extent that a Mill-kind is an eligible candidate for becoming a Peirce-kind. And candidates for Peirce-kinds are—in my terminology—those tracking groupings of mod-ally robust phenomena within and across several scientific perspectives, each displaying lawlike dependencies.

But in the presence of such lawlike dependencies, even in-the-making kinds enjoy the status of Peirce-kinds, I contend next. In-the-making kinds, by defini-tion, are hypothesized viable candidates for natural kinds. Take dark matter as an example. The 2019 Nobel Prize for Physics was given to James Peebles (along-side two other physicists, Michel Mayor and Didier Queloz, for their research in exoplanets) for

insights into physical cosmology [that] have enriched the entire field of re-search and laid a foundation for the transformation of cosmology over the last fifty years, from speculation to science. His theoretical framework, developed since the mid-1960s, is the basis of our contemporary ideas about the universe . . . . The results showed us a universe in which just five per cent of its content is known, the matter which constitutes stars, planets, trees—and us. The rest, 95 per cent, is unknown dark matter and dark energy. This is a mystery and a challenge to modern physics. (See https://www.nobelprize.org/prizes/physics/2019/press-release/)

It is clearly the job of physicists to find the evidence for and give the answer to the question about the nature of dark matter. For the purpose of my philosophical discussion here, dark matter is an instructive example of what I call *in-the-making kinds*. For despite all the good theoretical and experimental reasons for introducing it into the standard cosmological model, as I finish editing this volume (December 2021), scientists are still waiting to find dark matter particles through a variety of direct and indirect searches,[24] including work at the Large Hadron Collider.

What in my philosophical account makes dark matter an example of *in-the-making kinds* are the laws of nature that enter into a number of phenomena for which dark matter is required, and the perspectival data-to-phenomena inferences at play across a range of intersecting scientific perspectives. Very briefly, here are some of the cosmological details relevant to my philosophical discussion here (drawing on Massimi 2018d; for a historical retrospective, see de Swart et al. 2017; Peebles 2017).

The term 'dark matter' was originally introduced in the 1930s by the Swiss cosmologist Fritz Zwicky (1933) to account for why large numbers of galaxies cluster together much closer than one would expect from gravity alone (cf. Bradley et al. 2008 for a recent study). But the notion of dark matter did not take off until the 1970s, when the idea resurfaced to explain another puzzling phenomenon: namely, how spiral galaxies retain their distinctive shape over time (see Ostriker and Peebles 1973).

---

[24] And when it comes to its nature, a variety of hypotheses are available. One of the current candidates are hypothetical WIMPs (or weakly interacting massive particles), whose weak interaction with ordinary matter could lead to the recoils of atomic nuclei detectable using large liquid xenon chambers located underground. One such possible WIMP candidate is the so-called neutralino, the 'lightest supersymmetric particle' (LSP), whose searches at the Large Hadron Collider (CERN), among other experiments, have given null results as of today. Similarly, direct detection searches for dark matter candidates at two of the largest experiments, LUX in South Dakota and PandaX-II in China's JinPing underground laboratory, have produced null results so far (see Akerib et al., LUX Collaboration 2017; Tan et al., PandaX-II Collaboration 2016). Alternative possible candidates for dark matter are hypothetical particles called axions (see Di Vecchia et al. 2017), gravitinos (see Dudas et al. 2017), self-interacting dark matter (SIDM), and hypothetical superheavy and super-weakly interacting particles called WIMPzilla (see Kolb and Long 2017).

The hypothesis of a dark matter halo surrounding galaxies was introduced to explain the phenomenon, and the later measurements on spiral galaxies' rotational velocities by Vera Rubin and collaborators (Rubin et al. 1980) corroborated Zwicky's original idea. The rotational velocity of spiral galaxies, instead of decreasing with distance from the centre of the galaxy, remains fairly stable. This is taken as evidence for the existence of dark matter halos surrounding galaxies, and inside which galaxies would have formed (the same massive halos are necessary to guarantee dynamic stability to galactic discs). The current standard cosmological model, the so-called ΛCDM model, postulates dark energy in the form of Λ to explain the accelerated expansion of the universe,[25] in addition to Cold Dark Matter.

Some of the best evidence for dark matter comes from phenomena at a scale much larger than that of individual galaxies or even clusters of galaxies. Since the 1990s, scientists have known that out of the total gravitating mass density of the universe as a whole (Ω_m), only a small fraction is made up of baryons (the heaviest elementary particles—see White et al. 1993). The baryon density is measured from the baryon-to-photon ratio. Data from the WMAP and Planck Collaboration about the cosmic microwave background (CMB) provide an accurate indication of the photon energy density at the time of the last scattering after the Big Bang, while Big Bang Nucleosynthesis (BBN) provides constraints on the abundance ratios of primordial elements (hydrogen, helium, etc.—see Steigman 2007 for a review). From data such as these, cosmologists infer that Ω_m far exceeds Ω_b. This modally robust phenomenon in turn provides strong evidence for an additional kind of non-baryonic (maybe weakly interacting) matter yet to be experimentally detected: dark matter.

---

[25] The universe has long been known to be expanding, with the Hubble constant H_0 measuring the rate of the accelerated expansion. Ade et al., Planck Collaboration (2015) performed an indirect and model-dependent measurement of the Hubble constant based on ΛCDM and the cosmic microwave background (CMB). More recent measurements using Supernova Ia calibrated by Cepheids (see Riess et al. 2016) have led to an estimated measurement value for H_0 of 73.24 ± 1.74 km s^{-1} Mpc^{-1}. This value is in 3.4σ tension with Aghanim et al., Planck Collaboration (2016). Recent research has further increased the 'tension' between the value of the Hubble constant from Planck's model-dependent early-universe measurements, and more model-independent late-universe probes. In particular, members of the H0LiCOW 2019 (H_0 Lenses in COSMOGRAIL's Wellspring) collaboration—using a further set of model-independent measurements of quasars gravitationally bending light from distant stars—have recently measured the Hubble constant at 73.3 ± 1.7. Wendy Freedman et al. (2019) from the University of Chicago used measurements of luminous red giant stars to give another new value of the Hubble constant at 69.8 ± 1.9 km s^{-1} Mpc^{-1}, which is roughly halfway between Planck and the H0LiCOW values. More data on these stars from the James Webb Space Telescope, which launched in December 2021, will shed light on this controversy over the Hubble constant, as will additional gravitational lensing data. See Verde et al. (2019) for a comprehensive overview of the state of the art in this debate.

Other phenomena evinced from different kinds of cosmological data support dark matter as a kind-in-the-making. The angular power spectrum of the CMB shows initial density fluctuations in the hot plasma at the time of last scattering. Over-dense regions in these maps show the seeds that led to the growth of structure, and the gradual formation of galaxies and rich galaxy clusters, under the action of gravity over time. Cosmologists infer the existence of a non-baryonic (weakly interacting) dark matter that must have been responsible for the early structure formation.

The matter power spectrum inferred from data about baryon acoustic oscillations (BAO) is yet another piece of evidence. BAO are the remnants of original sound waves travelling at almost the speed of light shortly after the Big Bang and before the universe started cooling down and atoms formed. BAO measurements are used to probe the rate at which the universe has been accelerating at different epochs (and hence as a probe for dark energy). But BAO are also important for dark matter because they are related to the shape of the matter power spectrum, which diverges in a dark matter model and in a no-dark-matter model of the universe (see Dodelson 2011).

Thus, evidence for dark matter as an in-the-making kind accrues through a number of data-to-phenomena inferences:

(1) from Zwicky's data about the radial velocities of eight galaxies in the Coma cluster to the more general phenomenon of *galaxy clusters*;
(2) from Ostriker and Peebles's data about the *N*-body computer model simulation to the phenomenon of the *stability of the galactic discs*;
(3) from Rubin et al.'s spectrographic data to measure rotation velocities to the phenomenon of galaxies' *flat rotation curves*;
(4) from BAO data to the phenomenon of the shape of the *matter power spectrum*;
(5) from CMB data to the phenomenon of large-scale structure formation of the universe via computer simulations;
(6) from data about Big Bang Nucleosynthesis to the phenomenon of $\Omega_m \gg \Omega_b$.

Each of these inferences is perspectival in a distinct way. The data are gained from experimental and technological resources that are an integral part of the ΛCDM cosmological model. The inferences from data to phenomena tend to be very much model-dependent (with the caveat presented in footnote 25 for the measurement of the Hubble constant). The methodological and epistemic principles that guide and justify the reliability of the knowledge claims so advanced (e.g. Bayesian statistics with a well-motivated choice of priors for theoretical parameters and nuisance parameters—see Massimi 2021 for a review of the dark energy case) are also part of a distinctive scientific perspective.

Laws of nature enter into every one of the data-to-phenomena inferences above.[26] Despite the increasing group of phenomena at different scales which point to dark matter, dark matter particles have not yet been detected as I finish editing this book (December 2021). The current status of dark matter as a powerful kind in-the-making can be explained in terms of its *nomological resilience* across a number of perspectival data-to-phenomena inferences at different scales.

We are looking for a kind that allows us to make successful inferences about an identified and open-ended group of phenomena at different scales. Laws of nature play a key role in turning this hypothetical Mill-kind into a fully fledged Peirce-kind, to use Hacking's terminology. The final chapter of this story still needs to be written. New physics beyond the Standard Model might hold the key to this puzzle. Whether and when this in-the-making kind will become an evolving kind depends on how the scientific perspectives of contemporary cosmology and astrophysics may intersect with the current and future perspectives of particle physics.

That concludes my two main points in this section. First, in the absence of laws of nature, a once successful Mill-kind turn out to be an empty kind. And, second, laws of nature underpin the nomological resilience of in-the-making kinds and make them enjoy the status of Peirce-kinds.

Lesson no. 4: The lawlikeness of natural kinds is not downstream from some prior holding of microstructural essential properties and relations. But it is not a disposable add-on to Mill-kinds either. For without laws, Mill-kinds turn out to be empty kinds. And with laws, even in-the-making kinds enjoy the status of Peirce-kinds.

In summary, in this chapter I have made the point that a perspectival realist view on natural kinds should be able to accommodate the aforementioned four functions of natural kinds, suitably revised in light of the examples discussed. In particular, a perspectival realist view on natural kinds should be able to accommodate:

* how *engineered* kinds count as just as natural as more familiar kinds;
* how *evolving* kinds offer an unexpected platform for unanimous judgements;

---

[26] Just to mention a few (non-exhaustive) examples here, the virial theorem enters into the calculation of the dynamic mass of the Coma cluster and related inference to the possible presence of dark matter in (1). Force laws entered into Ostriker and Peebles's (1973) N-body model and estimate for the dark matter halo in the data-to-phenomena inference (2). The relativistic Doppler effect for light entered into Rubin and collaborators' use of data about optical emission lines to establish the discrepancy between the surface brightness of the luminous mass of galaxies vis-à-vis their mass density in (3). The ΛCDM model, and hence Friedmann equations, are assumed in the inference from the data for BAO from the Sloan Digital Sky Survey to the phenomenon of the power spectrum of matter in (4) and the relativistic Doppler effect in (5) to go from the CMB data to the phenomenon of large-scale structure. Measurements of the light elements' abundances (especially deuterium measurements) underpinning the phenomenon of the baryon density in (6) are typically compared with CMB-inferred constraints (see Burles et al. 2001, Steigman 2007). I thank Alex Murphy for helpful discussions on this point.

* how the unreasonable (temporary) projectibility of *empty* kinds has less to do with theories and predicates and more to do with machines and instruments designed with them in mind;
* how kinds *in-the-making* are eligible candidates for evolving kinds as long as they remain nomologically resilient.

In the next three chapters, I articulate the details of this perspectival realist view of Natural Kinds with a Human Face.

8

The inferentialist view of natural kinds

8.1. Introduction

In this chapter, I begin to unpack the various elements in my definition of Natural Kinds with a Human Face (NKHF), which I stated in Chapter 7 as follows:

> Natural kinds are (i) historically identified and open-ended groupings of modally robust phenomena, (ii) each displaying lawlike dependencies among relevant features, (iii) that enable truth-conducive conditionals-supporting inferences over time.

My attention in this chapter is on (i) *historically identified and open-ended groupings of modally robust phenomena*. Phenomena do not make up kinds via any part–whole relation. A natural kind is not just any set of phenomena. Only interrelated phenomena in a well-defined historically identified grouping qualify as natural kinds—though they form an open-ended, malleable and revisable grouping, under the view I articulate.

I see our encounter with natural kinds as akin to how evidence is gathered to make inferences about what *might have been the case* in a detective story. If we knew from the beginning who the criminal was and how the events happened, we would not need any forensic science and there would be no story either. Similarly, if we knew from the beginning the natural joints of nature, we would not need the natural sciences, and all the experimenting, model-building, theorizing that go with them. That is why I see our encounter with natural kinds as *navigating our ways in the space of possibilities* as a guide to actuality. This is something human beings have learned to do over millennia. It is the historical-cultural achievement of epistemic communities working within and across scientific perspectives on a variety of modally robust phenomena. This chapter lays out the inferentialist story underlying NKHF.

8.2. Neurath's Boat and the inferentialist turn

Thinking of natural kinds is first and foremost thinking about what a bunch of things have in common. Consider two examples from botany and chemistry.

Perspectival Realism. Michela Massimi, Oxford University Press. © Oxford University Press 2022.
DOI: 10.1093/oso/9780197555620.003.0011

The eighteenth-century Dutch botanist Jan Frederik Gronovius classified with the name of *Linnaea borealis* flowers that had a distinctive Y-shaped stem, campanula-like petals, a white or pink colour, and whose geographical distribution we now know ranges from the mountains of Alaska to the Dolomites. Hugo Erdmann classified as 'noble gases' those with the least chemical reactivity. And Ramsey's discovery of argon forced Mendeleev to add a special column for them in his periodic table.[1]

Thinking about natural kinds as 'given', part of nature's furniture, has engendered a metaphysical exercise of asking which *properties* are constitutive for each kind. One can think of flowers as coming with defining morphological properties (say, a Y-shaped stem, particular shapes for their foliage, petal colour), the properties one would find listed in a pocket-size guide to Alpine flowers. Similarly, chemical elements come with the properties systematized in the periodic table, with atomic numbers, electronic arrangements, and associated chemical properties. Particles in physics equally come with defining properties such as their mass, their electric charge, and spin values.

That is how one often learns about natural kinds. School textbooks imply that we live in a world of properties and some of them define the difference between angiosperm and gymnosperm, alkalis and bases, vertebrates and invertebrates, and so on. Such kind-constitutive properties are usually regarded as *essential* properties of the kind. This tradition begins with Aristotle and his theory of predicates, with how he saw genus–species relations originating from the way adjectives like 'rational' vs 'irrational' are brought to bear on nouns such as 'animal', demarcating the division between, say, man qua 'rational animal' and the rest of the animal kingdom, as Aristotle saw it.

In contrast, I see natural kinds as the outcome of humankind's scientific and cultural history. We come to *historically identify* them among a bewildering array of empirical regularities in nature. They are the end products of concerted efforts of generations, who have successfully identified relevant groupings of phenomena in nature. But what counts as a relevant grouping?

One can imagine Gronovius going about collecting specimens of *Linnaea borealis* on his mountain treks and observing some properties ('one Y-shaped-stem white flower, two Y-shaped-stem white flowers, three Y-shaped-stem white flowers, . . . ') and on that basis concluding 'All Y-shaped-stem white flowers belong to the natural kind *Linnaea*'. But this is a caricature. For the minimal units of identification for natural kinds cannot be properties—too many to count, and too diverse for kind identification. Is a slightly more pinkish petal still passable as a *Linnaea borealis*? What about other flowers with Y-shaped stems?

---

[1] See Gordin (2018) for Mendeleev's response to the discovery of argon by Ramsey and the inclusion of noble gases in the periodic table.

Considerations of this nature have traditionally pushed discussions of properties from observable macroscopic properties to unobservable microstructural essential properties—one is here reminded of Putnam's discussions about jadeite vs nephrite, molybdenum vs aluminium. One would not define water as a colourless transparent liquid, but instead as having a certain microstructural composition as a molecule with two atoms of hydrogen and one of oxygen.[2] And for *Linnaea borealis*, looking at the genome might give a more clear-cut definition of kindhood than looking at the colour of the petals or shape of the stems.

However, as indicated by isomers and the story of DENDRAL and chemoinformatics in Chapter 7, there is more to chemical compounds than just atomic composition. Topological considerations about structure and bonds are equally crucial. And AI-led techniques help to differentiate among isomers in a swarm of possible combinations of atoms, situating *engineered kinds* in a continuum with our most familiar natural examples.

Here, then, is an alternative approach to natural kinds. The minimal units of classification for natural kinds are not properties but phenomena, as I have defined them. To identify *Linnaea borealis*, one has to be able to distinguish a variety of phenomena first. Some concern leaves and stem-shapes: this is the task of plant morphologists. Others concern the geographical distribution of the plant (is the *Linnaea* of the Dolomites the same kind of plant as the similar one found in Alaska?)—a task for phytogeographers. The various chemical processes going on in the plant fall under the remit of phytochemists, while the relation between the plant and its biome is a specialty of plant ecologists. Equally important are phenomena concerning how poisonous the plant might be, whether it might be usable for medicinal purposes. Here, often knowledge of local communities proves important (be they the Ladin community of Süd Tirol for *Linnaea*, the Scottish Gaelic communities of the Hebrides with their knowledge of seaweeds, or the Malagasi community for the rosy periwinkle, or any other similar examples—I shall return to the latter two in Chapters 10 and 11, respectively).

In what follows, I use the expression 'local knowledge' in a specific sense following Suresh Canagarajah (1993, 2002): namely, to denote knowledge that is 'context bound, community specific, and nonsystematic because it is generated ground up through social practice in everyday life' (Canagarajah 2002, p. 244).[3]

---

[2] For a recent defence of microstructural essentialism that is at a distance from the Putnam–Kripke tradition, see Tahko (2015, 2020) and Hendry (2006, 2019).

[3] The term 'local knowledge' must be handled with care. As Canagarajah (2002, p. 244) points out, the term in modern science has often been the target of a 'systematic and concerted campaign to denigrate local knowledge at the global level' in the name of the exacting standards of 'universality, standardization, and systematicity, all for the end of predictability' of empirical science in the Western world. A wealth of anthropological, sociological, and postcolonial studies have shown how knowledge of local communities—especially minority ones and often colonized ones—was systematically suppressed or appropriated in Western science (see Mignolo 2000). Only in recent times has there been a rediscovery of local knowledge (see, among many others, Santos and Meneses 2020).

I shall expand on this point in Section 8.3 and show its relevance to my discussion of NKHF.

To know the natural kind *Linnaea borealis* involves grouping a variety of phenomena very different in nature. Each community uses different data as evidence for their phenomena. Hence the data-to-phenomena inferences are perspectival as each epistemic community resorts to data, experimental techniques, modelling resources, and methodological-epistemic principles that belong to their own situated practices.

That human practices condition how we carve the world's joints has become common currency since the work of Hacking (1991, 1999, 2007a) and Dupré (1981, 1993). More recently, human epistemic practices behind each natural kind have been put centre-stage by Kendig (2016a) in what she calls 'kinding' processes, and by Reydon (2016), who refers to the 'co-creation' of categories by merging empirical properties with human cognition. Ludwig's practice-dependent kinds have further strengthened this approach by looking at examples in ethnobotany and ethnozoology (see Ludwig 2017, 2018a, 2018b, and Ludwig and Weiskopf 2019). And Bursten's work on kinds in nanotechnology (2016, 2018, 2020a) has offered further material for reflecting on what natural kinds really are. In what follows, I clarify how I see the relevance of this scholarly tradition to the perspectival realist view and where the differences lie.

Traditionally, appeal to a variety of epistemic communities at work in parsing natural boundaries has often been combined with brands of realism: realism about individuals, or entities, or dispositional properties, combined with pluralism about taxonomic classification. Dupré's (1993) promiscuous realism combines realism about individuals with promiscuity in the semantic cross-classification. Hacking's experimental realism (1991, 2007a) combines realism about entities with nominalism about kinds.[4] Chakravartty's defence of what he calls 'sociable properties' combined with 'manifestation-based pluralism' is yet another realist view that in this case takes the properties of scientific interest as

---

It goes well beyond my expertise and goal to cover this literature here. But in the rest of this chapter I show how perspectival realism, with its emphasis on the historically and culturally situated nature of perspectives, is a natural ally of the movement of rediscovering local knowledge. For, like standpoint epistemology, perspectival realism too embraces in full the *locality* and *situatedness* of all scientific knowledge. Moreover, as I show in Section 8.3, perspectival realism takes the vantage point of local communities as an important lens for recalibrating discussions of natural kinds.

[4] For Hacking, the grounds for experimental realism are causal properties that an entity like the electron has—properties that epistemic communities learn how to use, and manipulate so as to create new phenomena in a lab: 'There are an enormous number of ways in which to make instruments that rely on the causal properties of electrons in order to produce desired effects of unsurpassed precision. . . . There is a family of causal properties in terms of which gifted experimenters describe and deploy electrons in order to investigate something else. . . . The "direct" proof of electrons and the like is our ability to manipulate them using well-understood low-level causal properties' (Hacking 1983, pp. 265, 272, 274).

dispositional.[5] For these different realist views, one can be a realist about tigers, lemons, hellebores, or electrons in believing that all these entities or biological individuals or their dispositional properties are 'out there' in the world.

Perspectival realism, while sharing the pluralism of other varieties of realism, eschews discussions about properties, causal-dispositional roles, or ways of packaging properties together. But how, then, to make sense of the idea of natural kinds as historically identified groupings of modally robust phenomena?

Consider again chemical elements. Identifying elements as natural kinds involves being able to identify phenomena such as atomic spectra, chemical reactions, melting and boiling points, oxidation, and so forth. Each of them typically occurs in a particular domain: spectra are the fingerprints of atoms; oxidation is a chemical reaction; melting and boiling points mark phase transitions. All these phenomena are modally robust in that epistemic agents infer what *could, would,* or *should* happen. What would happen to the atomic spectrum if the sodium atom were placed in a weak magnetic field? How does oxidation occur in iron? In which way can atmospheric pressure affect the boiling point of water? Examples multiply endlessly.

This shift from properties to phenomena-first reflects the central role played by *epistemic communities occupying a plurality of situated perspectives* in the identification of groupings of phenomena candidates for natural kinds: the variety of scientific practices they engage with, the inferences drawn from them and the phenomena they accordingly model and encounter. We encounter kinds by identifying relevant groupings of phenomena. They are as *historical* as our historically and culturally situated perspectives are. Can an account of natural kinds that centres on epistemic communities and perspectival vantage points qualify for the label 'realism'? What will it take to replace the marble-solid metaphysical foundations of natural kinds with something that looks as flimsy as Neurath's Boat?

---

[5] 'The idea of manifestation-based pluralism begins with a particular understanding of the nature of many properties of scientific interest: *viz.*, that such properties are dispositional. . . . [M]y intention is simply to illuminate one potential consequence of the position for the prospects of realist-compatible pluralism. Thus, begin with the idea that properties of scientific interest—those whose patterns of sociability underwrite practices of scientific classification—are generally (if not always) dispositional. That is to say, they dispose the things that have them to behave in certain ways in specific circumstances. On this view, *inter alia*, the sciences yield knowledge of the modal features of their target systems in the world. . . . Consider a simple, uncontroversial example. The molecular structure of a compound disposes it to behave in a number of different ways, depending on the ambient circumstances. It may dispose the compound to change phase (from solid to liquid, or liquid to gas) at different temperatures depending on variations in other environmental conditions (ambient pressure, the presence or absence of other chemical agents, and so on). In this way, different stimulus conditions may elicit different causal processes involving the compound, and thereby elicit different contributions of its molecular structure to its behaviour. And so, one and the same property can dispose an entity to manifest different behaviours in different contexts' (Chakravartty 2011, pp. 176–177).

That Boat is indeed the inspiration for my inferentialist view of NKHF. As Nancy Cartwright, Jordi Cat et al. have argued, 'What propelled Neurath was an *idea*: the idea not simply that our stock of knowledge claims keeps on changing forever, but that a decisive revision of our concept of knowledge is required if reason is to fulfil its Enlightenment promise' (Cartwright et al. 1996, p. 92, emphasis in original). Neurath's anti-foundationalist programme in philosophy was a reaction both against the presumption of first foundations (*pace* Descartes and Kant) and against 'the unbridled relativism supposedly encouraged by the absence of foundations' (*pace* Spengler) (p. 136).

Neurath's suggestive metaphor was put to varied uses.[6] Quine (1969), for example, adopted it as an emblem of naturalism in epistemology. But there is more to the metaphor's aptness for the view of natural kinds canvassed here than mere naturalism. The metaphor points to the importance of *communication* among epistemic communities as a guard against methodological solipsism (and Spengler's type of relativism).[7] Replacing metaphysical foundations with a view of scientific knowledge that gives epistemic agents their due has long been central to a family of philosophical views that style themselves as 'inferentialist'.

Take Brandom's inferentialism, where in 'calling what someone has "knowledge" one is doing three things: *attributing a commitment* that is capable of serving both as premise and as conclusion of inferences relating it to other commitments, *attributing entitlement* to that commitment, and *undertaking* that same commitment oneself. Doing this is adopting a complex, essentially *socially* articulated stance or position in the game of giving or asking for reasons' (Brandom 1998, p. 389, emphases in original). Or consider Huw Price's project of rethinking the notion of representation in language (Price et al. 2013); or Richard Healey appealing to Brandom's inferentialism in his interpretive reading of quantum mechanics as mentioned in Chapter 6 (Healey 2017).

---

[6] The metaphor of the boat appears a number of times in Neurath's writings across a span of 30 years, as Cartwright et al. (1996, p. 92) have extensively documented. They call it 'the first Boat' in 1913, the 'second Boat' in 1921, the 'third Boat' in 1932, and 'the fourth Boat' in 1944. And on each occasion the context and the envisaged interlocutor are different. Kant's and Descartes' foundationalism in philosophy are the intended interlocutors of the first Boat in 1913. 'Spengler's claim that truth exists only relative to certain types of humans' (p. 139) is the recipient of the second Boat in 1921. Carnap's protocol-sentences are the target of the third Boat in 1932. Finally, in 1944, the context is politically charged against the backdrop of World War II and the metaphor is largely seen as a nod to Max Weber and Georg Simmel, among others, in the prescient warning that 'A new ship grows out of the old one, step by step—and while they are still building, the sailors may already be thinking of a new structure, and they will not always agree with one another. The whole business will go on in a way we cannot even anticipate today. That is our fate' (Neurath 1944, p. 47—quoted in Cartwright et al. 1996, p. 165).

[7] 'Against cultural relativism Neurath sets a rudimentary sketch of the hermeneutics of communication that are presupposed even by science—and that support its claim to objectivity. Communication with members of our own communities—with whom we share many beliefs—does not differ in principle from communication with members of alien cultures' (Cartwright et al. 1996, p. 140).

In philosophy of science, Mauricio Suárez (2004) has spearheaded an influential inferentialist account of scientific representation whereby 'a source *s* represents a target *t* only if (i) the representational force of *s* points to *t* and (ii) *s* allows an informed and competent agent to draw specific inferences regarding *t*' (Suárez 2004, p. 773). And Suárez (2015a, 2015b) has offered a sustained defence of this inferentialist approach to scientific representation against alternative approaches that have often emphasised a range of relations holding between the source and the target system (from isomorphism to similarity) to explain scientific representation.

Surprisingly, though, discussions about natural kinds have so far escaped the inferentialist turn. I aim to remedy that, in pursuit of perspectival realism. Here, I begin to unpack the NKHF strategy by going back to its first notion, that natural kinds are (i) historically identified groupings of modally robust phenomena (I will have more to say about their being 'open-ended' in Chapter 9). The next section offers examples for rethinking the metaphysics of natural kinds as downstream from the epistemology of science—as aids for navigating the Neurath's Boat of natural kinds.

8.3. Historical naturalism and situated knowledge

If the 'naturalness' of natural kinds is not derived from the joints where nature is carved, how should one understand it? If we can have an engineered synthetic DNA with the potential of reproducing DNA's main properties, what does this teach us? I think it teaches us how to think of naturalism in an anti-foundationalist and historicized way.

The assumption that kinds track natural divisions in nature often precedes debates about realism or nominalism about kinds. Quine advocated a minimal naturalism whereby natural kinds are the scientific, discipline-specific outcomes of what he saw as our 'innate subjective spacing of qualities' (Quine 1969, p. 126). His main challenge was to explain how innate standards of similarity 'have a special purchase on nature and a lien on the future' (p. 126). Quine went Darwinian in his answer: '[S]pacing that has made for the most successful inductions will have tended to predominate through natural selection' (p. 126). His Darwinian approach was inspired by Neurath:

I see philosophy and science as in the same boat—a boat which, to revert to Neurath's figure as I so often do, we can rebuild only at sea while staying afloat in it. There is no external vantage point, no first philosophy. . . . For me then the problem of induction is a problem about the world: a problem of how we, as we now are (by our present scientific lights), in a world we never made, should stand better than random or coin-tossing chances of coming out right when we predict by inductions which are based on our innate, scientifically unjustified similarity standard. Darwin's natural selection is a plausible partial explanation. (Quine 1969, p. 127)

Natural selection seems a plausible but only partial explanation, in my view. Cognitive neuroscience and developmental psychology show how children learn the concepts for object-kinds following specific constraints. For example, developmental psychologist Ellen Markman and collaborators (Markman 1989, Ch. 5; Markman and Hutchinson 1984; Markman and Wachtel 1988) have studied how children learn the meaning of a new word from a single labelling event, guided by two main constraints: the 'whole object constraint' and the 'taxonomic constraint'. Children aged 18 to 24 months have a preference for the word to refer to the whole object (rather than parts of the object or attributes of the object—the rabbit, rather than the rabbit's ears or tail); by age 3 to 4, children prefer to generalize the meaning of the new word to taxonomically similar objects (e.g. rabbits, mammals, animals) rather than thematically related ones (e.g. rabbits, carrots, and burrows). The developmental advantage of doing so is clearly expressed by Markman (1989, p. 111):

> By expecting unforeseen nonperceptual properties to be common to members of a kind, children could go beyond the original basis for grouping objects into a category and discover more about the category members than they knew before. Children might start out assuming that categories will have the structure of natural kinds. With development, they would then refine these expectations, limiting them to properties, domains, and category types that are appropriate.[8]

But Darwinian considerations are only a partial explanation. For a gulf separates the basic similarity standards at work in, say, children's pre-scientific classificatory abilities from the complex and sophisticated taxonomies of scientific disciplines. Distinguishing, say, brown edible mushrooms from brown poisonous ones is one thing. Plotting scientific taxa is quite another. The University of Oslo's Nordic mycological herbarium features 14,695 currently accepted names of fungi (taxa and genera).[9] Accounting for this exceedingly complex system must call on our cultural–scientific history, which Quine's naturalism did not pay attention

---

[8] For a criticism of the view, see Callanan et al. (1994). More recent work on Bayesianism has aimed to explicate the inferential mechanism that might be at work in such economical concept learning acquisition by thinking of constraints as Bayesian priors at work in ruling out logically possible alternatives and increasing the efficiency of the learning process from a few examples (see Tenenbaum et al. 2011; and Xu and Tenenbaum 2007).

[9] See http://nhm2.uio.no/botanisk/sopp/tax-list/index.htm for details.

to. A distinctive variety of *historical naturalism* is at work in my Neurathian approach to kinds. Our subjective spacing of qualities has a purchase on nature only insofar as situated epistemic communities *historically* learn how to classify relevant and open-ended groupings of phenomena into evolving kinds. But how should this historical naturalism be characterized?

Appeal to historical considerations is not entirely new. As already mentioned, the historicity of natural kinds is emphasized by philosophers of biology reacting against 'eternal natural kinds'. Ruth Millikan (1999) has objected to the Kripke–Putnam view and argued that biological kinds are identified not by properties but by their lineages. These lineages are traceable to a clade having a common ancestor. Other philosophers of biology have appealed to the notion of homologues (see Ereshefsky 2012) to explicate phylogenetic continuity with a common ancestor. Many physical and chemical kinds can also be regarded as historical in some relevant sense. Muhammad Ali Khalidi, for example (2013, pp. 139ff.), has observed how the history of many chemical kinds (say, gold) coincides with the history of our universe and the formation of such elements inside stars.

A common theme of some of these approaches is the identification of historicity with *causal history*. Biological lineages are causal lineages of mating, breeding, and survival-adaptive evolution to a changing environment. Similarly, the causal history of our universe since the Big Bang—interspersed with supernova explosions, and the formation of stars and galaxies—underpins the historicity of chemical elements formed inside the stars.

But there is another, equally important, non-causal sense in which natural kinds are historical. Their historicity is also the outcome of how real epistemic communities across a plurality of situated scientific perspectives have come to *historically* identify a group of modally robust phenomena as candidates for natural kindhood. This kind of historicity goes to the heart of human epistemic practices.[10] It is not about locating a natural kind within a causal network of events, a phylogenetic lineage, or similar. But it is about the natural kind being a historically identified grouping of phenomena that different historically and culturally situated epistemic communities have encountered over time. Historical naturalism so understood offers then an answer to the question left open by Quine's naturalism: why is it that we seem so good at navigating nature and encountering 'functionally relevant groupings'?

---

[10] An important step in this direction has been taken by Kendig in her edited volume (2016a), where she points out how natural kinds are the outcome of a number of activities of 'kinding' (as she calls it). And in the same volume Hasok Chang (2016) has, for example, explicated the rise of chemical natural kinds through practices of what he calls 'epistemic iteration', building from Chang (2004). In what follows, I explicate how I see historical naturalism as a fellow traveller of these views on natural kinds.

In previous work (Massimi 2014), I sketched an answer through a version of *naturalized Kantianism*. The idea behind it is that natural kinds

> latch onto stable empirical clusters evinced by robust experimental data, i.e., observable records of occurrences that cannot be ascribed to error or background noise. From the pre-scientific ability of children to cluster objects with same empirical properties (pears with pears, apples with apples), to the mineralogist's ability to cluster minerals, it is our human ability to identify and track recognisable patterns of empirical properties in nature that gave us the upper hand in the evolutionary gamble. Peaks in magnetometers, sparks in scintillation counters, bubble trails in cloud chambers that have proved genuine (i.e., not due to background noise or experimental error) are the sophisticated scientific counterpart of children and laymen's pre-scientific clustering ability. (p. 427)

What I then called 'recognisable patterns of empirical properties in nature' should have been better characterized as the modally robust phenomena evinced through perspectival data-to-phenomena inferences.[11] How any specific grouping could become eligible for the title of natural kind is a question ultimately for scientific practitioners. And as I stressed (p. 428), 'one would need to tell a very detailed, discipline-specific and context-specific story' for each of those kinds.

In the following two chapters, I look into this in more detail. But in the rest of this chapter, I focus on how historical naturalism explains the need for a variety of scientific perspectives to contribute to natural kind classifications. The perspectival pluralism at work in modelling practices is an aspect of a more general epistemological pluralism in ways of knowing. This plurality of ways of knowing has far-reaching consequences for realism in science. In particular, it calls attention to the key role of situated knowledge in natural kind classifications, under the Neurathian view of NKHF.

The emphasis I have placed so far on technological, experimental, and modelling resources to reliably advance claims of knowledge should not be misunderstood. Scientific perspectives—as I have been using the term—are not akin to Kuhn's paradigms: the Newtonian perspective, Lavoisier's chemical perspective,

---

[11] In Massimi (2014), I was primarily concerned with a reply to Richard Boyd's argument that Kantianism about kinds is a form of constructivism and that a naturalized Kantian would have to rely on Boyd's brand of realist accommodationism to make sense of the projectibility of kinds. Hence the emphasis in that article was on showing how some Kantian intuitions might play out in this debate on natural kinds vis-à-vis Boyd's own view. Accordingly, most of the discussion was couched in terms of Boyd's view of homeostatic cluster kinds as clusters of properties. Moving beyond the specific details of that article and its internal dialectic, in what follows I expand on its main insights that I am still committed to.

and so on (more on this in Chapter 11). For I have stressed all along the social, collaborative, inferential nature of the epistemic exercise that underwrites perspectival realism. In what follows, I return to this point and highlight a few implications for the view of natural kinds I put forward:

1. First, and this is almost a platitude, there are many ways of making knowledge from a plurality of historically and culturally situated scientific perspectives.

2. Therefore, scientific knowledge is always and necessarily local and situated knowledge, according to perspectival realism: knowledge originating from situated vantage points (or perspectival$_1$ representations).

3. Reading natural kinds through the lenses of perspectival realism means, then, acknowledging this historical plurality of perspectival encounters with different phenomena.

4. The open-ended grouping of phenomena that any natural kind gets identified with is the reflection of these perspectival encounters, which are historically situated, specific to communities, and often enough (albeit not always) embedded in epistemic practices informed by daily needs.

5. An important consequence of this historical naturalism is that the perspectival realist would not say that if one wants to know what 'gold' is, one must ask atomic physicists for the atomic number; or if one wants to know what a particular plant is, one must ask for the DNA sequence for it. Instead, the perspectival realist would insist that to be classified as a natural kind K is to satisfy an (open-ended) series of historically identified phenomena. In the case of plants, the phenomena in question include those that are morphological, karyological (i.e. concerning chromosomes), physiological, ecological, and ethnobotanical (e.g. concerning the toxicity or biodiversity role or similar), among many others (for an example concerning Alpine flora, see, e.g., Fischer 2018). Natural kind classification is not like distilling a pure prototype or identifying a historically and culturally deracinated archetype meant to be valid always and everywhere.

6. This is neither a restatement of Hacking's *homo faber* nor of Dupré's promiscuous realism (with vernacular kind terms and scientific taxonomic classifications), much as it shares the spirit of both. This is instead a distinctively perspectival argument to the effect that the vantage points of differently situated communities offer in different contexts the privileged standpoint for encountering particular phenomena as modally robust. Each phenomenon is inferred from perspectival data-to-phenomena inferences. Such encounters are made possible in virtue of occupying historically and culturally situated perspectives that allow different epistemic communities to sift through nature's stable events in some way rather than

others, and to identify relevant modally robust phenomena which are candidates for NKHF.

7. A further important aspect of historical naturalism is therefore the central role it gives to local knowledge of epistemic communities that—often socially oppressed and epistemically marginalised—have been at the periphery of traditional narratives and canons concerning scientific knowledge production. Historical naturalism takes local knowledge as an integral part of how to go about encountering nature as teeming with modally robust phenomena and grouping them into NKHF. Let me illustrate this last point with an example from ethnobotany.

8.4. *Gymnopodium floribundum* and *Ts'iits'ilche'* honey: an example from ethnobotany

Ethnobotany, and ethnotaxonomy more in general, are a good illustration of historical naturalism at work. Among the varieties of local knowledge, traditional ecological knowledge plays a special role when it comes to natural kind identification for plants and animals. The United Nations defines 'traditional knowledge' as 'the complex bodies and systems of knowledge, know-how, practices and representations maintained and developed by indigenous people around the world, drawing on a wealth of experience and interaction with the natural environment and transmitted orally from one generation to the next' (UN 2019a, p. 2/13).

The recent reappraisal of traditional knowledge so understood comes after a long history of denigration (see Canagarajah 2002), erasure, and appropriation. (I shall return in more detail to the topic of epistemic injustices in this particular context in Chapter 11.) Institutional efforts led by the UN, among others, to tackle this endemic problem led in 1992 to the establishment of the Convention on Biological Diversity, whose Conference of the Parties,

in its decision XIII/18 adopted the Mo'otz Kuxtal Voluntary Guidelines . . . for the development of mechanisms, legislation or other appropriate initiatives to ensure the 'prior and informed consent' . . . of indigenous peoples and local communities for accessing their knowledge, innovations and practices, for fair and equitable sharing of benefits arising from the use of their knowledge, innovations and practices relevant for the conservation and sustainable use of biological diversity, and for reporting and preventing unlawful appropriation of traditional knowledge. (UN 2019a, p. 5/13)

Indeed, in the 1992 Convention on Biological Diversity (CBD), article 8(j) the UN affirmed the need to

> respect, preserve and maintain knowledge, innovations and practices of in-
> digenous and local communities embodying traditional lifestyles relevant for
> the conservation and sustainable use of biological diversity and promote their
> wider application with the approval and involvement of the holders of such
> knowledge, innovations and practices and encourage the equitable sharing of
> the benefits arising from the utilisation of such knowledge, innovations and
> practices.

The 1992 UN text mentioned 'indigenous and local communities', and the acronym IPLC (Indigenous People and Local Communities) has entered the ensuing literature on CBD[12] and wider discussions at the Intergovernmental Science-Policy Platform on Biodiversity and Ecosystem Services (IPBES).[13] Mulalap et al. (2020, p. 2) have clarified the acronym along the following lines:

> "Local communities," unlike Indigenous Peoples, do not necessarily have
> a history of being invaded or colonized by external entities. However, like
> Indigenous Peoples, local communities have cultural values, practices, and sys-
> tems developed through multiple generations and poised to be passed to future
> generations. . . . We acknowledge, however, that conceptualizations of indige-
> neity are contested and highly context-specific.

With this important caveat in mind, the acronym IPLC has entered the legal and scientific literature on biodiversity to refer to holders of traditional knowledge broadly construed. It is against this institutional and legal backdrop that in what follows I urge for the need to expand the classical remit and boundaries of the literature on scientific perspectivism and realism too. My aim is to clarify the mechanisms through which a plurality of historically and culturally situated perspectives leads to knowledge production that is always inherently local and perspectival, and enables the relevant local epistemic communities to identify modally robust phenomena which are candidates for NKHF.

There is one particular aspect that is instructive about situated knowledge and my story here on historical naturalism. I am going to call it the *fine-graining* and *coarse-graining* of descriptions of natural kinds. The historical identifi-cation of groupings by real epistemic communities depends on their localized epistemic practices and needs. Coarse-graining or fine-graining means that,

---

12 See https://www.cbd.int/topic/indigenous-peoples-and-local-communities.
13 See IPBES: https://ipbes.net/glossary/indigenous-peoples-local-communities.

without necessarily changing the membership of the open-ended grouping (i.e. without necessarily adding or removing specific phenomena from it), one might nonetheless give a description of it that zooms out or zooms in on specific phenomena. Zooming in on specific phenomena involves varieties of situated knowledge that are often a prerogative of perspectival local practices.

Take the following botanical example. There is a species of melliferous flora in Mexico called *Gymnopodium floribundum*. It belongs to the family of buckwheat (Polygonaceae). It is one of the most common plants in the Yucatán peninsula and its Mayan name (and the name of the honey produced from it) is *Ts'iits'ilche'*. In the Kew Royal Botanic Gardens herbarium, specimen records for this plant come from Belize. In other databases, such as the National Centre for Biotechnology Information (NCBI), the same species is classified on the basis of a group of proteins and nucleotides. And if we switch from NCBI to the Biodiversity Heritage Library, we find a list of relevant bibliographical references where the role of the plant in ethnobiology is included.[14]

It would be thus restrictive to think that the historical classification of this species comes down to the identification of a specimen (along the lines of Putnam's 'archetype') like the one preserved at Kew Gardens. Several concurrent perspectival phenomena have historically entered into the identification of the natural kind known in Western botany as *Gymnopodium floribundum*. Some are morphological and phenotypic: these concern the comparative analysis of the anatomy of the plant, its reproductive organs, and development, for which herbaria specimens are a useful source of knowledge.

Others concern nucleotides and protein groups inferred from the perspective of cytogenetics. Yet others are karyological phenomena: phenomena concerning the structure of cells and chromosomes. Biodiversity and ecosystem phenomena are also important: these are phenomena concerning the inter-relations among living organisms in a certain environment.

For example, particularly interesting are the plant–pollinator interactions that pollination ecology studies. It has long been known that such plant–pollinator interactions are highly context-dependent and that 'the degree of specialization within a study system can depend not only on the perspective of interest (plant vs. pollinator) but also on the community context' (Rafferty 2013/16). Let us then take a quick look at the role of the community context when it comes to identifying the relevant pollination phenomenon for this particular plant of the Yucatán.

[14] Compare the specimen at Kew Royal Botanic Gardens with the record at the National Centre for Biotechnology Information and the Biodiversity Heritage Library entry.

Its flowering season peaks between February and April and the plant has traditionally played a key role in the api-botanical cycle. Rain and high temperature make the plant blossom in the spring months and its nectar proves a vital resource for the local bee species (see Quezada-Euán 2018, pp. 195–196). Beekeeping and honey production have historically been an important element in the Maya culture. Archaeological evidence of beekeeping goes back to the late Pre-Classical Maya period (see Crane 1999, p. 295). Local species of stingless bees, including *Melipona beecheii*, made hive keeping a popular practice among the Maya. Honey production in specific locations, known as *meliponarios* (see Bratman 2020), continue to be an integral part of the local economy of the communities of Xmabén, Hopelchen, and Campeche in Mexico (see Coh-Martínez et al. 2019).

It is the situated knowledge of these local communities about the api-botanical cycle rotating around the nectar of the plant that has the epistemic upper hand when it comes to identifying particular phenomena such as pollination peak, for example. In my philosophical idiolect from Chapter 6, the pollination of *Gymnopodium floribundum* is a modally robust phenomenon which involves (a) a stable event (i.e. the transfer of pollen from the anther to the stigma); (b) data that provide evidence for it (i.e. the buzzing of the bees around the scented flowers, the fruity clusters); and (c) the historically and culturally situated epistemic communities teasing out a network of perspectival inferences from the data to the stable event in question. The situated knowledge of local beekeepers is an unrivalled source of information for identifying the particular phenomenon of pollination peak in a way that plant morphologists, or cytogeneticists cannot offer. Let us see why.

The pollen being deposited on to a stigma is a stable event in that it follows lawlike dependencies that are independent of there being or not being any epistemic community or perspective. An example of lawlike dependencies at play in the pollination of flowers is pollinator performance, defined as the product of flower coverage (FC)[15] and pollen deposited (PD).[16] Recall from Chapters 5 and 6 that lawlikeness plays the realist tether in perspectival realism. It grounds a first-tier modality at play in, for example, whether a flower *would* be pollinated *if* a pollinator were to visit it (depending on the pollen being deposited).

In turn, a phenomenon is a stable (qua lawlike) event whose occurrence *can be inferred in many different possible ways*. In this example, the phenomenon pollination is modally robust in embedding the very many ways in which this stable event of pollen transfer might occur. For example, honeybees perform this act by

---

[15] 'Flower coverage' is defined as the product of how many pollinators of a given species are present on flowers and the number of flowers they visit in a certain interval of time.

[16] 'Pollen deposited' is defined as the number of pollen grains deposited on the stigma for each pollinator's visit (for a discussion see Pérez-Balam et al. 2012).

carrying pollen on their legs; wasps may carry it in their mandibles; flies in their abdomen. And of course teasing out all the ways in which this is done by various pollinator species and the impact it has on pollination peaks and flowering seasons for individual plant species is something studied by pollination ecology, entomology, among other disciplines. Modal robustness expresses the many ways in which epistemic communities infer the relevant phenomenon by connecting often diverse datasets to the occurrence of the stable event in question.

Different pollinators perform differently, more or less efficiently, and the presence of insecticides in nearby crops significantly affects the number of local honey bees in each particular area and the associated process of pollination. For example, melissopalynology studies the variety of pollen and pollen sources present in particular samples of honey. In so doing, it provides insights into the percentages and varieties of flower nectars visited by honey bees. Melissopalynological studies in the region of the Yucatán have found that *Ts'iits'ilche'* is under-represented at ca. 3% among the single-flower honeys of the region, despite the plant being common there. This finding has in turn suggested that *Gymnopodium* pollen production must be lower than that of other varieties of melliferous flora in the region (see Alfaro Bates et al. 2010, p. 60). An explanation for this under-representation might be sought in the reproductive biology of the plant.

Local communities and their situated knowledge play an integral role in trying to understand and explain this finding. Beekeepers know best how to protect their apiaries across seasons; how to control insecticides that have devastating effects on bees; and when the nectar peak for the local plants is so as to sustain honey production throughout the year. It is by virtue of their being historically, geographically, and culturally situated that local epistemic communities know best about the phenomenon pollination peak: they know how to identify this modally robust phenomenon among a swarm of stable events. This example of local situated knowledge about a phenomenon (call it P_k) enables in turn other epistemic communities (e.g. plant morphologists) to investigate related phenomena P_j (e.g. about the reproductive biology of the plant and the possible causes for the pollen under-representation in honey).

In my philosophical lingo, the *Gymnopodium* taxon is *all these phenomena*. The situated knowledge of different epistemic communities—melissopalynologists, beekeepers, plant morphologists, pollination ecologists, etc.—makes it possible to fine-grain or coarse-grain the description of the taxon by focusing on one phenomenon rather than another. For example, plant morphologists can describe the reproductive organs of the plant, but to gain insight into its reproductive performance, one needs to fine-grain the description at the level of the pollination peak. And it is here that the local knowledge of beekeepers and honey producers has the epistemic upper hand in better understanding what might be causing the under-representation of *Ts'iits'ilche'* pollen in the honey of the region as spotted by melissopalynologists.

Under historical naturalism, the identification of relevant groupings of phenomena qua candidates for NKHF is *humankind's collective historical and epistemic* achievement, something that geneticists at NCBI, botanists at Kew Gardens, and beekeepers in the communities of Xmabén, Hopelchen, and Campeche equally share and can reclaim as their own. This is what historical naturalism is ultimately about: a celebration of the social and cooperative nature of scientific knowledge where a variety of perspectival situated practices by specific epistemic communities at particular historical, geographical, and cultural locations get intertwined in delivering knowledge of natural kinds.

Each epistemic community contributes one or more phenomena to the grouping. And the open-ended groupings that the natural kinds get identified with can always be fine-grained or coarse-grained by focussing attention on *one* or *more* particular phenomena and associated descriptions. But it would be a mistake to conclude on this basis that therefore the natural kind *Gymnopodium floribundum* should be *primarily* or even *exclusively* identified with the herbarium specimen or the nucleotide sequence as if these gave us some privileged handle. It would equally be hasty to reach the opposite conclusion and defend some form of conventionalism about this natural kind. The different epistemic practices are not an invitation to pick and choose a phenomenon P_1 at perspective sp_1 instead of a phenomenon P_2 at perspective sp_2, depending on specific needs.

Historical naturalism does not make natural kinds *social constructs*. The phenomena that communities have learned to identify over time and across perspectives are as real as the effect of the dry season on the melliferous flora of the Yucatán; as tangible as the specimen at the Kew Gardens herbarium; as reliably inferred from data as nucleotide sequences held at the NCBI database.

There is nothing 'constructed' about these modally robust phenomena. Human construction is of course involved in creating instruments, making machines, and devising methods through which modally robust phenomena are inferred from a variety of data and get eventually historically identified as belonging together. But this does not in turn license the *metaphysical conclusion* that phenomena and their groupings are themselves a human construction.[17]

---

[17] The often-heard charge that epistemic moves of this nature about natural kinds fall prey to some kind of constructivism is based on a series of misconceived hidden assumptions, in my view. I have reviewed some of those in Massimi (2014), where I discuss Kantian kinds, and I will not repeat the arguments here. The notion of NKHF that I am articulating in this book shares some of the distinctively epistemic features of Kantian kinds (in the modally robust notion of phenomena). But the present discussion is meant to be more general and broader than the original intuition behind Kantian kinds (nor is it tied to any particular paraphernalia of Kant's own account).

Going back to engineered kinds, their effective naturalness is neither surprising nor mysterious on this account. The epistemic ability to *historically identify* relevant groupings of phenomena explains why our subjective spacing of qualities has a purchase on nature, no matter how engineered *hachimoji* DNA or DENDRAL-aided molecules might be. Thus, recalling Lesson no. 1 from Section 7.3 in Chapter 7, *the naturalness of kinds is not just the product of our 'subjective spacing of qualities' (as Quine maintained) but it is also the result of our perspectival scientific history*, where a genuine plurality of scientific perspectives—broadly understood—have historically defined the boundaries of what we think of as a 'natural kind'.

Yet the discussion so far is incomplete. What still remains to be done is to offer a philosophically more detailed account of how an identified group of phenomena *with relevant features and lawlike dependencies* become eventually a natural kind.

8.5. The inferentialist view behind NKHF

All natural kinds are born as *in-the-making kinds*. This might seem a bold assertion. Surely, either something is a natural kind or it is not. The way in which our beliefs, representations, or conceptions change should not affect basic metaphysical facts.

Yet, given the perspectivalist stance against the 'view from nowhere', starting *from somewhere*, namely local and situated epistemic practices, this is inevitable. Perspectival realism deepens the tradition that has emphasized the role of human practices in natural kind classifications as a way of keeping metaphysics in check. What are natural kinds if not a presumption humankind makes about nature and its *possible* joints given the robust phenomena one gets perspectivally acquainted with?

Some in-the-making kinds survive the ongoing and never-ceasing inferential work within and across scientific perspectives. These in-the-making kinds become *evolving kinds* resilient across scientific perspectives. They are effectively the 'natural kinds' we know and love. Other in-the-making kinds do not survive: they prove to be *empty kinds*. Nomological resilience plays an important role in the ability of in-the-making kinds to survive and become evolving kinds.

The natural kinds that we know and love evolve with our scientific history. Hence, a perspectival realist who takes seriously the situated nature of knowledge does not take natural kinds as placeholders for clusters of essential properties (discovered or still to be discovered), causal powers, categorial properties, dispositions, or similar. Natural kind is the name we give to what makes our presumptions of 'natural kindhood' for in-the-making kinds a little less presumptive.

In this final section, I return to the role of laws of nature and clarify why they provide the realist tether for natural kinds. Let me, then, go back to the second element in my definition of NKHF: that is, the reference to modally robust phenomena '(ii) each displaying lawlike dependencies among relevant features'. In what sense do phenomena manifest such dependencies?

As we saw in Chapter 5, perspectival models are an exercise in *physical conceivability*. They are an invitation to imagine something about the relevant target system that complies with the state of knowledge and conceptual resources of a community C and is *consistent with the laws of nature known* by C. As the analogy with inferential blueprints suggested, perspectival models involve by and large an inferentialist exercise of physically conceiving guided by the laws of nature and with an eye to delivering modal knowledge of what might be the case.

The laws endorsed by a particular epistemic community support the open-ended exercise of modelling what *might be the case*. I have discussed the various ways in which laws of nature enter into physical conceivability: by driving analogical reasoning between different modelling practices; by enabling non-causal explanations; or by fixing the general nomological boundaries within which the physical conceivability exercise takes place.

I also stressed the difference between lawhood and lawlikeness. The former is contingent on whichever series of perspectival Best Systems we happen to work with. The latter is displayed in nature among specific features of phenomena underpinning the stability of the event and not contingent on there being a perspectival Best System in place.

The events that are candidates for phenomena are 'stable' precisely because they display lawlike dependencies. The stable event associated with the phenomenon 'cathode rays bending' is the expression of the lawlike dependency between the electrical nature of cathode rays and the way they respond to a magnetic field. The stable event (structural loop in the kinase regulatory region) associated with the phenomenon 'phosphorylation' is the expression of the lawlike dependency between the ability to carry phosphate molecules and changes induced in proteins. The stable events (melting of glaciers, ocean heat uptake, etc.) associated with the multifactorial phenomenon of 'global warming' are all associated with various lawlike dependencies concerning increased GHG and associated retention of incoming radiative energy.

These lawlike dependencies can be causal, as with the electrostatic force that causes the bending of cathode rays; or the addition of phosphates to proteins. Other relevant dependencies are non-causal, as with Pauli's principle constraining nucleon structures. At yet other times, the lawlike dependencies fix general constraints within which the exercise of perspectival models takes place, as when R-parity conservation and consistent electroweak symmetry breaking are involved in the pMSSM-19 to physically conceive ways in which hypothetical candidate SUSY particles might manifest themselves (see ATLAS Collaboration 2015 and Section 5.5.3 in Chapter 5).

Thus, the nomological resilience of in-the-making kinds is a good indicator that the groupings of historically identified phenomena have (causal or non-causal) *lawlike dependencies among relevant features*. The relevant features might be empirical properties that manifest themselves in the relevant phenomena. I call them 'empirical' to avoid confusion both with the 'sparse natural properties' of Lewisian memory and with the metaphysical properties of the dispositional realist or dispositional essentialist. The charge-to-mass ratio is an example of lawlike dependency between two empirical properties at play in the phenomenon of cathode rays bending. The kinase inhibitors are another example of lawlike dependencies (well studied in pharmacology and drug discovery) among empirical properties (e.g. activation loop) of particular molecules at work in the phenomenon 'phosphorylation'.

At other times, the relevant features are not empirical properties but measurable quantities (e.g. the coefficient of viscosity for a fluid, the potential gradient, soil pH). Or they are physical or chemical constants (the Planck constant h, the elementary charge e, various thermal constants for inorganic, organic, and metallo-organic substances). These are just some illustrative and non-exhaustive examples.

But natural kinds talk is not so much concerned with specific lawlike dependencies among relevant features indexed to a particular domain. One wants to find out what is common to groupings of phenomena *across different domains*. For example, one might be interested in finding out whether the lawlike dependencies at play in the phenomenon of Moon–Earth alignment and the times of the tides are related to the lawlike dependencies observed among the speeds of different kinds of balls rolling down inclined planes. Or whether the phenomenon of the stability of matter at the level of stars is related to the phenomenon of the stability of matter at the subatomic scale. Or whether the lawlike dependencies observed in water electrolysis have anything to do with those in the bending of cathode rays. Or whether the phenomenon of pollen under-representation in *Ts'iits'ilche'* honey is related to other phenomena concerning, for example, the reproductive biology of the plant.

Are these *stable (lawlike) dependencies* identified in a number of phenomena, each indexed to a particular domain, indicative of how these phenomena might somehow belong together? Depending on how one answers, the in-the-making kind might turn out to be an empty kind, or prove to be an evolving kind. Identifying *which grouping of phenomena* speaks to an evolving kind and which one hides an empty kind is something that epistemic communities learn how to do over time through a network of perspectival inferences, which I will have to return in more detail in the next chapters.

The ether exemplifies a former in-the-making kind that turned out to be an empty kind. A number of phenomena with seemingly lawlike dependencies were identified by Newton, Boerhaave, and Kant, among other natural philosophers of the eighteenth century. The putative kind 'ether' was introduced to encompass a wide-ranging array of phenomena including bodies repelling each other;[18] calcination in chemistry;[19] and even the formation of planets in the solar system,[20] among others. Yet there is no natural kind 'ether' in this grouping of phenomena across different domains. Although there were lawlike dependencies underpinning each one, there was no genuine *bona fide* inferential link connecting them. The phenomena of metallic filings floating on liquids, calcination of metals, and formation of the solar system have nothing in common. It took almost a century and a half to downgrade the ether to an empty kind.

Consider, on the other hand, the grouping of phenomena ranging from water electrolysis to cathode rays bending, to black-body radiation. The lawlike dependencies at play in each of these phenomena were this time indicative of how this particular grouping of phenomena did in fact belong together under what—to the best of our knowledge still today—we have reasons for believing to be one of our evolving kinds: the electron. In Chapter 10, I shall return in detail to the nature of the inferential links among phenomena in this historical episode.

In sum, how epistemic communities come to identify which phenomena group together and which ones do not is not a matter of happenstance or convention. As I argue in Chapter 9, successful groupings of phenomena typically satisfy a *sort-relative sameness relation*. But for now, the main points to stress are the following:

- The identification of modally robust phenomena is effected by historically and culturally situated epistemic communities.

[18] The ether was, for instance, invoked in relation to phenomena such as those concerning melted lead which does not adhere to an iron vessel, or metallic filings floating on liquids, to give two examples taken from Newton's *De Aere et Aethere* (see Massimi 2011b for a discussion of these examples).

[19] 'As evidence for the existence of the ether, Newton referred to the experiments of Boyle on calcination, whereby "metals, fused in a hermetically sealed glass for such a time that part is converted into calx, become heavier" [Newton 1674/1962, p. 227]' (Massimi 2011b, p. 533).

[20] As I reconstructed in Massimi (2011b), in *Universal Natural History*, Kant, for example, built on Newton's speculative experimentalism about the ether and believed that a fine ethereal stuff filled the universe at the beginning of time and that through a mechanism of whirling according to Newton's principles of attraction and repulsion, the primordial ethereal matter gave rise to the different planets (p. 530).

- Knowledge of phenomena is always situated local knowledge. It is the specific vantage point occupied that gives communities their upper hand in teasing out modally robust phenomena from data.
- Lawlike dependencies at work in each phenomenon—and across different phenomena—play a major role in identifying which phenomena are genuine candidates for grouping into an evolving kind.
- Even if there might be phenomena each of which displays some lawlike dependencies, establishing whether particular phenomena hang together as a natural kind requires genuine (i.e. truth-conducive conditionals-supporting) inferences among the relevant types of phenomena.

That there are features and lawlike dependencies in phenomena qua stable events is a fact about nature, and ultimately it *is* the *realist* tether of perspectival realism. The *perspectival pluralist* aspect is that situated epistemic communities are able over time to rely on these and engage with epistemic practices and perspectival models—qua inferential blueprints—that allow them to tease out the relevant conditionals-supporting inferences linking genuine groups of phenomena into evolving kinds.

But how do we possibly come to know the world *as is* if all that is ever given to us is the world as *it appears to epistemic communities over time?* The question has a genuine bite. If our scientific knowledge is always situated knowledge of phenomena, no matter how reliable in the ways I describe, it still feels like something is amiss. The world as it *appears to us* is never going to be the world *as is*, unless we pass phenomena off as noumena, a critic might reply.

I need to show that our modelling—and more broadly epistemic—practices and their associated groupings of phenomena deliver indeed $perspectival_2$ representations, which offer a genuine window on reality, despite the situated nature of each representation. Let me give you a specific example which I shall return to in Chapter 10. Take J. J. Thomson's (1897) $perspectival_1$ data-to-phenomena inference about cathode rays bending, Max Planck's (1906/1913) $perspectival_1$ data-to-phenomena inference about black body radiation, and Theodor Grotthuss's $perspectival_1$ data-to-phenomena inference about water electrolysis around 1805–1806. This particular group of identified phenomena, over a span of a century, enabled conditionals-supporting inferences that still underpin our knowledge claims today about the electron as a natural kind, one of our best examples of evolving kinds. Evolving kinds are analogous to the unbounded space reflected by the mirror in the *Arnolfini Portrait*, a space that extends beyond what is visible on the canvas, beyond the $perspectival_1$ representation.

How can this work? If there is no God's-eye view from which one can access natural kinds as given, how can there be a God's-eye view from which to access relevant features in phenomena, and their lawlike dependencies, as somehow indicative of whether a grouping of phenomena belongs together? Are not any such groupings of phenomena ultimately at the mercy of historical contingencies?

J. J. Thomson, doing experiments on cathode rays and X-ray ionization in the late 1890s, was still couching his findings in the idiom of 'carriers of negative electricity', 'corpuscles', and even 'Faraday tubes', harking back to the nineteenth-century view of a strained state of the ether (see Falconer 1987, p. 260). My Neurath's Boat of natural kinds begins to look alarmingly leaky. How can NKHF resist the ever-present stresses and strains of historical changes and scientific revolutions?

The worry is genuine, but it conceals once more an invitation to hold epistemology in check, by offering a metaphysically more solid tether of some sort. Yet this fails to appreciate the epistemological pivot behind NKHF: the third condition in my definition—how lawlike dependencies among relevant features (iii) *enable truth-conducive conditionals-supporting inferences over time.*

I will spell out the full details of this point (iii) in the next two chapters, which explain the nature of the inferential exercise that joins the dots among types of phenomena and allows epistemic communities to navigate safely in the Neurath's Boat of natural kinds. These inferences supporting indicative conditionals over time lead epistemic communities to the identification of the relevant links

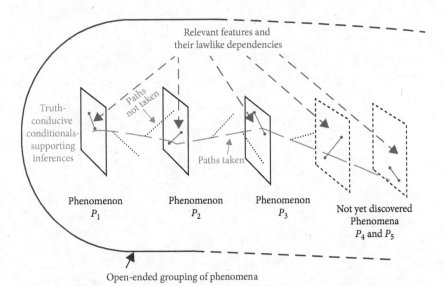

Figure 8.1 An evolving kind is an open-ended grouping (in purple) of *historically* identified phenomena P_1, P_2, P_3 (including not yet discovered phenomena, e.g. P_4 and P_5), (ii) each displaying lawlike dependencies among relevant features (in blue), (iii) that enable truth-conducive conditionals-supporting inferences (in green) over time. (For interpretation of the references to colours in this figure, the reader is kindly referred to the Open Access eBook version). © Michela Massimi

among lawlike dependencies in different phenomena across different domains (see Figure 8.1).

Subjunctive conditionals in turn act as signposts in inviting epistemic agents to walk in the inferential garden of forking paths. Ultimately, albeit always provisionally, they tell us which among the *very many* features and lawlike dependencies present in each type of phenomenon are related to which others across a multitude of perspectives. The next two chapters take a closer look at the remaining steps in this journey navigating the Neurath's Boat of NKHF.

9

Sorting phenomena into kinds

'From now on, I'll describe the cities to you', the Khan had said, 'in your journeys you will see if they exist'.

But the cities visited by Marco Polo were always different from those thought of by the emperor.

'And yet I have constructed in my mind a model city from which all possible cities can be deduced', Kublai said. . . .

'I have also thought of a model city from which I deduce all the others', Marco answered. 'It is a city made only of exceptions, exclusions, incongruities, contradictions. If such a city is the most improbable, by reducing the number of abnormal elements, we increase the probability that the city really exists'.

Italo Calvino (1972/1997) *Invisible Cities*, Cities & Names, p. 61[1]

9.1. Navigating Neurath's Boat of natural kinds

Natural Kinds with a Human Face (NKHF) are equidistant from essentialism and conventionalism. In this section, I briefly review this philosophical landscape, before I introduce the idea of natural kinds as sortal concepts in the rest of this chapter.

These days, essentialists about natural kinds are keen to defend a suitable notion of *necessity* that accompanies kind membership. Necessity can be delivered in many ways: by appealing to microstructural essences like atomic numbers for chemical elements (see Hendry 2008, 2019),[2] by pointing at essential properties

---

[1] Copyright © Giulio Einaudi editore, s.p.a. 1972. English translation copyright © Harcourt Brace Jovanovich, Inc. 1974. Reprinted by permission of The Random House Group Limited. For the US and Canada territories, *Invisible Cities* by Italo Calvino, translated by William Weaver. Copyright © 1972 by Giulio Einaudi editore, s.p.a. Torino, English translation © 1983, 1984 by HarperCollins Publishers LLC. Reprinted by permission of Mariner Books, an imprint of HarperCollins Publishers LLC. All rights reserved.

[2] Hendry (2019) has argued for microstructural essentialism by departing from traditional approaches such as the semantic route of Kripke and Putnam, and the scientific essentialism of Ellis. Hendry defends instead what he calls 'immanent microstructural essentialism' on the grounds that 'microstructure is the basis of chemistry's own classification of and naming of chemical substances: the International Union of Pure and Applied Chemistry (IUPAC) has developed systematic nomenclatures for chemical substances, referring exclusively to microstructural properties' (2019, p. 5). Hendry's second argument is that 'microstructural properties and relations are

for biological species and taxa (see Devitt 2008, 2010), or by highlighting properties that are super-explanatory and therefore metaphysically necessary (see Godman et al. 2020).

For all of them, natural kinds are not haphazard. There is an underlying *necessary* order that explains why lemons cannot be blue, why water cannot have atomic structure H_4O, nor gold atomic number 9. Unanimity seems to demand a notion of *necessity* in kind membership. And necessity in turn seems to cry out for metaphysical foundations in terms of essential properties: why things *ought to be* the way they are and cannot be otherwise by virtue of some core properties.[3] To reject essentialism as a metaphysical view is, then, to face an epistemic conundrum: what else might secure the necessity of kind membership? If necessity goes, what to make of unanimity?

Cluster accounts of natural kinds (from Boyd 1991, 1992, 1999b, to Slater 2015, more recently) endeavour to bypass this conundrum by swapping essences with homeostatic clusters of properties, or 'cliquish' properties (to use Slater's terminology). The necessity of kind membership does not, after all, require marble-solid metaphysical foundations in essential properties. Sufficiently stable clusters of empirical properties would do, as long as there is some underlying mechanism to warrant their stability over time.

Cluster accounts are fascinating. They seem to achieve the best of both worlds in securing necessity for kind membership while also accommodating for the historical vagaries of conceptual change and fuzzy boundaries. This seems an attractive middle ground between essentialism and conventionalism. In Boyd's

---

involved indispensably in understanding the physical properties, chemical reactivity and spectroscopic behaviour of chemical substances' (p. 5). And 'The third argument for microstructuralism is that there is no alternative. There is no conception of the sameness and difference of chemical substances that is both independent of microstructure and consistent with the ways in which chemistry in fact classifies substances, and the ways in which it explains their behaviour. How might one attempt to individuate substances independently of microstructure?' (p. 6). I share with Hendry the emphasis placed on scientific practice in the first argument. Where I would depart from him is in the conclusion (argument number 3 here) that there is no alternative and that is all there is to say about chemical elements (or any other examples of natural kinds). My defence of historical naturalism in Chapter 8 should have made clear already why I see a plurality of perspectival data-to-phenomena inferences at work in historically identifying the relevant phenomena, and why the impression that 'there is no alternative' is the product of coarse-graining or fine-graining the plurality of descriptions contributed by various scientific perspectives, rather than a methodological fiat.

[3] This sentiment is well captured by Godman et al.: '[T]here is a widespread intuition that chemical substances have their molecular structure *necessarily*. The intuitive consensus is that there are no metaphysically possible worlds where water is not H_2O and moreover that, in any possible world, anything that is H_2O is water' (Godman et al. 2020, p. 327, emphasis added). Along similar lines, see Hendry (2019, p. 13): 'Solid, liquid and gaseous ethane each have different geometrical structures. But what is common to all three is the ethane molecule, so if we ask "What is the structure of ethane?" independently of any particular state of aggregation, the appropriate answer is the ethane molecule, which is common to all three. Being composed of molecules of this kind is necessary and sufficient for being ethane. So what remains of the task of arguing for microstructural essentialism is explaining why having this microstructure is what makes ethane the particular substance that it is'.

(2010, p. 220) own words, 'Natural kinds are social constructions'. Over time communities have reworked and expanded those fuzzy boundaries for kind membership by dropping old properties from or adding new properties to any semi-stable cluster of them. Yet cluster accounts too face epistemic quandaries. As Kendig and Grey (2021, pp. 372–373) have argued,

[T]he explanatory work that cliquish stability does in picking out the property of natural kindness is (at least in part) grounded in our knowledge that the entity has already been discovered to be a natural kind and so—by virtue of that knowledge—is the appropriate sort of entity that we can attribute natural kindness to. This of course is not only a problem Slater must face, but one that must be faced by all who seek a metaphysically neutral approach to explain the nature of natural kinds (for example, Khalidi [2013]; Ereshefsky and Reydon [2015]; Magnus [2014]). In order to secure a metaphysically neutral account of natural kinds that does not rely on kinds that have already been determined to be natural and stable, we must be able to identify mistakes that arise when the wrong conditions of stability for putative natural kinds are prescribed.

Often enough problems like these have tipped the balance of the debate towards more decisively conventionalist views. If natural kinds are indeed the workmanship of people, why bother with homeostasis or other underlying mechanisms to explain why they form a cluster at all? Why not just conclude that what holds a cluster of properties together is a conventional label that people have decided to attach to things that look sufficiently similar on some ground?[4]

This sentiment resonates with historically motivated criticisms of scientific realism (see, e.g., Chang 2012a and Stanford 2015, p. 406). Stanford has drawn attention to the role of 'interpretive decisions concerning past speakers and

---

[4] See for example LaPorte (2004). I discussed LaPorte's view in Massimi (2012b, p. 532): 'Against Putnam's causal theory of reference, LaPorte maintains that possessing underlying properties and relations does not guarantee sameness of kind to paradigm samples. For example, being H_2O is neither necessary nor sufficient for being water; nor did we discover that water is H_2O. On the contrary, if confronted with XYZ, we might conclude that "XYZ is water"; or vice versa, if confronted with H_2O having unusual characteristics, we might conclude that "Some H_2O is not water". LaPorte concludes that the kind term 'water' is vague and changes its meaning over time, so that XYZ is neither clearly in nor clearly out of the extension of the term until a decision is made by speakers to include or exclude it. Such stipulation on behalf of speakers would amount to a precisification of the term; it would not be a scientific discovery. To illustrate this point, LaPorte revisits Putnam's Twin Earth story with the fictional scenario of Deuterium Earth. The story is functional to LaPorte's point that scientists stipulated that deuterium oxide (D_2O), i.e. one of the most relevant isotopic varieties of water, is a kind of water (usually called 'heavy water'). But they could have chosen otherwise, and they would not have been wrong in doing so, anymore than we are right in including deuterium oxide in the extension of water.'

linguistic communities' over and above any metaphysical and semantic fiat.[5] Currie has expanded on Collingwood (1936/1976) and defended the view that

> in the natural world, fixed forms are sometimes instantiated, sometimes not; in the world of history, the forms themselves arise. . . . The European Union was not merely instantiated, but was founded with the Maastricht Treaty in 1992. And the particular shape, properties and—fate—of that institution are intimately linked to those historical conditions. (Currie 2019, pp. 29–30)

At other times, the sentiment finds its counterparts in pragmatist accounts of kinds (see Franklin-Hall 2015; Khalidi 2013; Magnus 2012; for examples on protein folding, see Mitchell 2009 and Mitchell and Gronenborn 2017). In Kitcher's words (2001, p. 59):

> Our ways of dividing up the world into things and kinds of things depend on our capacities and interests . . . [T]he aim of the sciences is to address the issues that are significant for people at a particular stage in the evolution of human culture . . . Whatever language, or compendium of languages, is apt for this large purpose will mark out privileged divisions in nature. It will identify the *real* natural kinds.

Or in Franklin-Hall's (2015, p. 940) *categorical bottleneck account*:

> [N]atural kinds correspond to those categories that are metaphorical bottlenecks . . . : they reflect the categories that both ourselves and a large array of scientific inquirers with epistemic aims and cognitive capacities differing from our own would sanction in common, thereby converging on a single set of categories and kinds from multiple, distinct starting positions or points-of-view.

Present-day anti-essentialists raise an important contingentist point: things could have gone otherwise. Zoologists could have decided to classify the platypus as a non-mammal on the ground of egg-laying. The International Astronomical Union could have decided for Pluto to remain as a 'planet' if the property of clearing-its-neighbourhood had not been added to the definition. And the Royal Society back in the 1930s could have decided not to include isotopic varieties in the definition of 'water' on the ground of their macroscopically very different chemical properties.

---

[5] 'And we would do better to deliberate about the conditions under which our own uses of terms such as "atoms" and "gene" will be *held* to be referential (and our claims that "atoms exist" or "there are genes" will be *held* true) by the members of future linguistic communities who interpret *us* than about whether such referential status or truth is straightforwardly established by even the sum total of facts that are presently established' (Stanford 2015, p. 407, emphases in original).

Necessity and contingency are difficult masters for any account of natural kinds. No matter where one sits in this spectrum of views, the tension between the two intuitions is tangible. Cluster accounts sit uncomfortably between them. They seem to be forced to either concede with the essentialists that some properties of the cluster are more core than others, or nod to the conventionalist that indeed the whole cluster is on wheels. Reports of twilight for accounts of natural kinds (Hacking 2007a) reflect on this philosophical stand-off.

In this chapter, I argue that navigating between necessity and contingency requires natural kinds to be flexible and malleable enough to accommodate contingentism without necessarily opening the door to conventionalism. As I've already outlined, phenomena (rather than properties) should be the minimal unit for natural kind classification, and natural kinds are nothing but historically identified and open-ended groupings of phenomena. A phenomena-first ontology can avoid being caught up in debates about *which properties* are natural and fundamental, and which ones are not; which properties are necessary for kind membership, and which ones can be dropped without loss.

In Section 9.2, I review some motivations behind so-called 'deep essentialism' about kinds, and show how they trade on an ambiguity between explanatory necessity and metaphysical necessity for some properties. More to the point, I outline a number of reasons why a perspectival realist might want to do away with deep essentialism.

9.2. Doing away with deep essentialism

That there is more to water than just a molecular composition of one atom of oxygen and two atoms of hydrogen or more to lemons and hellebores than their genetic codes, as Putnam had it, is not news. It has been abundantly pointed out in the literature of the past twenty years on various metaphysical, semantic, and scientific grounds, which it is not my intention to repeat here (see Häggqvist and Wikforss 2018; Havstad 2018; Needham 2000, 2011; Tobin 2010; and the vast literature in philosophy of biology). Yet essentialism still figures in many contemporary realist accounts of natural kinds: from Chakravartty's (2017) dispositional realism to Bird's (2018) dispositional essentialism;[6] from the

---

[6] Bird has offered an argument for natural kind essentialism. The argument is put to use to explain why phlogiston for example cannot be a natural kind: 'Let us imagine that current chemistry is a horrible mistake, and that phlogiston theorists were right all along. There is a substance which explains combustion by being given off in that process; that substance is phlogiston. But does that substance have an essence that includes being emitted in combustion? It would be a matter of scientific investigation to discover whether that is the case' (Bird 2018, p. 1409).

aforementioned Godman et al. view on super-explanatory essential properties to Hendry's microstructural essentialism. Why is this?

As mentioned already in Chapter 3, following Shoemaker and Swoyer Chakravartty (2007), for example, defends the 'dispositional identity thesis' whereby what makes a property the property that it is (as opposed to some other property)—in other words, what constitutes its 'essence'—is the set of dispositions associated with it. And in his latest book (2017, p. 111) Chakravartty offers an insight into the interrelations among properties, laws, and kinds that is often at work in varieties of realism, including dispositional realism:

> [A] neatly unified account of kinds and laws follow immediately from disposi-
> tional realism and the concomitant view of causation. . . . Of course sometimes
> laws are stated more directly in terms of the behaviours of members of kinds
> (e.g. things having opposite electrostatic charges attract) rather than in terms of
> relations between properties as such, but for the dispositional realist this works
> just as well, because once again the properties shared by the members of these
> kinds dispose them to behave in the shared ways they do.

Two main factors are at play behind the ongoing promise of essentialism for re-alist accounts of science:

(1) an ontology of properties-first seems to be *metaphysically explanatory* for a great variety of phenomena;

(2) laws of nature have long been regarded as *instantiations of properties* (be these causal properties identified by their dispositions or be they irreduc-ibly dispositional properties), and this view—in conjunction with (1)—naturally explains the projectibility of natural kinds.

Some authors defend (2) but not (1); and others defend (1) without being committed to the dispositional view of laws in (2). For example, Hendry's microstructural essentialism endorses a metaphysically explanatory view for microstructural-essential properties, without defending a concomitant view of laws as dispositional manifestations of such properties. Chakravartty (2007, pp. 135ff.), on the other hand, has articulated a dispositional account of laws of nature, without necessarily taking dispositional properties as metaphysically ex-planatory. For his account of 'sociable' properties is designed to combine disposi-tional realism with ontological pluralism.

I do not aim to offer any specific argument against (1) or (2). As I said at the outset, I see perspectival realism as making room for different ways of thinking about realism in science rather than winning arguments about who is right and who is wrong on metaphysical matters. However, I do want to make some observations.

A leitmotiv of this book is that one might consider replacing an ontology of properties with one of phenomena. I have not given any knock-down argument for this claim, but offered instead a series of localized moves to this effect. My characterization of perspectival$_2$, where the perspectivity of the representation should *not* be regarded in terms of property attribution to the same target system, was functional to my take on the problem of inconsistent models in Chapter 3. There I pointed out how the tension between plurality of models and realism holds only under readings of realism that commit to *representing-as-mapping* and *truth-by-truthmakers*. These in turn appealed to essential properties.[7] Those unduly restrictive realist readings, I further argued, were behind the HYCAEI argument in Chapter 3, Section 3.6.

An epistemology of science that gives perspectival pluralism its due forces us to rethink the realist ontology that should accompany it. An ontology of phenomena-first is the congenial companion of perspectival pluralism, where phenomena should be reconsidered beyond the strict empiricist view and the strong realist one that sees them as teeming with properties. Thus, the following question arises: can the *necessity* of kind membership be retrieved in some form within my phenomena-first ontology, without having to endorse (1)?

I think the answer to this question is positive. But to understand the reluctance of essentialists to even consider a switch from properties-first to phenomena-first, one needs to understand how traditional discussions about laws of nature have got us here. In particular, one needs to understand how the notion of lawlikeness that I have taken as a primitive in my earlier discussion departs from notions of laws of nature typically at play in the essentialist ontology of properties-*with*-laws.

The projectibility of natural kinds has often been read as the expression of how some properties (but not others) can be predicated of the kind (e.g. 'green' can be predicated of emeralds, but not 'grue'). Realists with essentialist leanings tend to see the projectibility of natural kinds as linked with an ontology of property-first and laws as instantiations/dispositional manifestations of properties. Let us take it as a law that 'All emeralds are green', and that 'All samples of water boil at 100 degrees Celsius'. 'Being green'

---

[7] Recall how in Chapter 3, Section 3.6, I defined the two notions: '*Representing-as-mapping.* The true model is the one that offers an accurate, partial, *de re* representation of relevant essential features of the target system. Offering an accurate, partial, *de re* representation means to establish a one-to-one mapping between relevant (partial) features of the model and relevant (partial)—actual or fictional—states of affairs about the target system'. And '*Truth-by-truthmakers.* States of affairs ascribe essential properties to particulars, and, as such, they act as ontological grounds that make the knowledge claims afforded by the model (approximately) true'.

and 'having a boiling point of 100 degrees Celsius' are typically regarded as essential properties that pertain to emeralds and to samples of water, under a number of realist approaches.

For example, for a dispositional essentialist the properties that feature in statements of laws are essential properties conferring dispositions to behave in certain way in the presence of the right stimuli. For a categoricalist, these essential properties can be ascribed to particulars and enter into laws understood as necessitation relations among universals (think of Armstrong's view on laws among others).[8] For a Humean follower of David Lewis (what Beebee 2013 refers to as a 'shallow essentialist'), these are sparse natural properties (not to be identified with either dispositions or categorical properties). For my purpose here, the difference between the categoricalist view and the dispositional view of properties can be left aside. I also leave aside here the 'shallow essentialist' and concentrate my attention on what Beebee (following the terminology of L.A. Paul 2006) calls 'deep essentialist'. Thus, henceforth, whenever I say 'essentialist' I mean 'deep essentialist':[9] namely, the view that some essential properties determine the nature of a certain natural kind by explaining why the kind has the unique and distinctive features it has.

Laws of nature play a key role for essentialism. In Armstrong's categoricalist version, the law statement 'All samples of water boil at 100 degrees Celsius' captures a state of affairs whereby a particular (i.e. *this* sample of water *here*) partakes in the universal categorical property of 'being H_2O'. The latter, in turn, is regarded as being in a necessitation relation with another universal categorical property, that of 'boiling at 100 degrees Celsius'.

In the dispositionalist version, the law 'All samples of water boil at 100 degrees Celsius' is regarded as emergent from the dispositional essence of water: boiling at 100 degrees Celsius is to manifest the disposition to boil in the presence of the right stimuli and environment. In either case, the fact that water has property H_2O is regarded as *metaphysically explanatory* for other physical and chemical properties that can be attributed to the same kind.

---

[8] One example is given by Bird (2018, p. 1418): 'Hawley and Bird (2011) do argue that natural kinds are indeed complex universals. . . . For example, according to this view, the universals *yellow coloured, metallic, dense, being constituted of atoms with atomic number 79*, etc. form a complex which is the natural kind gold. One of these, *being constituted of atoms with atomic number 79*, explains why this complex exists, and so will be part of the complex wherever it exists. Hence it will be a necessary and plausibly also essential feature of the kind'.

[9] In L.A. Paul's terminology, deep essentialists 'take the (nontrivial) essential properties of an object to determine its nature—such properties give sense to the idea that an object has a unique and distinctive character, and make it the case that an object has to be a certain way in order for it to *be* at all. . . . Shallow essentialists oppose deep essentialists: they reject the view that objects can be said to have essential properties independently of contexts of description or evaluation, and so substitute context-dependent truths for the deep essentialist's context-independent ones' (2006, p. 333, emphasis in original).

As I see it, the case for essentialism about natural kinds rests on a series of assumptions:

(i)   Some properties are *metaphysically explanatory* for natural kind K.

(ii) These properties are *essential* properties of K.

(iii) Essential properties *that enter into law statements* explain why a range of other properties holds for members of the same kind K.

(iv) Therefore, essential properties are kind-constitutive in explaining why particular *a* belongs to natural kind K, and why other properties of *a* inferred on the basis of its belonging to K are in turn projectible.

I have already offered some reasons for resisting the assumption of essential properties (ii) in Chapter 3, Section 3.6, and Chapter 6, Section 6.6. What now of (i) and (iii)? I take it that key to the essentialist view are two common intuitions. One is about the metaphysically explanatory role of some properties. The other is about the role of laws of nature in acting as 'ladders' to go up and down different domains of inquiry, in the hope of reducing a plurality of phenomena to a set of core essential properties regarded as metaphysically explanatory.

Genetic codes, chemical microstructure, atomic numbers, and clades can be regarded as metaphysically explanatory properties: they *enter into law statements* that seem to explain why a range of other properties hold for members of the same kind. Given atomic number 79, and given the laws of chemistry, one can explain a range of other properties for the element gold. Or given the genetic code for hellebores and Mendel's laws, one can explain other properties such as the dominance of some petal colours over others.

The essentialist would concede that there is a division of labour: the chemist is better equipped to answer questions about chemical reactivity for gold than the nuclear physicist; the botanist is better equipped to answer questions about morphological properties of hellebores than the cytogeneticist. But, the essentialist would also insist, if one wants to know what gold *really* is, one goes to the nuclear physicist; and if one wants to know what hellebore *really* is, one goes to the cytogeneticist. This is because once one knows what the essential properties are—by using laws of nature—one can avail oneself of explanatory ladders to connect with other properties in other domains (morphology, chemical reactivity, etc.). The projectibility of natural kinds follows effortlessly.

Yet, in my view, essentialism about natural kinds is inadequate on two main grounds. First, it outsources the *explanatory role* of particular phenomena to *metaphysical properties*. Second, it treats laws of nature as if they were ladders to go up and down different domains of inquiry in a strongly reductionist exercise to try and reduce a plurality of phenomena to an allegedly fundamental one.

That the explanatory role of some phenomena should not be outsourced to *metaphysical properties* emerges clearly from the story about the ice calorimeter and the unreasonable projectibility of empty kinds (Chapter 7, Section 7.5). Lavoisier outsourced the *explanatory role* of phenomena such as state transitions from ice to water to *metaphysical properties* of a putative kind caloric (e.g. its ability of being released from the pores of ice). But this unwarranted leap from explanation to metaphysics underpins the famous case studies that historicist critics of scientific realism typically appeal to. While the phenomena revealed by the ice calorimeter could help Lavoisier *explain* why ice and water were two physical states of the same kind water, they did not nonetheless license any conclusion about caloric and its putative metaphysical properties.

One might reply that in the case of caloric it proved an empty kind because there was no law of nature at stake. After all, there turned out to be no conservation of caloric. Still, the essentialist would rejoin, when we do have confidence that there are laws in place for the relevant phenomena, the leap to metaphysically explanatory properties of a kind is warranted. Being H_2O is metaphysically explanatory for the kind water and all its phenomena. This is so because there are laws about boiling points, freezing points and similar, no matter how different communities at different historical times might have *explained* water-related phenomena.

I have never found the essentialist story very convincing, for a variety of (a) *epistemic*, (b) *historical*, and (c) *nomological* reasons that I shall briefly outline in what follows. From (a) an *epistemic* point of view, appeal to metaphysically explanatory properties is dubious. Epistemically, it is debatable whether atomic composition is for example metaphysically more explanatory than molecular structure and thermodynamical considerations—underpinned by intermolecular lawlike dependencies—to *explain* a variety of macroscopic phenomena such as the liquidity of water (as Needham 2000 has long pointed out).

Indeed, even if being H_2O is a *conditio sine qua non* for these phenomena, this is so only by virtue of a presupposition that phenomena P_1 at the atomic scale and phenomena P_2 at the molecular scale and P_3 at the thermodynamic scale *belong to the same grouping* that the natural kind 'water' has been historically identified with. This is *not* so by virtue of phenomena at the atomic scale P_1 being *metaphysically more explanatory* for phenomena at the molecular scale P_2 and at the thermodynamic scale P_3; any more than, say, phenomena about individual galaxies can be said to be metaphysically more explanatory for phenomena at the large-scale structure of the universe. More in general, phenomena indexed to a particular domain cannot be taken to be metaphysically more explanatory of others in different domains unless one assumes some strong form of reductionism. Deep essentialism requires a strong form of explanatory reductionism that is often hard to maintain in scientific practice.[10]

But one might insist that while explanatory reductionism among phenomena across different domains sounds epistemically dubious, the advantage of thinking in terms of properties rather than phenomena is precisely that they lend themselves more easily to this kind of explanatory reductionism. Thus, one might reply, let us grant for the sake of the argument premise (i). Let us assume there are indeed properties as the first ontological unit, and that on the basis of some scientific investigation some of these properties are found to be *metaphysically explanatory* for others. Imagine the chemists telling us on the basis of their scientific research that microstructure is indeed one such property and that there are well-founded empirical grounds for taking some properties as metaphysically explanatory for a natural kind K.

So be it. Yet from a (b) historical point of view, the shift from explanation to metaphysics remains dubious. Certainly, the chemists or the chemistry textbooks are advancing *empirical* claims when they state that microstructure is the explanatory *conditio sine qua non* for other properties of chemical compounds. I doubt they also intend to make a *metaphysical* claim about microstructure being an essential property *explanatory* for all the remaining ones of a given natural kind K.

Historical considerations speak against such a leap from an empirical claim to a (presumed) metaphysically explanatory claim concerning some properties. Indeed, such a leap—and associated division of linguistic labour[11]—has historically lent itself to scientific narratives that are often exclusionary. Natural kind essentialism has often been associated with a single-sided historical take on who can legitimately reclaim the epistemic upper hand, be it the nuclear physicist at the cost of excluding the chemist; the cytogeneticist at the cost of leaving out the botanist; or the systematic botanist at the cost of neglecting the ethnobotanist.

To rectify such historical narratives about divisions of labour and metaphysical-explanatory priority, let us go back once more to the example of water. What is the natural kind 'water'? And what properties are metaphysically explanatory for it (hence good candidates for essential properties)? Whilst much attention has been given to the Chemical Revolution with the discovery of oxygen and atomic composition (and later the discovery of isotopes in the twentieth century), water being one of the most common natural kinds our species has been acquainted with, it might be instructive to dig a little bit further back in

---

[10] I am grateful to Wilson Poon for constructive comments on this point.

[11] One is here reminded of traditional ways in which the division of linguistic labour has often been portrayed in the literature: namely, if one wants to know what gold really is, one should ask the physicist about the atomic number, and not the metallurgist who studies its properties as a metal, or the jeweller who knows how to make golden rings, or the assayer who knows how to determine the quality of the metal.

time than Lavoisier and the Chemical Revolution. This will also be instructive to reinforce my case as to why a shift from properties to phenomena might be helpful in addressing debates about natural kinds.

For anyone like myself born in central Italy, just outside Rome, where the clergy and aristocratic families of D'Este and Farnese owned large estates in the Renaissance, one of the most common phenomena water displays are majestic garden fountains. Anyone skipping secondary school in Tivoli to spend a lazy morning in Villa D'Este would associate water with fountains before studying chemistry and learning that water is H_2O, that water molecules are subject to very specific kinds of intermolecular forces and so on. And if one wants to know how water forms such spectacular jets, the answer has nothing directly to do with it being H_2O.

The gardens of Villa D'Este have 51 fountains, 364 water jets, 64 waterfalls, 220 basins, and a remarkable water organ fountain. These monuments offer testimony to the fact that from a historical point of view which properties count as *explanatory* for others depends very much on *which phenomena* an epistemic community is accustomed to and *care for* in their own scientific perspective.

The construction of the Villa started in 1550 and went on for almost 20 years, including acquiring land, excavating grottos, and diverting the river Aniene. The hydraulic engineers who worked on the project needed to use the river Aniene to feed the remarkable number of fountains and jets in the garden (see Berger 1974). This was a problem that ancient Roman hydraulic engineers had already faced in building tanks of aqueduct water (so-called *castelli*) that distributed water for Roman baths. It was mastered by Islamic engineers in Granada in the thirteenth century, building a system of underground galleries, waterwheels, and reservoirs to bring water from the river Darro up to the Alhambra and Generalife, 778 metres above sea level (see García-Pulido 2016).

The relevant engineering phenomena (e.g. jetting, gushing, cascading down, overflowing) in all these cases involved features such as surface tension in pipes, pressure difference in communicating vessels (as in the Villa D'Este water organ fountain), flow under the action of gravity, and so on. From Vitruvius to the Islamic engineers of the Alhambra, whose work reported by the Venetian Andrea Navagero in 1563 seems to have inspired in turn Pirro Ligorio for his designs of the fountains of Villa D'Este in Tivoli (Berger 1974, p. 304). From the perspective of hydraulic engineering in its very many historically situated versions, one gets a very different impression about which (i) properties are *metaphysically explanatory* for the natural kind water. The disarming but inescapable answer is: 'it depends on the context'.

If the context is that of hydraulic engineering within a variety of historically situated perspectives—where water played a key geopolitical role for building empires or for asserting authority in sixteenth-century Tivoli—surface tension and pressure difference are some of the relevant *explanatory* properties for the phenomena of interest.

From a (b) historical point of view, which properties count as *explanatory* for a natural kind *K* depend very much on *which phenomena* a particular epistemic community is accustomed to and care for in their own scientific perspective. Phenomena come first (not properties) as my phenomena-first ontology recommends. Without belabouring this issue any further, from a (b) historical point of view, the emphasis on metaphysically explanatory properties flies in the face of the very many ways in which different epistemic communities have come to encounter nature *through a plurality of phenomena*, not all of which can be explained in terms of a *single* selected group of (metaphysically explanatory) essential properties.

I come to my last point: how the essentialist view faces difficulty also for (c) nomological reasons. Laws of nature are not 'ladders' to access different domains of inquiry in the hope to reduce a plurality of phenomena to a set of essential properties regarded as metaphysically explanatory. If anything, laws or lawlike dependencies are tied to specific phenomena. The engineers of Alhambra could succeed without any knowledge of chemical laws about water being H_2O. They knew instead how to exploit other relevant lawlike dependencies, such as flow under the action of gravity and surface tension in ceramic pipes, among others. Each phenomenon is imbued with lawlike dependencies. As I have explained in Chapter 6, these are primitive relations among relevant features of events that enter into what I have called 'empowered phenomena'. For example, the aforementioned hydraulic engineering phenomena rely (among others) on Darcy's law[12] which expresses a simple causal relation between the total discharge of water flow and the area of the cross-section where water flows.

The anti-essentialist move in my view of NKHF is that the natural kind water cannot *just* be identified with phenomena at the atomic scale. There is more to it: the droplet formation studied by Henry Cavendish[13] and the phenomenon of ground-water motion and jet formation mastered by hydraulic engineers of Alhambra and Villa D'Este have an equal right to be considered as 'kind-defining'. Moving away from properties to phenomena means recognizing the

[12] Darcy's law (from the French engineer Henry Darcy, who was studying problems about ground-water flow for public fountains in Dijon) states that $\frac{Q}{A} = -K\frac{dh}{dl}$, with K being a constant of proportionality, h the height and l the length of the manometer tubes (see Darcy 1856; and for a discussion Hubbert 1940).

[13] As early as 1766, Cavendish had come to identify 'inflammable air' (i.e. today's hydrogen) as a separate kind of air, and building on the phlogiston tradition of Joseph Priestley in England and experiments by Warltire, in 1781 he used an electric spark to fire a mixture of inflammable air and common air (containing what we now know to be oxygen) to obtain dew (water) and heat. Cavendish (1784/2010) concluded that the dew was produced by the combination of inflammable air and one-fifth of the common air (which he believed to be dephlogisticated air) losing 'elasticity' and condensing into 'dephlogisticated water'.

plurality and variety of phenomena that historically and culturally situated epistemic communities have encountered and learned how to historically group
under natural kind concepts and terms.

Water concerns all these phenomena at once: ground-water motion no less
than droplet formation or chemical bonds. Some of the laws involved—like
Darcy's law for ground-water motion—are not reducible to, say, the Navier–
Stokes equations in fluid dynamics. Laws are not ladders. A phenomena-first
approach is congenial to the pluralism inherent in perspectival realism, and
is anti-reductionist in denying any foundations. In a genuinely Neurathian
spirit, we are rebuilding the Boat of Natural Kinds while at sea. The *historically*
identified grouping of phenomena does not take any particular phenomenon
(and its associated lawlikeness) as more essential, more foundational, or more
metaphysically explanatory than any other.

Even if H_2O appears in all these phenomena across different physical scales,
invoking the property of being H_2O as 'kind-constitutive' (as the essentialist
would have it) *in and of itself* is irrelevant to answering questions about other
phenomena such as ground-water motion or jet formation. Insisting that water
is H_2O, while true, does not begin to answer the question as to why all these phenomena have historically been grouped together under the natural kind concept
of water.

But there is more. Phenomena that we have historically grouped together
into kinds are as open-ended as is scientific inquiry itself. More phenomena
are encountered as new experiments are devised and new models conceived
of. Novel instruments and technologies pave the way to expanding and refining our historically identified groupings of phenomena. Blossoming
new practices and techniques bring to the fore previously unidentified phenomena. We thought that DNA consists of four basic letters, and now, thanks
to new technology, we know that *hachimoji* DNA of eight letters is possible.
We thought that caloric was released from ice melting in a calorimeter, and
with the phenomena observed by Davy, Rumford, and Joule we came to conclude that it was impossible.

That historically identified open-ended groupings of phenomena belong together tells us only that there is something *shared among* them, and that what
is shared is not some 'kind-constitutive' microstructural essential property.
However, if phenomena do not encode essential properties, the projectibility
and unanimity of kinds might seem mysterious. How and why do historical and
open-ended groupings of phenomena hang together? How can a perspectival realist consistently maintain that although it is necessary for water to be H_2O, we
should leave open the possibility of discovering new phenomena that might be
added to the existing grouping? How to square necessity with historical contingency, the realist view that there is a way the world is with the perspectivalist

insight that our scientific knowledge is always from a human point of view? What glues the planks of Neurath's Boat together?

The task ahead for perspectival realism is to explain why it is necessary for water to be H_2O in a way that is consistent with the open-endedness of the historical grouping of phenomena. Make no mistake, though. This is no re-enactment of the 'shallow essentialist' exercise whereby (in L. A. Paul's words), one can still maintain that there are essential properties for natural kindhood but they just so happen to be context-dependent. There are no essential properties—context-dependent or independent as they might be—in a phenomena-first ontology. There are instead kind-defining stable events, whose lawlikeness makes them candidates for phenomena eligible for the title of 'natural kinds'. At the same time, in order to explain why it is necessary for water to be H_2O in a way that is consistent with the open-endedness of the historical groupings, the perspectival realist must avoid the unbridled historical contingency of Goodmanian scenarios.

Imagine the following Goodmanian scenario where something is *phlater* if it is observed before 1774 and is found to consist of dephlogisticated air and inflammable air; or it is observed after 1774 and found to consist of oxygen and hydrogen. One might argue that, given the available evidence before and after 1774, the putative kind *phlater* behaves like Goodman's grue in licensing diverging inductive inferences after 1774. This could in turn act as the basis for a strong historicist argument against realism (and not just scientific realism) as the view that there are some phenomena that are necessary in the sense of being 'kind-defining'. If some radical underdetermination affects the phenomena themselves, should not we conclude that things could have indeed gone otherwise? Should not we say that it was a historical accident that after 1774 we ended up making inductive inferences about H_2O rather than about dephlogisticated air and inflammable air? 'Does not it look like happenstance that we ended up with the kind concept *water* rather than *phlater*?', the Goodmanian historicist critic might insist.

In what follows, I want to resist this strong historicist move that might be levelled against perspectival realism no less than scientific realism. I do not think it is a matter of historical accident and contingent happenstance that particular phenomena have been identified as belonging to the same grouping and as defining a given natural kind. I think of what holds a group of phenomena together in terms of a *sort-relative sameness relation* taking my cue very loosely from Delia Graff Fara and moving the insight in a rather different direction.[14] Natural kinds

---

[14] Delia Graff Fara (2008, 2012) was primarily interested in developing an alternative to David Lewis's counterpart theory, and her distinction between identity and sameness was functional to addressing open problems in Lewis's view. In brief, Fara's idea of relativizing *sameness relation* to a 'sortal' was designed to deliver on *de re* possibility, and to elucidate how things might have been or will be like when we ask whether some x is 'the same x as' something else either at some later point in time or in another possible world w (to use Lewis's possible worlds). The underlying intuition is that

are an invitation to ask whether a phenomenon P_1 is *of the same sort* as some other phenomenon P_2. Is the P_1 that runs from my tap, your tap, Elsa's tap, and so on, of the *same sort* as the P_2 whose atomic structure the physicists found to be H_2O? Is the P_1 that produces *ts'iits'ilche'* honey for the Yucatán beekeepers of the *same sort* as the P_2 morphological specimen in the Kew Gardens herbarium known as *Gymnopodium floribundum*? Natural kind thought and talk should allow one to explain why a bunch of historically identified phenomena are of the *same sort* (and necessarily so) while also exploring how they *could* (contingently) *be otherwise*. The idea behind this is a simple one.

9.3. Why I am a contingentist of the most unusual kind

So far I have focused on the anti-essentialist and anti-foundationalist aspect of my position, and not addressed an equally important challenge from historicist accounts. Have not anti-foundationalism and anti-essentialism already found expression in views that have stressed the importance of human agents and their epistemic practices in thinking about kinds? Contingentism has become common currency in history and philosophy of science.

We have learned from Dupré (1981) that whether we classify garlic and onions as *Liliacee* depends on the specific needs of the community of reference—be it botanists or chefs. Hacking (2007a, 2007b) has made the case for a nominalist account of natural kinds, independent of psychological or social facts about human beings. And LaPorte (2004) has advanced a robust plea for historical contingencies and conventional undertakings in tempering realist-essentialist views of natural kinds.

Moreover, historicist accounts such as Chang's and Stanford's have drawn attention to the historical foundations of our thought and talk about natural kinds. Things could have historically gone otherwise in the way epistemic communities have come to identify water (Chang 2012a), temperature (Chang 2004), or cells (think of the case study discussed in Stanford 2006), just to mention a few well-researched historical examples. How does a perspectival realist differ from this broad family of views that share the same anti-foundationalist and anti-essentialist spirit?

If NKHF were nominal labels or conventional groupings, there would be no way of telling apart an in-the-making kind that turns out to be an empty kind from one that turns out to be an evolving kind. Steering Neurath's Boat clear of

---

two things may count as the *same* even if they are not necessarily numerically identical. Fara's analysis is confined to sort-relative sameness of tokens. In what follows, I want to build and expand on her analysis by taking it in a novel direction because it is the sort-relative sameness relation among phenomena (qua *types* of phenomena) that matters to my present discussion on natural kinds.

strong contingentist arguments requires shedding light on why the boundaries of natural kinds are malleable but not in a conventional fashion; or at least not in a fashion that is entirely reducible to some stipulative decisions by someone at some point in time.

The view I am suggesting is *contingentist* in that natural kinds are not after all identical to fixed-once-and-for-all groupings of phenomena. I see all natural kinds ultimately as open-ended, revisable, and evolving kinds: there is no complete A–Z list of phenomena that are constitutive of any specific natural kind. Things could always go otherwise, as contingentism would have it.

For example, epistemic communities might not have devised the experimental resources to make the relevant data-to-phenomena inferences. Imagine if Joule had never experimented with his paddle-wheel machine in the 1840s and measured the exchange rate between mechanical energy and thermal energy. Or, imagine if the Laplace–Lavoisier ice calorimeter had proved a more reliable instrument than it ever was in measuring the amount of ice melting into water and feeding into conditionals-supporting inferences about caloric. Or things could have gone otherwise if other undreamt-of phenomena had been identified, perhaps via modelling practices that epistemic communities simply failed to historically conceive of.

Yet I am a contingentist of the most unusual type. For ultimately I do not believe we could have lived in a world in the year 2021 where caloric was an evolving kind still going from strength to strength, whereas I do have fallible confidence (not God's-eye certainty) that in the year 2100 the electron will still be an evolving kind. My realism is born out of this unusual brand of contingentism.[15] The situation can be summarized as follows:

(1) A natural kind K is identical to some historically identified open-ended grouping of phenomena $G = \{P_1, P_2, P_3, \dots\}$.

---

[15] In a way, the view I am going to defend and articulate chimes with a recent analysis by Helen Beebee (2013), where she argues that one can legitimately see Kripke–Putnam essentialism as *consistent* with some suitable form of what she calls 'Kuhnian relativism' in situations where natural kinds cross-cut different scientific paradigms (e.g. *water* before and after the discovery of isotopes): 'The Kuhnian relativist will characterise this situation as one in which the natural kinds have *changed*: it is not that the *same* kind of substance or entity has been discovered to have different essence to the one we thought it had; rather, the old kind has genuinely ceased to exist, and has been replaced with another kind. After all, if it is the *same* kind of substance or entity, *which* kind of substance or entity, exactly, would that be? Not the same *chemical* kind, because the old chemical kinds are no longer a part of the classificatory framework of chemistry. Albumin and rennin would not both be enzymes, and hydrogen and deuterium would not both be isotopes of the same element, because such categories, we are imagining, have ceased to apply. My claim here is not that relativism is a defensible position. It is rather that relativism—which is to say, interparadigm crosscutting—is *consistent* with KP: we simply relativise the *same-NK* relations to the classificatory framework enshrined by a given paradigm. Hence KP, on its own, does not have the resources to defeat relativism' (pp. 161–162, emphases in original). I am no Kuhnian relativist; nor do I defend KP essentialism. But the middle ground between necessity and contingency that I discuss here resonates with Beebee's analysis.

(2) A natural kind *K could* include phenomena *different from* those actually included in the historically identified open-ended grouping of phenomena
$G = \{P_1, P_2, P_3, \dots\}$.

(3) Yet there is a sense in which the phenomena defining the natural kind *K*
could not have been *that* different from the historically identified open-ended
grouping of phenomena $G = \{P_1, P_2, P_3, \dots\}$ that *K* is actually identified with.

There is a clear tension in holding points (1), (2), and (3). While (1) is a restatement
of historical naturalism, (2) gives historicity its due by allowing for the possibility
of other kind-defining phenomena, different from those (historically) actually included in the grouping. Point (2) is meant to take care of historical contingencies
and open futures. Point (3) in turn conveys a different notion of necessity from the
traditional (deep essentialist / kind-constitutive) one that accompanies natural
kinds: a necessity about which particular phenomena (qua stable events and their
lawlike dependencies) are kind-defining for particular natural kinds.

Take again the example of water. Point (2) invites us to consider other *possible*
phenomena from those we have actually already identified water with. For example, we might not have included deuterium oxide in the list $G = \{P_1, P_2, P_3, \dots\}$.
After all, in the 1930s when Harold Urey discovered it, scientists debated whether
or not it was indeed a kind of water. At the same time, point (3) suggests that water
could not have been *that* different from the historical grouping $G = \{P_1, P_2, P_3, \dots\}$
that it has been identified with. There is a sense in which for a scientist in 2021 it is
necessary that isotopic varieties are part of the natural kind water. But there is also
a contingentist intuition that things could have gone otherwise. Back in the 1930s,
scientists might have decided not to take deuterium oxide as a kind of water: after
all, it is poisonous to all forms of life, has different boiling and melting points, and
so on (see LaPorte 2004). But the contingentism that I see as relevant to my NKHF
account is different from LaPorte's stipulative conventionalism.

The tension among (1), (2), and (3) can be mitigated using the strategy of
replacing identity with a sort-relative sameness relation. In asking whether *K* is indeed defined by the grouping $G = \{P_1, P_2, P_3, \dots\}$, we are not asking whether the
two are *numerically* identical. Instead, we are asking whether there is a *sort-relative
sameness relation* both within the phenomena already included in the grouping G
and between them and other *possible* phenomena (either past uninferred ones or
future yet-to-be inferred ones). Consider the following analogy. Suppose I want to
define my family (F): the Massimi–Sprevak family (F_{MS}).[16] To do so, suppose I proceed by identifying a group of people p (namely, individuals):

$$F_{MS} = \{p_1, p_2, p_3, p_4, \dots\}$$

[16] I am very grateful to Helen Beebee for helpful comments and for suggesting this analogy in private correspondence.

Some individuals are obvious candidates for this group: in addition to myself, my husband, and our son, our respective parents, grandparents, and great-grandparents are clear candidates. What about first-order and second-order cousins? What about the branch of the family who migrated to the United States in the early twentieth century, got naturalized, and we met only once in our life-time? What about family members who remarried and those who will be adopted in the future? Surely, while there is a sense of necessity that seems to accompany the inclusion of some family members in F (like my mother) in this grouping, there is also a sense in which the grouping is open-ended and malleable, and rightly so. It would be unduly restrictive to say that the Massimi–Sprevak family consisted of a set of three individuals or five or seven individuals. It would also be unduly exclusionary to identify the Massimi–Sprevak family only with members sharing blood relationships.

Natural kinds qua evolving kinds—I want to suggest—are groupings of phe-nomena that share something like the 'same-family relation'. When asking what holds an open-ended grouping of phenomena together into a natural kind, one is not asking for a closed and fixed once-and-for-all set of things, entities, or indi-viduals, nor for something akin to blood relations or genetic lineages. One is in-stead simply asking whether this type of phenomenon P_1 is *of the same sort* as that type of phenomenon P_2 and that other type of phenomenon P_3, and so forth. Let me expand on this point in the next section.

9.4. Spinozian sortals for natural kinds

Discussions about natural kinds start with a deceptively simple question: is this type of phenomenon P_1 *of the same sort* as that type of phenomenon P_2? Is this type of phenomenon (droplet formation) observed by Henry Cavendish *of the same sort* as that type of phenomenon (ice melting) in Lavoisier's calorimeter?[17] Is this type of decay in a four-lepton channel *of the same sort* as that type of decay in the two-photon channel? Is this type of flowering in the mountains of Alaska *of the same sort* as that type of flowering in the Dolomites? The presumption of a *sort-relative sameness relation* invites two further questions:

  (I) The same *what*?
  (II) *In which respect* can two types of phenomena be *of the same sort*?

---

[17] One is here reminded of how Kuhn (1990) insisted against Kripke-Putnam essentialism that a hypothetical Oscar living in 1750 before the Chemical Revolution would not have concluded that ice is *of the same sort* as liquid water or steam.

On the face of it, droplet formation (like dew drops on a blade of grass, say) does not necessarily look like *of the same sort* as ice melting; and a four-lepton decay channel does not necessarily look like *of the same sort* as a two-photon decay channel; nor are the morphological features of this flowering in one region of the globe exactly *of the same sort* as those in another region of the globe. These differences matter. Indeed, they are the reasons why different decay channels are explored by particle physicists, different species of flowers are introduced by botanists under the same genus, and phase transitions are identified by physicists.

Thus, coming back to my questions, it is important to clarify the sense in which these wide-ranging types of phenomena can be regarded as being *of the same sort*. In daily parlance, instead of saying 'of the same sort', one uses natural kind terms and concepts and asks whether the same sort of particle (the *Higgs boson*) is at play in this type of decay channel as in that other type of decay channel; or whether that type of phenomenon concerning ice melting is the same sort of chemical substance (*water*) as that other type of phenomenon concerning droplet formation, and so on. Natural kinds are *sortal concepts* (henceforth I shall use italics to denote concepts and single quotes for the corresponding natural kind terms).

Sortals have traditionally been used in philosophy precisely to take care of questions like (I) and (II). John Locke is usually credited with having coined the term in his *Essay Concerning Human Understanding*, in a discussion of the distinction between nominal essences and real essences:

> *The common Names of Substances,* as well as other general Terms, *stand for Sorts:* which is nothing else but the being made signs of such complex *Ideas,* wherein several particular Substances do, or might agree, by virtue of which, they are capable to be comprehended in one common Conception, and be signified by one Name. (Book III, Chapter VI, 1. 28–30 and 1–3; Locke 1689/1975, pp. 438–439)

The neologism got entangled with Locke's distinction between real essence and nominal essence and for a long time mention of the word 'sortal' evoked some form of Lockean nominalism about essences. But in fact Baruch Spinoza had already introduced the notion of sortal sixteen years before Locke. And this time not in the context of a discussion of nominal vs real essences (although arguably Spinoza too was rather concerned with essences, in particular God's essence and why—he claimed—one is speaking improperly when one says that God is 'one'). In a letter to Jarig Jelles in 1674, Spinoza writes

> For we do not conceive things under numbers unless they have first been brought under a common genus. For example, someone who holds a penny

and a dollar in his hand won't think of two unless he can call the penny and the dollar by one and the same name, either 'coin' or 'piece of money'. For then he can say that he has two coins or two pieces of money, since he calls not only the penny, but also the dollar, by the name 'coin' or 'piece of money'.

From this is evident that nothing is called one or unique unless another thing has been conceived which (as they say) 'agrees with it'. (Spinoza 2016, p. 406)

Spinoza's notion has far-reaching implications: in order to count something as *two* rather than one, one needs to know *what sort of thing* one is counting. A sortal concept like *coin* is necessary to be able to count as two the penny and the dollar and to individuate them as the *same sort of* thing.[18]

Spinozian sortals are not Lockean nominal essences. They are not names standing for an 'abstract idea' qua *de dicto* essence (rather than *de re* essence) for a group of particulars. Spinoza's choice of the example of coin is telling. What makes *coin* a sortal is the fact that it is a proxy for a bunch of rather diverse items (the penny, the dollar, and nowadays the euro, the Swiss franc, the Japanese yen, the Chinese yuan, the Ghanaian cedi, the Malagasy ariary, and myriad others).

These items are very different in nature: in the metal each coin is made of, in their sizes, their designs and inscriptions, their subunits, geographical, historical, and cultural provenance, and so forth. But they are also *of the same sort* in one major respect: they are a 'medium of exchange' (in the words of Kiyotaki and Wright 1989) with a purchasing power on goods and well-defined exchange rates. The sortal concept *coin* denotes neither a real essence, nor a nominal essence. It is not just a name, an abstract idea, a convenient label for a bunch of items that are found to agree with the idea either. Despite the absence of necessary and sufficient conditions for membership, there are nonetheless clearcut rules about what counts or does not count as a coin. For example, my son's chocolate coins are *not* coins even if they might be conveniently called as such ('chocolate coins'). The importance of the sortal *coin* lies, then, not in what it designates but in what one *can do with it.*

Coin is a sortal concept that allows human beings to count a penny, a dollar, a yen, a cedi, and an ariary as *five* items that are *of the same sort* in the sense that—in their respective geopolitical contexts—each performs exactly the same function: to allow people to trade goods and complete transactions in *exchange* for them. *Coin* as a sortal simply stands for a 'medium of exchange' that encompasses tokens of exchange value that happen to be very different on a number of counts (denomination, metal, weight, inscriptions, provenance, etc.).

---

[18] Fast forwarding a few centuries from Spinoza, counting requires a principle of individuation at work in 'count nouns', to use Quine's (1969) terminology. 'Coin' is a count noun, but so also is 'rabbit' or 'Linnaea'. Mass terms like water, by contrast, refer to an uncountable thing, although one can (and typically does) speak of one, two, or three *samples* of water.

Yet not everything might count as a coin, and telling aside counterfeit from genuine coins has a long history in numismatics.

Coin making presupposes a governance system that legislates the unit of account, the cost of minting, the revenues from transforming silver bullions into coins, and so forth. The various tokens of different denomination ought to be convertible into one another to facilitate trades: rather than exchanging spices for fabrics, it proved more expedient to exchange spices for coins and coins in turn for fabrics. The problem of figuring out how to trade spices with fabrics became the problem of figuring out how to trade goods in, say, dinars with goods in florins, or any other denomination.

Exchange rates among different denominations in turn presuppose that the coin tokens in each relevant denomination are a trustworthy transactional medium. They assume some monetary authority (e.g. the Crown, the Bank of Italy) that fixes the legal tender value at which the mint buys the metal to make the coins. Ultimately the value of each monetary denomination (and associated exchange rates) depends on a complex, highly variable, and context-dependent nexus of relations involving minting costs, the cost of the bullion, how many coins can be struck out of the bulk metal, the relative percentages of metals used, whether or not two metals are used (bimetallism), and whether the gold standard is used (for an interesting historical discussion of these issues, see Desan 2014 and Redish 2006). But it equally depends on how willing merchants and traders are to exchange goods across different geographical areas and cultures by trusting foreign coins.[19]

Likewise, I suggest the value of natural kinds qua sortal concepts is dictated by a network of relations among the phenomena in the grouping. At some point epistemic communities learned to call by one and the same name (e.g. 'water', 'Linnaea', 'Higgs boson') different types of phenomena, which stand in particular relations to one another and can be exchanged for one another. *Gymnopodium floribundum* is the botanical name for the plant whose nectar produces *Ts'iits'ilche'* honey and whose morphological features are displayed in the Kew Gardens herbarium, among other types of phenomena. Spinozian sortals capture the way in which natural kinds do indeed encompass a range of types of phenomena that stand in particular 'exchange rate' relations; phenomena that in some sense are *of the same sort* (and necessarily so) and in some other sense are not *exactly* the same (as contingentism would have it).

In a phenomena-first ontology, the phenomena $P_1, P_2, P_3, \ldots$ in any open-ended grouping are therefore clustered under a natural kind K as a sortal concept. The sortal is there to guarantee that we have a principle for counting types

---

[19] For an analysis of some troublesome aspects in the monetary governance of the early nineteenth-century Dutch empire, see, e.g., Weber (2018).

of phenomena $P_1, P_2, P_3, \ldots P_n$ as n samples of the *same sort in the appropriate contexts* and in some *specific respect*. In other words, I do not see phenomena as *instantiating* a sortal universal K.[20] Nor are $P_1, P_2, P_3, \ldots P_n$ temporally indexed phases of the same developing sortal.[21] They are not linked to one another by an arbitrary Lockean nominal essence either, which can be stripped and substituted easily in the course of scientific history and whose main function is to offer well-defined necessary and sufficient conditions for kind membership. I'd like to use the term 'sortal' in this distinctively Spinozian sense to denote a metaphysics-lite proxy for historically identified groupings of phenomena. Recent studies on concept learning and classification coming from experimental philosophy and cognitive science are instructive to shed light on the idea of natural kinds as Spinozian sortals.

Natural kind concepts and terms are of great interest to developmental psychologists. As already mentioned in Chapter 8, studies since the late 1980s have shed light on how children learn how to constrain hypotheses during the naming process. Ellen Markman et al. (2003) have demonstrated how before the onset of the naming explosion children tend to assume mutual exclusivity of object labels. When exposed to a novel term, children tend to associate the term with the whole object (rather than with a part of the object), and prefer taxonomic (rather than thematic) clustering (e.g. *rabbit* and *animal*, rather than *rabbit* and *carrot*). But they also tend to resist second labels for objects for which they have already a name—this is mutual exclusivity.

For example, consider a child presented with two objects: a familiar one (say, a ball) and an unfamiliar one (say, a whisk). If the child hears 'Can you hand me the whisk?', they will tend to associate the new word with the new unfamiliar object rather than use it as a second label for the already familiar object (see Markman et al. 2003, p. 243). Mutual exclusivity as a reasonable assumption in children's language acquisition was introduced by Susan Gelman and Ellen Markman

---

[20] Sortals have been at the centre of an important literature in metaphysics and philosophy of language after Spinoza and Locke, much of it beyond my scope here. Suffice to mention that Strawson (1959) put forward an influential notion of 'sortal universals'. He saw 'dog' as a sortal universal, which particulars (such as say Fido, Bobby, Rosy) are instantiations of.

[21] Wiggins (1980, 2001) has developed a sophisticated theory of sortal concepts. Instead of having one single concept applying at all times, one can capture continuity over time via the succession of *phase* sortal concepts (see Wiggins 1980, pp. 64ff.; Wiggins 2001, pp. 64ff.), as when we want to capture questions of identity and existence that involve phases: something ceases to exist and something new emerges (from tadpoles to frogs; from maggots to houseflies). Recently, Wiggins's account has attracted renewed attention in the philosophy of biology with DiFrisco (2019), who has approached the problem of biological individuality as a problem of finding a certain type of biological sortal or kind. Lowe (2009) considered whether changes such as from water to ice, or from wood to ashes in combustion, amount to substantial changes (changes of substance sortals) or to phase change within a single substantial kind. Lowe sought an a priori metaphysical framework of ontological categories that could offer a principled distinction between substantial change and phase change. I depart from both Strawson's and Lowe's treatment of sortals because I do not see natural kind sortals as *substance sortals*.

(1987) and again by Markman and Gwyn Wachtel (1988) while studying how children avoid redundant hypotheses about the meaning of category terms.[22] Natural kind terms are an interesting class of category terms.

Gelman and Markman (1987) documented how children aged 3 and 4 rely more on category membership than on perceptual appearances to infer properties that members of a kind share. In one experiment, 4-year-olds were shown a tropical fish, a dolphin, and a shark. The shark was given the label 'fish' like the tropical fish and children were asked whether the shark was breathing more like a dolphin or more like a tropical fish. They tended to liken it to the latter (p. 1533).

Developmental psychology offers further evidence for the view that by capturing deep causal-structural properties that promote inductive inferences, children acquire natural kind terms and other category terms within the constraint of mutual exclusivity. The 'essentialist bias'—as it has become known (see Gelman and Wellman 1991 and especially Gelman 2003)—captures children's tendency to learn names and generalize knowledge via inferences that rely on category membership and assumptions about the 'natures' of things hidden from appearances.

The dichotomy between hidden causal structure or essence and observable superficial properties has resurfaced in studies that have contrasted natural kind concepts, like *tiger*, with 'dual-character concepts', such as *musician* or *scientist*. Johsua Knobe et al. (2013) have defined dual-character concepts as those whose members have a dual set of criteria for membership including both a concrete feature (e.g. playing the electric guitar for *musician*) and the abstract value that the feature serves to realize (e.g. making rock music). The authors see an analogy between dual-character concepts and natural kind concepts. More recently, Newman and Knobe (2019) have reviewed varieties of psychological essentialism at play both in natural kinds (what they call *causal essentialism*, with reference to Gelman 2003) and in dual-character concepts involving value judgements.

For example, a recent paper by Kevin Tobia et al. (2019) applies 'dual-character concepts' to natural kind concepts like *water* in experiments where adults were exposed to the Twin Earth scenario and asked to judge whether the liquid from Twin Earth was or was not water. The tendency was to go for causal essentialism, whenever participants were told that the entity in question lacked the underlying causal properties of water but shared some of the superficial properties. However, results changed in a further experiment, where the participants

---

[22] 'A single object cannot both be a cow and a bird or a dog. Thus, in order for categories to be informative about objects, they will tend to be mutually exclusive, especially at the basic level of categorization. Of course there are exceptions: categories can overlap, as in "dog" and "pet", and they can be included as in "poodle" and "dog". The point here is that mutual exclusivity is a reasonable, though not infallible, assumption to make' (Markman and Wachtel 1988, p. 123).

were again exposed to the Twin Earth vignette but this time by placing it in three different contexts: scientific, legal, and neutral. The answers this time round varied depending on the context. The legal context (e.g. a town with legislation about approved liquid for pools) seemed to invite answers that prioritized superficial properties of the liquid. The scientific context elicited answers with a focus on deeper causal properties. And the neutral context attracted answers that were somewhere in between these two. The conclusion was unequivocal:

> People's ordinary judgments do not conform to the standard philosophical intuition that the deeper causal properties are the *sole* criterion of category membership. Instead, we find that people's actual judgments display a more complex dual-character pattern. Entities are categorized into natural kinds according to two different criteria. According to one, the Twin Earth liquid is water, but according to the other, it is not water. (Tobia et al. 2019, p. 205, emphasis in original)

What makes these psychological studies of interest for my story here is the role that *context* clearly plays in people's judgments about natural kind concepts. These experiments tell us something important, I think. Namely, that natural kinds as sortal concepts can hide conflicting intuitions, as when *different contexts* select *different phenomena* as the *most relevant ones* to answer question (I), i.e. the same *what*? For example, the sortal concept *water* allows us to count Earth's water (protium oxide by and large) and a hypothetical Twin Earth's water (deuterium oxide in LaPorte's 2004 story) as *two* isotopic varieties of the *same chemical kind*.

But when we ask (II) *in which respect* these samples are indeed *of the same sort*, similarities and differences become important. NKHF qua Spinozian sortals provide malleable *value-based sortal concepts* for fine-graining the description of *which* phenomenon is the *most relevant* to identifying the natural kind with in any given context of inquiry. For example, in the context of atomic physics, where values are placed on atoms and atomic composition, a phenomenon such as H_2O is the most relevant. But in the context of thermodynamics, where one cares about large ensembles of molecules, boiling points and freezing points are the most relevant. And in the context of hydraulic engineering, where fountains, ponds, and water jets become an expression of geopolitical power, ground-water motion and jet formation are the most relevant phenomena.

We have good (Spinozian) reasons for presuming that these are all phenomena *of the same sort* although in each context of inquiry distinct values are at play in the way one might be tempted to answer questions (I) and (II). The early 'essentialist bias' of children should not translate into adults' biases of reading off philosophical essences out of malleable Spinozian sortal concepts. Nor should malleable Spinozian sortal concepts for natural kinds collapse into Lockean nominal essences: for there is no presumption of essence in how phenomena P_1, P_2, P_3, ... are deemed *of the same sort* under a natural kind K. But then, in the absence of such a presumption, what holds open-ended groupings of phenomena together?

9.5. What holds open-ended groupings of phenomena together?

The sortalism I have described is congenial to perspectivism: whether a phenomenon P_1 is *of the same sort* as P_2 depends to a large degree on how one answers questions (I) 'The same *what?*' and (II) '*In which respect* can two types of phenomena be *of the same sort?*'

Thus, in asking whether something is a natural kind K, and in particular if *this* type of phenomenon P_1 is of the same sort as *that* type of phenomenon P_2 there and that other type of phenomenon P_3, and so forth, one is not asking whether P_1 bears the same microstructural kind relation to some archetype sample: that is, whether the tap water in the glass here bears a H_2O microstructural relation to *that* sample of water in some original baptism, as Putnam had it. Likewise, in asking whether Urey's first sample of deuterium oxide is indeed a sort of water, one is not asking whether $water_1$ and $water_2$ (or, if you like, the extension of the term 'water' before and after the discovery of isotopes) overlap completely or only partially so (*pace* Kuhn). Instead, one is asking for what holds together an open-ended grouping G of historically identified phenomena: G = $\{P_1, P_2, P_3, ... \}$.

Something might be a natural kind sortal K even in the absence of one or more phenomena in the grouping G =$\{P_1, P_2, P_3, ... \}$. Or something might be K even if other possible yet-to-be-found phenomena $\{P_i, P_{ii}, P_{iii} ... \}$ were added to the original grouping G to transform it into G* =$\{P_1, P_2, P_3, ..., P_i, P_{ii}, P_{iii} ... \}$. The types of phenomena are the planks in Neurath's Boat: they can be changed one by one. The Spinozian sortal concept K acts as the non-binding glue that holds the planks together. Let me clarify this analogy.

Natural kinds qua Spinozian sortal concepts perform two main functions:

(a) they create a standard, a unit of account (in analogy with coin) that brings with it the presumption of being usable by everyone;

(b) they act as a 'medium of exchange' across situated epistemic communities that might (or might not) be at some historical distance from one another (think again of *water* before and after the discovery of oxygen, or before and after the discovery of isotopes).

Spinozian sortals perform these two functions by circling (and re-circling) around open-ended groupings of phenomena that have been historically identified and traded as being *of the same sort* over time and across communities. The value of each grouping—their being regarded as types of phenomena *of the same sort*—depends on their 'purchasing power', so to speak: that is, how well they serve the needs of different situated communities over time. Communities often put in place governance systems that legislate the value of the grouping and tend to single out specific types of phenomena as being representative of the natural kind K (think again of water = H_2O or gold = atom with atomic number 79). Equally, communities might take the view that some groupings need be expanded and might legislate that oviparous platypuses count as mammals, or that *hachimoji* DNA counts as DNA.

Thus, do not think of NKHF as the spontaneous or serendipitous convergence of different communities on a particular grouping. Behind each NKHF there is a long history of trades and negotiations among communities and debates about what to include and what to exclude within broad nomological constraints fixed by the stable events and their underpinning lawlike dependencies. A faulty or structurally unstable *hachimoji* DNA with hydrophobic nucleotides slipping onto one another and distorting the double-helix structure would not be counted as DNA.

The necessity that NKHF typically bring along with them is expressed by lawlike dependencies. The contingency of NKHF is the effect of ongoing negotiation among communities as to what path to go down while walking in the inferential garden of forking paths at each point and juncture. In some cases, communities might negotiate ways around when the standard unit of account does not apply.

Consider again the example of *hachimoji* DNA. In asking whether this novel type of eight-nucleotide phenomenon P_i emerging from within synthetic biology is of the same sort as the familiar type of four-nucleotide DNA (P_1), one is asking whether P_i in synthetic biology reliably translates into RNA for protein formation as much as P_1 ordinarily does, thanks to important lawlike dependencies at work in both P_1 and P_i (in this case Etter's hydrogen bonding rules). And a positive answer means that a novel phenomenon such as *hachimoji* DNA can be included in the expanded grouping G* of phenomena that we historically identify with the natural kind *genetic code* (even if P_i lacks the essential property of having four nucleotides and P_1 and P_i are different in other relevant respects).

To return once more and unpack Spinoza's example of the coin, what makes the natural kind *genetic code* a sortal concept is precisely the ability to 'trade' types of phenomena such as P_1 and P_i. The exchange rate in this case is fixed by lawlike dependencies: for example, the hydrogen bonding rules that regulate how nucleotides can form stable double-helix structures.

Such lawlike dependencies and associated stable events are independent of us, our scientific perspectives and perspectival inferences. What does depend on us is what in Chapter 8 I called the fine-graining or coarse-graining in the description of the natural kind K, the emergence of new scientific perspectives (e.g. chemoinformatics or synthetic biology), and the novel modally robust phenomena that can be encountered via them. In other cases, epistemic communities can debate and legislate the purchasing power of the presumed natural kind in relation to a wide range of phenomena: how well does the sortal capture a broad range of phenomena? Let us return to the question of whether Urey's deuterium oxide is the same kind as the protium oxide that fills rivers and lakes. In asking whether Urey's new phenomenon P_j is indeed *of the same sort* as the more familiar phenomenon P_1, one is not seeking a privileged level of description that can furnish the answer. For no level of description is privileged under the NKHF account, where at best one can fine-grain or coarse-grain the multi-layered description of the kind by selecting one among many available descriptive levels, depending on which vantage point matters most.

In asking whether P_j is *of the same sort* as P_1, one is in fact asking how well the sortal concept *water* captures:

(1) the familiar (pre-scientific, if you like) type of phenomenon P_1 familiar from dripping from the tap or forming fountain-jets in Alhambra and Villa D'Este at the macroscopic level;

(2) the type of phenomenon P_i of chemical bonding at the intra- and inter-molecular level;

(3) Urey's new phenomenon P_j, whose boiling and freezing points are different from P_1;

and the list goes on (these are just three examples). But make no mistake. There is no conventional decision behind all this. It is not as though someone at some point decided that some phenomena rather than others were part of the necessary-and-sufficient conditions for defining the nominal essence of water. There are reasons—independent of any specific epistemic communities—as to why some phenomena go together while others do not. After all, for example, not all existing isotopic varieties of hydrogen and oxygen are candidates for isotopic varieties of water, depending on how stable the isotopes are and their ability to form stable bonding.

Or consider as another example dark matter as a kind in-the-making. One line of reasoning for taking dark matter as a candidate for a natural kind as of today resides in the fact that one can ask whether the phenomenon P_j of a higher-than-expected total gravitating mass density over the baryon density ($\Omega_m \gg \Omega_b$) is *of the same sort* as the more familiar phenomenon P_1 (galaxies' flat rotation curves). Here again, the sortal concept *dark matter* holds for (1) phenomena P_1 (at the level of galaxies); (2) phenomena P_i (angular power spectrum of the cosmic microwave background) at the large-scale structure; and (3) phenomena P_j at the Big Bang nucleosynthesis, among others mentioned in Section 7.6, Chapter 7. It is the 'purchasing power', so to speak, of the sortal concept for a range of phenomena that originally motivated the uptake of dark matter. Its ongoing fate as a kind in-the-making lies in its nomological resilience as explained in Chapter 7.

In yet other cases, a sortal might lose its 'purchasing power' and no longer be regarded as a trustworthy 'medium of exchange'. Consider, once again, caloric. This was a seemingly successful Mill-kind for a few decades. One might say that the sortal *caloric* was once used as a medium of exchange for trading phenomena such as P_1 (e.g. ice melting in the ice calorimeter) and P_i (e.g. the hot–cold reservoir in Carnot cycle), till eventually it was found to be a non-trustworthy medium of exchange. These phenomena (among others) cannot be grouped or 'exchanged' via it. Its 'exchange rate' was fixed by lawlike dependencies that turned out to be false. For there is no law of conservation of caloric at work either in the ice calorimeter (P_1) or in Carnot cycle (P_i).

It is time to take stock. My goal here was to show in broad terms how starting with the modest historical naturalism of Chapter 8, it is possible to develop an anti-essentialist and anti-foundationalist approach to natural kinds that is also distinct from traditional forms of conventionalism or nominalism. Things could have historically gone otherwise, but not in the drastically different ways the conventionalist story might suggest. That deuterium oxide is a kind of water, dark matter is an in-the-making kind of non-baryonic matter, and caloric is not a kind of hot matter—all these knowledge claims are grounded in stable events and their lawlike dependencies. The relevant phenomena enter into sort-relative sameness relations that explain what holds each open-ended grouping of phenomena together. This holding together is not contingent on *who* decided *what, where, when,* and *why*.

The way in which natural kinds get refined by constantly circling and re-circling the open-ended groupings of phenomena is ultimately a matter for empirical inquiry. This is down to the work of hydraulic engineers working on viscosity and pipes; synthetic biologists creating *hachimoji* DNA; ethnobotanists working on biodiversity in the Yucatán; colleagues in chemoinformatics working on AI methods for dialling new molecules; cosmologists and particle physicists devising searches for dark matter; and so forth.

Yet the way candidate phenomena get stock-piled in groupings is not chancy, random, or whimsical. The grouping of phenomena, loosely held together like the planks in Neurath's Boat, cannot be changed *ad libitum*. There are specific constraints. Some are fixed by what I referred to as lawlike dependencies; others by truth-conducive conditionals-supporting inferences over time.

Whether they are empirical rules such as Etter's hydrogen bonding rules (see Etter 1990), phenomenological laws such as Darcy's law for groundwater motion, or fundamental equations such as the Navier–Stokes equations in fluid dynamics, lawlike dependencies undergird epistemic communities' ability to establish sort-relative sameness relations across phenomena in different domains.

In Chapter 7, I discussed nomological resilience and pointed out that without laws, Mill-kinds might turn up empty kinds, and with laws, even kinds in-the-making might enjoy the status of Peirce-kinds. The further discussion in this chapter shows that we may think of the nomological resilience of NKHF as resilience under a number of lawlike dependencies inherent in each phenomenon in its own domain and providing an indication of how any grouping of phenomena might or might not hold together in the metaphysics-lite way I have just described. This is as much necessity as can be packed into my Neurathian account to explain why some phenomena (but not others) belong together.

At the same time, there is a sense in which the contingentist is right in saying that things could have gone otherwise. Other perhaps undreamt-of past scientific perspectives could have led to different data to phenomena inferences from the ones we are familiar with. Or maybe future scientific perspectives will bring new inferential blueprints and their practices will infer new phenomena that are currently beyond epistemic reach.

Spinozian sortals for open-ended groupings of phenomena allow us to understand why these historical groupings are candidates for natural kinds. There is no need to privilege one particular phenomenon at one particular domain at the expense of the others. Nor is there any presumption that what defines natural kindhood among phenomena across different domains is some special slicing of microstructural essential properties that cuts across domains. The value of a natural kind is the 'exchange rate' that governs the relations among the phenomena. And the exchange rate is in turn the sort-relative sameness relation that allows epistemic communities to consider and adjust as they go along the list of phenomena within broad nomological constraints.

The peculiar variety of contingentism I see at play in NKHF becomes evident in how some epistemic communities have over time enforced the use of particular natural kinds qua Spinozian sortals to assert some research programmes. We will see one such example in the history of the electric charge and Planck's quantum programme in Chapter 10.

But there are also more concerning manifestations of contingentism. They affect how communities in some cases decide to police the use of a particular NKHF as a transactional medium, to assert scientific power, and to configure intellectual ownership —a story I tell in Chapter 11 when I talk about the rosy periwinkle of Madagascar, the English vernacular name for the plant known in botany as *Catharanthus roseus*. Natural kinds qua Spinozian sortals are a powerful medium of exchange: they are the embodiment of the social and cooperative nature of scientific inquiry no less than of the epistemic risks associated with it.

10

Evolving natural kinds

Arriving at each new city, the traveler finds again a past of his that he did not know he had: the foreignness of what you no longer are or no longer possess lies in wait for you in foreign, unpossessed places. . . .

Futures not achieved are only branches of the past: dead branches.

'Journeys to relive your past?' was the Khan's question at this point, a question which could also have been formulated: 'Journeys to recover your future?'

And Marco's answer was: 'Elsewhere is a negative mirror. The traveler recognizes the little that is his, discovering the much he has not had and will never have'.

Italo Calvino (1972/1997) *Invisible Cities*, pp. 24–25[1]

10.1. *Die kleine h*

20 May 2019 marked World Metrology Day, celebrated by news headlines around the world: 'The International System of Units—Fundamentally better'.[2] It was the day that saw *die kleine h* replace *le grand Kilo*. For the first time in 130 years, one of the seven base units in the International System of Units (SI)—the kilogram—was forced into retirement and replaced with what is regarded as a much better and more fundamental unit: Planck's constant h.

It was a unanimous decision of representatives of more than sixty nations gathered in Versailles at the General Conference on Weights and Measures. The French *grand Kilo*—or, better, its 1889 built prototype made of platinum and iridium and kept in a vault at the Bureau of Weights and Measures in Paris—had to give way to a more reliable unit introduced by Max Planck back in 1900.

---

[1] Copyright © Giulio Einaudi editore, s.p.a. 1972. English translation copyright © Harcourt Brace Jovanovich, Inc. 1974. Reprinted by permission of The Random House Group Limited. For the US and Canada territories, *Invisible Cities* by Italo Calvino, translated by William Weaver. Copyright © 1972 by Giulio Einaudi editore, s.p.a. Torino, English translation © 1983, 1984 by HarperCollins Publishers LLC. Reprinted by permission of Mariner Books, an imprint of HarperCollins Publishers LLC. All rights reserved.

[2] See http://www.worldmetrologyday.org.

Perspectival Realism. Michela Massimi, Oxford University Press. © Oxford University Press 2022.
DOI: 10.1093/oso/9780197555620.003.0013

For a long time, metrologists had worried that the prototype in Paris was dissimilar to copies distributed around the world, with imperceptible and yet increasingly important changes occurring to its mass over time. Fundamental constants for their part do not need prototypes: they are unchanging over time, and, most importantly, they are accessible always and everywhere. What better way to secure the reliability of metrological practices than to have a unit of measure for mass based on one such fundamental constant, namely Planck's h?

Metrology is not the only domain where the quest for stable, unchanging, and universally accessible fundamental units matters. Such a quest permeates science more broadly. I argue in this chapter that this very same quest is often at play in how we talk and think of natural kinds in relation to historically identified and open-ended groupings of phenomena. It indeed explains the perceived unanimity and projectibility surrounding the natural kinds we know and love.

Consider the natural kind electron. Physics textbooks tell us that the electron is defined by a number of relevant features: its negative electric charge is the physical constant e; it has a spin of ½ in units of $h/2\pi$, and a rest mass of 0.511 MeV. Moreover, the dependencies among these relevant features—between charge and mass, or between half-integral spin and Fermi–Dirac statistics, among others—are lawlike. Some of these lawlike dependencies are causal in nature, as when one observes cathode rays bending by increasing the strength of the electric or magnetic field. Others are non-causal in nature, as in the relation between half-integral spin and Fermi–Dirac statistics, as already discussed in Chapter 5.

These lawlike dependencies are at work in a number of modally robust phenomena that over time epistemic communities have learned to identify through reliable data-to-phenomena inferences: from the phenomenon of the bending of cathode rays to spectroscopic phenomena about alkali,[3] from electrolysis to black-body radiation, just to give a few examples.

My realism about modally robust phenomena goes hand-in-hand with the Neurathian strategy on 'Natural Kinds with a Human Face' (NKHF). It does not eliminate natural kinds. But it does not take them either as the metaphysical seat of essences or as conventional labels. They are instead Spinozian sortal concepts that stand for open-ended groupings of phenomena. And Spinozian sortals are nothing but proxies for the 'exchange rate' among phenomena.

In this chapter, I complete my account of NKHF. That calls for a return to the inferences upon which kinds-in-the-making eventually become evolving kinds. Appealing to a sort-relative sameness relation sheds light on the mechanism underneath NKHF. But something is still missing: how is it that one can evaluate

---

[3] For the role of alkali doublets in the discovery of the electron spin, see Massimi (2005, Ch. 2).

as *veridical* claims of knowledge concerning kinds-in-the-making that *evolve* through historical journeys across scientific perspectives? Recall my definition:

(NKHF)

Natural kinds are (i) historically identified and open-ended groupings of modally robust phenomena, (ii) each displaying lawlike dependencies among relevant features, (iii) that enable truth-conducive conditionals-supporting inferences over time.

This chapter clarifies the last condition here. How can we veridically maintain that there are indeed electrons if our perspectival representations of them have changed radically? How can science ever be expected to offer a 'window on reality' if, at best, scientific representations reflect the agent's situated point of view?

I complete my answer here by placing centre-stage the perspectival$_2$ nature of our scientific representations and by showing how we do get a window on natural kinds, not in spite of but by virtue of our perspectival$_1$ data-to-phenomena inferences. After all, recall from Chapter 2 how perspectival representations in science, despite being always *from a specific vantage point* (perspectival$_1$) can nonetheless give us a 'window on reality' (perspectival$_2$).

I illustrate this point by recounting one particular episode out of the history of the electron. The electron is probably the best understood particle in contemporary physics. Much as we are all realists about the electron today, the story of the electron is ongoing. The electron is a paradigmatic example of what I'd like to call *evolving kinds*. And so much could be written about its puzzling quantum mechanical aspects that they deserve a book of their own.[4]

But I have a more modest philosophical goal. If you think that quantum mechanics is baffling, the earlier history of the electron is even more so. Physics textbooks teach that the electron is an elementary particle defined by a series of kind-constitutive properties: charge, mass, and spin. But a quick glance at the history of our coming to know about the negative electric charge *e* reveals the deeply perspectival nature of our scientific representations. Our veridically maintaining that there is an electron with charge *e* is—as my definition (iii) has it—the outcome of *truth-conducive conditionals-supporting inferences* enabled by a historically identified open-ended grouping of modally robust phenomena.

In the next sections, I recount how Planck's constant *h* played a role in our coming to know *that* there is an electric charge and about *what* it is.[5] I illustrate

---

[4] Luckily, such books already exist: see Baggott (2000) and Ball (2018), among others.
[5] The material presented in this chapter is reproduced in expanded and adapted form from Massimi (2019c) with permission from Springer.

how the realist commitment to the electric charge crystallized around a number of perspectival data-to-phenomena inferences between 1897 and 1906. These inferences involved three main scientific perspectives broadly construed: the Faraday–Maxwell field-theoretical perspective, in which J.J. Thomson was working (Section 10.2); the electrochemical perspective, to which Grotthuss and Helmholtz contributed (Section 10.3); and the emerging quantum perspective championed by Max Planck (Section 10.4).

Evidence for the electric charge appeared independently in each of these perspectives, no matter how diverse the data and data-to-phenomena inferences were in each case. The unexpected unanimity of natural kinds is not our convergence on a metaphysics of essential properties. It is a long and painstaking process of negotiation. Natural kinds as evolving kinds are the products of our perspectival scientific history and our collective willingness to engage in 'giving and asking for reasons' (to echo Brandom 1998) in a conditionals-supporting space of inferences, to which I return in Section 10.7.

10.2. Hebridean kelp, glass-blowing, and electrical research: J.J. Thomson's perspective around 1897–1906

In 1906, J.J. Thomson was awarded the Nobel Prize for his 'theoretical and experimental investigations on the conduction of electricity by gases'.[6] The award did not mention the electron as such because Thomson's experiments with cathode rays in 1897–1898 did not lead him to the conclusion that 'the electron exists'. The Presentation Speech by J.P. Klason, President of the Swedish Academy, is telling when read in conjunction with Thomson's own acceptance speech. Klason mentioned Thomson's work with H.A. Wilson (building on C.T.R. Wilson's method) on the discharge of electricity through gases, and presented Thomson as following in the footsteps of Maxwell and Faraday, especially Faraday's 1834 discovery of the law of electrolysis, which had shown

that every atom carries an electric charge as large as that of the atom of hydrogen gas, or else a simple multiple of it corresponding to the chemical valency of the atom. It was, then, natural to speak, with the immortal Helmholtz, of an elementary charge or, as it is also called, an atom of electricity, as the quantity of electricity inherent in an atom of hydrogen gas in its chemical combinations. Faraday's law may be expressed thus, that a gram of hydrogen, or a quantity equivalent thereto of some other chemical element, carries an electric charge of $28,950 \times 10^{10}$ electrostatic units. Now if we only knew how many hydrogen

---

[6] The Nobel Prize in Physics 1906. https://www.nobelprize.org/prizes/physics/1906/summary/.

atoms there are in a gram, we could calculate how large a charge there is in every hydrogen atom.[7]

Having presented Thomson as the scientist who 'by devious methods' was able to answer this puzzle, Klason added (almost as a caveat) that 'even if Thomson has not actually beheld the atoms, he has nevertheless achieved work commensurable therewith, by having directly observed the quantity of electricity carried by each atom. . . . These small particles are called electrons and have been made the object of very thorough-going researches on the part of a large number of investigators, foremost of whom are Lenard, last year's Nobel Prize winner in Physics, and J.J. Thomson.'[8] The qualification 'even if Thomson has not actually beheld the atoms' is important. For the fact that Thomson did not refer to his particles as 'electrons' was not just a terminological matter: he did not quite see them as genuine particles having inertial mass,[9] and believed that there were positive and negative electric charges whose field-theoretical behaviour was captured by what elsewhere he had modelled as a 'Faraday tube'.

But today Thomson has gone down in history as the discoverer of the electron. And for good reasons too, thanks to his precise experiments on cathode rays. Exhausted glass tubes had become a tool for electrical research since the time of Faraday in the 1830s.[10] Later on, William Crookes developed an active interest in producing high-quality exhausted tubes that were pivotal for his research on cathode rays, radiometry (the latter was deeply entangled with Crookes's spiritualistic beliefs), and, last but not least, the commercial manufacture of light bulbs.[11] Crookes went as far as training his research assistant Charles Gimingham in glass-blowing in the 1870s (see Gay 1996, p. 329) in addition to the expertise of two professional women glass-blowers in his lab.

And Crookes was not the only one to have in-house assistants trained in glass-blowing. J.J. Thomson himself at the Cavendish Lab had an assistant, Ebenezer Everett, who specialized in producing bespoke glass tubes for Thomson's experiments (see Crowther 1974). As Jaume Navarro (2012, p. 51) reports: 'Ebenezer Everett . . . became the Cavendish glass blower in 1887, after training in the Chemistry Dept. . . . The task of blowing glass was crucial for the kind of experiments that Thomson was performing on the discharge of gases in

---

[7] Presentation Speech by Professor J.P. Klason, President of the Royal Swedish Academy of Sciences, on 10 December 1906. https://www.nobelprize.org/prizes/physics/1906/ceremony-speech/.

[8] Ibid.

[9] In the rest of the Presentation Speech, Klason remarks that 'From experiments carried out by Kaufmann regarding the velocity of β-rays from radium, Thomson concluded that the negative electrons do not possess any real, but only an apparent, mass due to their electric charge' (ibid.).

[10] See Faraday's Bakerian Lecture (1830).

[11] Crookes took out a patent for light bulbs in 1881 as Director of the Electric Light and Power Company (see Gay 1996, p. 319 fn. 41).

tubes, and Everett proved to be very successful at this job, as Thomson always acknowledged'.

An industry of glass-blowing developed in the nineteenth century around optical and electrical researches. British glass manufacture at the time resorted to lead oxide, which hampered electrical conductivity. Hence British scientists preferred the use of what were known at the time as German or French glasses, which instead of lead used soda (see Gray and Dobbie 1898, p. 42). Increasingly, glass manufacturers such as Powell and Sons of Whitefriars in London were requested by scientists like Crookes to produce lead-free glass to conduct their experiments (see Powell 1919). Such requests intensified during World War I when the lines of supply of German-made glass were cut off. To reduce the melting temperature of the glass, increasingly the British glassware industry relied on alkali flux obtained from the ashes of seaweeds (kelp).[12]

The practice of using kelp for glass-making had been part of the economy of local communities on the West Coast of Scotland and also in Ireland since the eighteenth century (see McEarlean 2007). Samuel Johnson reported such a practice already back in 1775 in 'A Journey to the Western Islands of Scotland', where the kelp trade is said to have sparked litigation on the Isle of Skye between the Macdonald and Macleod clans for a ledge of rocks rich in seaweed.[13] While the kelp trade proved lucrative for local clans (see Gray 1951), the local population of the Western Isles did not enjoy similar fortunes.[14] It is against this socioeconomic backdrop of kelping that glass manufactures for scientific research took place.

The manufacture of kelp-fluxed glass continued throughout the eighteenth and into the nineteenth century and played an important role in the development of chemical research by Scottish-based Joseph Black and Lyon Playfair, with the glassware laboratory in Leith (Edinburgh) producing the glass used in University of Edinburgh lab (see Kennedy et al. 2018). Some of the glass tubes in the electrostatic induction section of the Playfair Collection at the University of Edinburgh, for example, reveal a high-calcium percentage with a 'presence of strontium indicating that kelp was used as the alkali flux' (p. 260).

---

[12] I thank Craig Kennedy for helpful comments on this.

[13] 'Their rocks abound with kelp, a sea-plant, of which the ashes are melted into glass. They burn kelp in great quantities, and then send it away in ships, which come regularly to purchase them. This new source of riches has raised the rent of many maritime farms; but the tenants pay, like all other tenants, the additional rent with great unwillingness; because they consider the profits of the kelp as the mere product of personal labour, to which the landlord contributes nothing. However, as any man may be said to give, what he gives the power of gaining, he has certainly as much right to profit from the price of kelp as of any thing else found or raised upon his ground' (Johnson and Boswell 1775/2020, p. 66).

[14] 'The kelper, gripped by his dependence on the landlord both for land and for employment, was paid at a piece rate which failed to rise with the price of the product.... His labour, then, brought him in not more than £10, sometimes considerably less' (ibid., p. 203).

And it was not just chemistry that benefited from lead-free glassware but also and especially the blossoming field of electrical research. J.J. Thomson's research on electrical conductivity in gases is deeply rooted in this long-standing scientific practice of producing high-quality (ideally lead-free) exhausted glass tubes, following in the footsteps of Faraday and Crookes but also Black and Playfair. While glassware was a key component of the experimental and technological resources available to Thomson's scientific perspective, its modelling assumptions too deserve a closer look. In brief, using cathode ray glass tubes and relying on classical laws of electrostatics and magnetism, Thomson could measure the displacement of the cathode rays in the presence of an electric or magnetic field. From these experiments, he was able to establish the charge-to-mass ratio (e/m) at work in the modally robust phenomenon of cathode rays bending (or m/e, as Thomson still referred to it in 1897).

The charge-to-mass value was found to be *stable* under a range of background conditions: it was independent of the velocity, the kind of metal used for the cathode and anode, and the gas used in the tube.[15] Most interestingly, under an additional range of interventions, the same lawlike dependency between charge and mass was observed to hold *stably* across a number of other phenomena in different domains, including electrolysis in chemistry and X-ray ionization in nuclear physics.[16]

Since the beginning of his career at the Cavendish Laboratory in Cambridge in the 1880s (and still visibly in the 1893 book *Notes on Recent Researchers in Electricity and Magnetism*), Thomson's research on electric discharge in gases took place within a well-defined scientific perspective, still popular in Cambridge at the turn of the twentieth century, that I am going to call—for lack of a better term—the *Faraday–Maxwell perspective*, with the caveat that such a perspective is not of course confined to or centred on the works of Faraday and Maxwell, and stretches well beyond those.[17] This was primarily concerned with electromagnetism: the interconversion of electricity and magnetism observed by Ørsted in Denmark and Faraday in England in the 1820s, and to explain which Maxwell in the early 1860s produced mechanical models of the ether. This perspective

---

[15] Thomson ran a series of experiments using air, hydrogen, and carbonic acid as different gases, and as cathode he used different materials from aluminium to platinum from which he concluded that 'the value of m/e is independent of the nature of the gas, and that its value 10^{-7} is very small compared with the value 10^{-4}, which is the smallest value of this quantity previously known, and which is the value for the hydrogen ion in electrolysis' (Thomson 1897, p. 310).

[16] The discovery of X-rays (or Röntgen rays) revealed interesting new phenomena about gas conductivity: gases exposed to X-rays conduct electricity at low potential. The phenomenon, as I mention below, could be modelled using Grotthuss's chain model from electrolysis with so-called Faraday tubes connecting positive and negative charges in gas molecules.

[17] See Falconer (1987, 2001) for an excellent historical account of the historical context in which Thomson operated.

centred on the field-theoretical analysis of the electromagnetic field (what Faraday had originally called 'magnetic lines of force').

But—as always with scientific perspectives—the perspective was not limited to (or exclusively centred on) a particular theoretical body of knowledge claims about electromagnetic phenomena. It equally involved the experimental and technological resources to advance them, including the aforementioned Hebridean kelping industry and glassware manufacture behind cathode ray tubes. But it also involved what I called second-order epistemic-methodological principles that justify the reliability of the knowledge claims so advanced. In this example, specific modelling assumptions concerning the so-called Faraday tubes, which physically conceived of 'tubes of electric force, or rather of electrostatic induction, . . . stretching from positive to negative electricity' (Thomson 1893, p. 2).[18]

Faraday tubes were a way of modelling what we would now call 'electric flux' as a measure of the electric field strength, with the two charges (positive and negative) at the two ends of the tube. In the nineteenth century, this was a semi-classical way of conceiving the electric field as a collection of ethereal vortex tubes, carrying electrostatic induction. Thomson toyed with the model of Faraday tubes in 1891 as they allowed him to reconcile claims about the discrete nature of electricity emerging from electrochemical experiments with Maxwell's electromagnetic field (whereby electricity was analysed primarily as electric displacement in a continuous field). Atoms of opposite electric charge connected by a Faraday tube could serve to represent molecules of electrolytes—polarized with the passage of electric current as in Grotthuss's chain model of electrolysis.

Yet the *Faraday–Maxwell perspective* gave a perspectival representation of the electric charge in stark contrast with the one emerging from the electrochemical perspective (as we shall see in the next section). Thomson (1891) had made it clear from the outset that Faraday tubes were not just an expedient to visualize mathematical equations, but they had 'real physical existence' and that the contraction and elongation of such tubes could explain the passage of electricity through metals, liquids, and gases.[19]

---

[18] See Smith (2001) for an excellent historical account of Thomson's experiments and intellectual background in 1897–1898.

[19] 'If we regard these tubes as having a real physical existence, we may . . . explain the various electrical process . . . as arising from the contraction or elongation of such tubes and their motion through the electric field. . . . As the principal reason for expressing the effects in terms of the tubes of electrostatic induction is the close connexion between electrical and chemical properties. . . . We assume, then, that the electric field is full of tubes of electrostatic induction, that these are all of the same strength, and that this strength is such that when a tube falls on a conductor it corresponds to a negative charge on the conductor equal in amount to the charge which in electrolysis we find associated with an atom of a univalent element . . . the tubes resemble lines of vorticity in hydrodynamics' Thomson (1891, pp. 149–150).

Twelve years later, just three years before the Nobel Prize, Thomson returned to the topic in the Silliman Lectures in May 1903 at Yale University (Thomson 1904). These lectures provide an instructive example of his long-standing ontological commitment to electric charge at the dawn of the new century (just when Planck was ushering in quantum physics). Four aspects of Thomson's methodological commitment to Faraday tubes deserve comment:

a. *Thomson's treatment of the electric charge is still deeply rooted in the nineteenth-century Faraday–Maxwell tradition of lines of force and mechanical ether models for electromagnetic induction in analogy with hydrodynamics.* Thomson refers to the Faraday tube as a 'tube of force' or a tubular surface marking the boundaries of lines of force so that 'if we follow the lines back to the positively electrified surface from which they start and forward on to the negatively electrified surface on which they end, we can prove that the positive charge enclosed by the tube at its origin is equal to the negative charge enclosed by it at its end' (p. 14). In this way, he explained the old ideas of positive and negative electricity with 'each unit of positive electricity in the field . . . as the origin and each unit of negative electricity as the termination of a Faraday tube' (p. 15).

b. *The boundary between Thomson's corpuscles and the Faraday tube is a lot more subtle than it might seem.* The mass of the Faraday tube is nothing but the mass of the bound ether, or, as Thomson puts it, 'the mass of ether imprisoned by a Faraday tube' (p. 39). The term 'corpuscle' is introduced to refer to 'those small negatively electrified particles whose properties we have been discussing. On this view of the constitution of matter, part of the mass of any body would be the mass of the ether dragged along by the Faraday tubes stretching across the atom between the positively and negatively electrified constituents' (p. 50). Thus, Thomson's corpuscle is effectively nothing but a 'concentration of the lines of force on the small negative bodies' so that 'practically the whole of the bound ether is localised around these bodies, the amount depending only on their size and charge' (p. 52).

c. *The electric charge is presented as a natural unit and its atomicity is explained in terms of Faraday tubes.*[20] However, by contrast with the electrochemical

---

[20] 'Hithertho we have been dealing chiefly with the properties of the lines of force, with their tension, the mass of the ether they carry along with them, and with the propagation of the electric disturbances along them; in this chapter we shall discuss the nature of the charges of electricity which forms the beginning and ends of these lines. We shall show that there are strong reasons for supposing that these charges have what may be called an atomic structure; each charge being built up of a number of finite individual charges, all equal to each other. . . . [I]f this view of the structure of electricity is correct, each extremity of the Faraday tube will be the place from which a constant fixed number of tubes start or at which they arrive' (ibid., p. 71).

perspective, the reasoning leading to Thomson's conclusion that the electric charge is somehow atomistic does not rely exclusively on electrolysis but also on Wilson's experiments on the conductivity of the vapour obtained from metallic salts (the so-called electron vapour theory).[21]

d. *An explanation of Röntgen rays is given in classical terms of corpuscles* and via Faraday tubes and with no reference to the quantum hypothesis and electrons losing part of their quantized energy as in Planck's contemporary treatment of the topic (Section 10.4).

Thomson started with data from cathode ray tubes, which were made possible by a century-long history of kelping behind the glassware industry for scientific instruments. He built on Faraday's lines of force and Maxwell's honeycomb model of the ether, and resorted to a field-theoretical model that made it possible to infer what *might happen, under the supposition that the Faraday tubes were stretched and elongated.* For this conditionals-supporting inference to be truth-conducive, other perspectival data-to-phenomena inferences had to be brought to bear on it across a network of inferences that eventually guided epistemic communities to the correct identification of the electric charge. This was indeed what happened.

10.3. Grotthuss's and Helmholtz's electrochemical perspective ca. 1805–1881

The Faraday–Maxwell field-theoretical perspective on electromagnetic phenomena such as cathode rays bending was at some distance from what I call the *electrochemical perspective*. In 1874, G. Johnstone Stoney used Faraday's law of electrolysis to conclude that in the phenomenon of electrolysis, 'For each chemical bond which is ruptured within an electrolyte a certain quantity of electricity traverses the electrolyte which is the same in all cases' (Stoney 1874/1894, p. 419). Stoney introduced the term 'electron' to describe this minimal quantity of electricity.[22] In 1881, Hermann von Helmholtz in Germany championed the

---

[21] 'Wilson found that the saturation current through the salt vapour was just equal to the current which if it passed through an aqueous solution of the salt would electrolyse in one second the same amount of salt as was fed per second in the hot air. . . . Thus whether we study the conduction of electricity through liquids or through gases, we are led to the conception of a natural unit or atom of electricity' (ibid., p. 83).

[22] In a 1874 talk presented at the British Association meeting in Belfast and entitled 'On the Physical Units of Nature', Stoney presented this minimal quantity of electricity as 'one of the three physical units, the absolute amounts of which are furnished to us by Nature, and which may be the basis of a complete body of systematic units in which there shall be nothing arbitrary' (Stoney 1874/1894, p. 418). But Stoney believed that these electrons within each molecule or chemical atom were 'waved about in a luminiferous ether' and that in this motion through the ether the spectrum of each gas originated.

hypothesis that elementary substances were composed of what he called 'atoms of electricity'[23] (or 'ions', as Lorentz later called them). He motivated and justified this view in light of chemical studies of electrolysis going back to the German chemist Theodor von Grotthuss, who in 1805–1806 had published his influential chain model for water electrolysis.

The atoms of electricity were regarded here as the minimum quantity carried by electrolytes (or by the hydrogen atoms) when molecules decomposed with the passage of electricity. Helmholtz's argument originated from Faraday's first and second law of electrolysis, which had established that the electric charge of hydrogen atoms (or what we now know to be their valence electrons) was a fundamental unit not further divisible. Helmholtz's reasoning for taking the electric charge as a physical unit (and in Britain, Stoney's analogous reasoning) was entirely chemical, rooted in the well-known tradition of eighteenth- and nineteenth-century electrolytical experiments and a long-standing debate on the animal vs metal nature of electricity going back to Galvani's frogs and Volta's electric pile (see Pauliukaite et al. 2017). What made e a minimal unit under this perspective was the fact that it was the charge corresponding to chemical valence 1. Thus, a different data-to-phenomena inference was at play in this scientific perspective, one that fed into the indicative conditional

(E.1) If hydrogen and oxygen form a Grotthuss's chain, hydrogen is released at the negative electrode.

By physically conceiving of a minimal (positive and negative) electrical unit for the ions of electrolytes standing in a chain, Grotthuss's model could be used to explore what might happen in the well-observed phenomenon whereby water molecules decompose with the passage of electricity with oxygen at one end and hydrogen at the opposite one.

Bringing this kind of information to bear on J.J. Thomson's perspective proved key in this story. As Thomson himself recounted in his Nobel Prize speech, it became apparent that there was a disparity between the ratio E/M of the hydrogen atom (known from the phenomenon of water electrolysis) and the ratio e/m emerging from the phenomenon of cathode rays bending

---

[23] 'The most startling result of Faraday's law is perhaps this. If we accept the hypothesis that the elementary substances are composed of atoms, we cannot avoid concluding that electricity also, positive as well as negative, is divided into definite elementary portions which behave like atoms of electricity. As long as it moves about on the electrolytic liquid each ion remains united with its electric equivalent or equivalents. At the surface of the electrodes decomposition can take place if there is sufficient electromotive force, and then the ions give off their electric charges and become electrically neutral' (Helmholtz quoted in Stoney 1874/1894, p. 419).

within the Faraday–Maxwell perspective. A numerical discrepancy appeared of the order $e/m = 1,700 \ E/M$. This led Thomson to the following reasoning, which was pivotal for the identification of the 'corpuscle' as the first sub-atomic particle:

We have already stated that the value of e found by the preceding method [i.e. Wilson's][24] agrees well with the value E which has long been approximately known. Townsend has used a method in which the value e/E is directly measured, and has shown in this way also that e equal to E. Hence since $e/m = 1,700E/M$, we have $M = 1,700 \ m$, i.e. the mass of a corpuscle is only about 1/1,700 part of the mass of the hydrogen atom. (Thomson 1906, p. 153)

But the inferences that led to the electric charge were not confined to phenomena about water electrolysis and the bending of cathode rays in an external field (in addition to other phenomena that I do not have the space to cover here). On the other side of the Channel, German physicists were laying the foundations of a new scientific perspective, which was bound to have a lasting impact in the story so far.

10.4. Max Planck's quantum perspective: the electric charge as a 'natural unit'

In the Preface to the Second Edition of *The Theory of Heat Radiation*, Planck announced that his measured value for e lay in between the values of Perrin and Millikan. More importantly, he presented the idea of an 'elementary quanta of electricity' as the most important new evidence in support of his hypothesis of the quantum of action:

Recent advances in physical research have, on the whole, been favorable to the special theory outlined in this book, in particular to the hypothesis of an elementary quantity of action. . . . Probably the most direct support for the

---

[24] The equivalence between e and E was established thanks both to the work of C.T.R. Wilson, which in turn made possible H.A. Wilson's measurement of charged droplets, and to John S. Townsend's measurement of the charges of gas ions. As the historian of science George E. Smith points out, Townsend's experiment was 'predicated on Maxwell's diffusion theory. . . . Townsend inferred a magnitude for Ne, where N is the number of molecules per cubic centimetre under standard conditions. The uniformity of this magnitude for ions of different gases and its close correspondence to the value NE from electrolysis (where E is the charge per hydrogen atom), then allowed Townsend to conclude, *independently of any specific value of e or N*, that the charge per ion, when generated by X-rays, is the same as the charge on the hydrogen atom in electrolysis' (Smith 2001, pp. 74–75, emphasis in original).

fundamental idea of the hypothesis of quanta is supplied by the values of the elementary quanta of matter and electricity derived from it. When, twelve years ago, I made my first calculation of the value of the elementary electric charge and found it to be 4.69×10^{-10} electrostatic units, the value of this quantity deduced by J.J. Thomson from his ingenious experiments on the condensation of water vapour on gas ions, namely 6.5×10^{-10} was quite generally regarded as the most reliable value. This value exceeds the one given by me by 38 per cent. Meanwhile the experimental methods, improved in an admirable way by the labors of E. Rutherford, E. Regener, J. Perrin, E.A. Millikan, The Svedberg and others, have without exception decided in favor of the value deduced from the theory of radiation which lies between the values of Perrin and Millikan.

To the two mutually independent confirmations mentioned, there has been added, as a further strong support of the hypothesis of quanta, the heat theorem which has been in the meantime announced by W. Nernst, and which seems to point unmistakably to the fact that, not only the processes of radiation, but also the molecular processes take place in accordance with certain elementary quanta of a definite magnitude. (Planck 1906/1913, p. vii)

With these words, Planck established a tradition with far-reaching philosophical consequences. The idea of an elementary electric charge corroborated his quantum hypothesis and showed how it could be extended beyond the radiation of the black-body, into the nature of matter and electricity. And there was no better evidence for this than to identify e as a physical constant (along the lines of Planck's own constant h) and present the experiments of Thomson, Rutherford, Perrin, and Millikan as all dealing with the same task: to measure the value for the elementary charge. Planck's desire to find a connection between h and other physical constants was revealed by Max Klein in a letter to Ehrenfest of 6 July 1905, at a time where the existence of an elementary charge quantum e was only a conjecture. As reported by Klein, Planck was keen to find a 'bridge' between his quantum hypothesis h and the experimentally found values for e (see Holton 1973, p. 176 fn. 19).

In Chapter 4 of *The Theory of Heat Radiation*, Planck returned to the hypothesis of quanta and the temperature of black-body radiation from a system of stationary oscillators and embarked on what in my view is an illuminating journey into the nature of physical constants. After introducing Planck's constant, he went on to a discussion of the kinetic theory of gases and how to estimate the number of hydrogen molecules contained in 1 cm³ of an ideal gas at 0 Celsius and 1 atmosphere. He concluded that the 'elementary quantity of electricity or the free charge of a monovalent ion or electron' e in electrostatic unit is 4.67×10^{-10}, adding that 'the degree of approximation to which these numbers represent the corresponding physical constants depends only on the accuracy of the measurements of the two radiation constants' (Planck 1906/1913, p. 173).[25] In a single stroke, Planck effectively established:

    i. the theoretical equivalence between the 'free charge of a monovalent ion or electron' with the 'elementary quantity of electricity' (it is worth stressing Planck's ambiguous use of the double terminology of Lorentz's 'ions' as interchangeable with Stoney's 'electron');[26]

    ii. the identification of the 'elementary quantity of electricity' e with a 'physical constant' among others in the context of black-body radiation;

    iii. and the accuracy in the values of the physical constant e depending on the refined measurements of radiation constants.

The ambiguity in the terminology ion/electrons is, in my view, symptomatic of Planck's disengagement from the ontological debate about the nature of the minimal unit of electric charge (and of atoms more generally).[27] For Planck, electric charge helped establish the validity and universal applicability of the quantum hypothesis. And what up to that point had been just a hypothesis—the 'ion hypothesis', as the German physicist Paul Drude still called it—had become in Planck's hands a 'natural unit'.

Drude's electron gas theory was an important influence for Planck (see Kaiser 2001). Drude himself was working on metal optics, and how to explain phenomena such as dispersion of light and optical reflection from metal surfaces within Maxwell's electromagnetic theory. Building on van't Hoff's kinetic theory of osmotic pressure, Drude patterned electrical conductivity in metals on the model of the kinetic theory of gases, and used Boltzmann's equipartition theorem with the universal constant a to establish that 'If a metal is now immersed in an electrolyte in the case of 'temperature-equilibrium' [that is, thermodynamic equilibrium] the free electrons ['kernels'] in the metal would have the same kinetic energy as the ions in the electrolyte' (Drude 1900, quote from Kaiser 2001, p. 258).

---

[25] Among them, the constant that features in the Stefan–Boltzmann law for the black-body, which takes black-body radiation as proportional to the fourth power of the absolute temperature. Planck took its numerical value from Kurlbaum's original measurements, although Kurlbaum's results were soon rectified and improved by a series of measurements performed by others.

[26] See Arabatzis (2006, p. 79) for a historical reconstruction of Lorentz's 'ions' vs 'electrons' as they were called by Stoney, Larmor, and Zeeman.

[27] I refer the reader to the excellent historical reconstruction of this episode by Arabatzis (2006, Ch. 4).

Planck did not speculate on the nature of the minimal unit of electricity. He was more interested in identifying *e* as a physical constant and in establishing accurate values of various inter-related physical constants. What makes some units natural, according to Planck, are two features that we still identify with physical constants.

(1) Physical constants are *objective*. Planck maintained that their holding does not depend on us qua epistemic agents: it is not meant to cater to our epistemic needs, or to our research interests. Physical constants are thus set aside from metrological considerations that typically apply to other units, for there is no conventional element presumably affecting their validity.

(2) Physical constants are *necessary*. They are part of the fabric of nature: they exist and would have existed even if humankind had not existed (or had not developed our particular scientific history). The naturalness of these constants is tied to laws of nature, according to Planck. Their 'natural significance' is retained as long as the relevant laws 'remain valid; they therefore must be found always the same, when measured by the most widely differing intelligences according to the most widely differing methods' (Planck 1906/1913, p. 175).[28]

The introduction of the elementary quantity of electricity *e* in this context, then, marks an important shift in the debate about the nature of electric charge. It signals that ontological discussions about what the electron *really* is do not matter, because the fundamental unit is not the electron (or hydrogen atom or gas ion or corpuscle), but the electric charge. And electric charge is a physical constant. It is entrenched in laws of nature, whose validity—Planck insisted—holds always and everywhere.

10.5. Walking in the garden of inferential forking paths

How it is possible for different epistemic communities to reach the same conclusion (e.g. *that* something is and *what is*), perspectival representations

[28] 'All the systems of units which have hitherto been employed . . . owe their origin to the coincidence of accidental circumstances, inasmuch as the choice of the units lying at the base of every system has been made, not according to general points of view which would necessarily retain their importance for all places and all times, but essentially with reference to the special needs of our terrestrial civilization. . . . In contrast with this it might be of interest to note that, with the aid of the two constants *h* and *k* which appear in the universal law of radiation, we have the means of establishing units of length, mass, time, and temperature, which are independent of special bodies or substances, which necessarily retain their significance for all times and for all environments, terrestrial and human or otherwise, and which may therefore be described as "natural units"' (Planck 1906/1913, pp. 173–174).

notwithstanding? Natural kind thought and talk depends on explaining why a bunch of phenomena are *of the same sort* while also making room for the possibility that things could have gone otherwise. While the sort-relative sameness relation sheds light on the mechanism, if you like, behind NKHF, there are still gaps to fill in. I anticipated that truth-conducive conditionals-supporting inferences ultimately explain *how and why* epistemic communities come to historically agree that a certain open-ended grouping of phenomena is a natural kind.

In this historical episode, the modally robust phenomena, each in its own domain, were beyond anyone's doubt: the bending of cathode rays, the electrolysis of water, Röntgen rays, optical reflection in metals, and others. How did this coming to historically agree happen?

Here is a classical realist way of thinking about this. There is a world out there packed with natural kinds (e.g. the electron) having some distinctive properties (e.g. negative electric charge). Over time and with great experimental and theoretical efforts, scientists come to know the kinds and their properties. They might have some approximately true beliefs and other false beliefs about them. Thomson might be said to have had approximately true beliefs about the charge-to-mass ratio of his object of study but false beliefs when it came to Faraday tubes and all that. Over time, these false beliefs are rectified and eliminated as we get more true beliefs.

But here is another realist way of thinking about our coming to agree, which I am now putting to test with this episode. Let our fiat be not some granted metaphysical picture of the world out there but our *ways of knowing* the natural world. As long as we are ready to make this switch from a metaphysical fiat to an epistemological one, our starting point becomes the plurality of historically situated scientific practices—scientific perspectives—through which humankind has encountered the natural world as teeming with modally robust phenomena.

Different perspectives produce different perspectival representations. Helmholtz's, Thomson's, and Planck's each operated with different perspectival$_1$ representations of what we call the electric charge e in that they availed themselves of a variety of *situated* scientific practices pertinent to their respective scientific perspectives. Helmholtz identified the electric charge as a fundamental unit corresponding to chemical valence 1 via the Grotthuss chain model and decades of experiments on electrolytes. Thomson resorted to the Faraday tube to model electric flux, and to a century-long tradition of kelp-making and glass-blowing to run his experiments. And Planck deployed h as a way of programmatically rethinking units of measure and identified the electric charge with a fundamental constant, ushering in a tradition that continues to these days with *die kleine h* entering the SI.

What makes these representations perspectival$_1$ is not therefore that they each represent a *given* property—electric charge—from different points of view. They are not *representing* a *given* content *as* portrayed from vantage point *a* rather than *b* or *c*. For establishing the existence of such a property and its nature was precisely what was at stake in all these investigations—the existence of the electric charge was *not* epistemically given. It was not the *starting point* but the *end point* of these scientific endeavours. If one knew in advance that there is indeed such a constant in nature, it would not be necessary to go through such a century-long painstaking experimental effort to measure it, to model it, to theorize about it.

It is in this specific sense that these representations can therefore also be said to be perspectival$_2$ in the language of Chapter 2, where the two notions of perspectival$_1$ and perspectival$_2$ were presented as Janus-faced and complementing each another. They are perspectival$_2$ in being *directed towards* establishing that there is indeed an electric charge and finding out its nature. In so doing they open for us a 'window on reality' thanks to *methods, experimental tools,* and *modelling practices* that were the expression of genuinely different scientific perspectives at the time— the electrochemical, the electromagnetic, and the quantum one—through which a plurality of data-to-phenomena inferences were reliably delivered.

In *Philosophy and the Mirror of Nature* (1979, pp. 330–331), Richard Rorty famously concluded about the controversy surrounding Galileo's new discoveries that 'Galileo won the argument, and we all stand on the common ground of the "grid" of relevance and irrelevance which "modern philosophy" developed as a consequence of that victory'. One could similarly be tempted to claim that Planck won the argument in 1906 and we all stand on the common ground of the 'grid' that quantum physics has developed as a consequence of that victory.

But the story I have told differs from the classic realist one as much as from its Rortian counterpart. Our unanimous agreement is neither the result of uncovering 'hidden goings on', nor the outcome of converging towards some final reality. Equally, *pace* Rorty, our unanimously coming to agree about the electric charge is not a matter of winners or losers.

It is instead the unpredictable, unforeseeable, and extraordinary epistemic feat of a plurality of epistemic communities in their historically and culturally situated scientific perspectives, and their sophisticated inferential game between 1897 and 1906 of 'giving and asking for reasons' (to echo once again Brandom 1998, p. 389) as to why a particular grouping of phenomena belong together. Progress takes place not *in spite of* but *thanks to* a plurality of scientific perspectives. I will return to the importance of a plurality of perspectives in my final chapter. But for now, let me clarify how I see the inferential reasoning at play here through:

(a) the use of perspectival models as inferential blueprints to identify modally robust phenomena;

(b) the conditionals-supporting nature of the inferences linking various phenomena;

(c) their being truth-conducive.

Let me unpack each in turn.

10.6. Inferential blueprints *encore* and modally robust phenomena

Willingness to engage with other epistemic agents occupying different scientific perspectives (synchronically and diachronically) is key to perspectivism as a pluralist view. I see our coming to unanimously agree *that* something is and about *what* it is as the outcome of conditionals-supporting inferences linking phenomena across different domains so that they come to be historically identified as being *of the same sort*. In the historical example I have briefly examined, the inferential game becomes the game of considering a number of phenomena (let us call them P_1, P_2, P_3) in their respective domains that *at the time* had been historically identified via a plurality of perspectival data-to-phenomena inferences.

Recall how in Chapter 5 I defined perspectival models as inferential blueprints. The key idea was that

> Perspectival models model possibilities by *acting as inferential blueprints* to support a particular kind of conditionals, namely *indicative conditionals with suppositional antecedents*.

The representational value of a blueprint consists in its ability to enable the relevant users to make *relevant and appropriate inferences* over time. The perspectival models offer instructions to an often diverse range of epistemic communities for making relevant and appropriate inferences about the phenomena of interest within broad constraints. Just as architectural blueprints offer a sketch of a building's shape, proportions, and relations among the relevant parts, perspectival models sketch the lawlike dependencies among relevant features of the phenomena at stake.

Faraday tubes and Grotthuss's chain model are examples of perspectival models qua inferential blueprints. Consider again Grotthuss's 1805–6 model—still conceived within the electrochemical perspective which at the time featured the Galvani–Volta controversy and a plurality of models about the nature of animal vs metallic electricity. The model supposed that water formed a chain of positive and negative charges that would be released at the two ends of the electrodes as a way of exploring how electricity might affect water and other fluids.

The model acted as an inferential blueprint for a series of experimental observations run by Michael Faraday in London in the 1830s, observations that eventually revealed stable events in the form of lawlike dependencies across a wide array of electrolytic substances. Grotthuss conceived his model using Volta's pile, which he took to have poles with opposite attractive and repelling forces. Thirty years later, Faraday did not believe that electrochemical decomposition was the effect of the powers between opposite poles.

Yet just as architectural blueprints give teams of different craft workers instructions about how to build a house, Grotthuss's model gave Faraday and other scientists of the time helpful instructions for experiments. In Faraday's case, the experiments were designed to show that '*for a constant quantity of electricity, whatever the decomposing conductor may be, whether water, saline solutions, acids, fused bodies . . . the amount of electrochemical action is also a constant quantity*' (Faraday 1833/2012, vol. I, p. 145, emphasis in original). Grotthuss's model equipped Faraday with an inferential blueprint for thinking about the outcome of his experiments with a variety of oxides, chlorides, and salts. Faraday concluded that 'many bodies are decomposed directly by the electric current, their elements being set free; these I propose to call *electrolytes*' (vol. I, p. 197, emphasis in original). The *stability* of the *events* observed by Faraday concerning the decomposition under the action of electricity was due to their inherent lawlikeness.[29]

Faraday went on to call it 'the general law of *constant electro-chemical action*' (Faraday 1833/2012, vol. I, p. 225, emphasis in original). The lawlike dependency that he saw as inherent in 'the chemical decomposing action' made the events stable under a number of changes in background conditions: in the intensity of the electricity, the location of the electrodes, or the conductivity or non-conductivity of the medium.

The associated phenomenon—the electrolysis of water, of saline solutions, acids, and so forth—was modally robust in the sense I explained in Chapter 6, Section 6.7.3. A triadic relation linked the data observed, the stability of the event (qua lawlike chemical decomposing that is constant for a constant quantity of electricity), and the perspectival inferences from the data to the stable event. The phenomenon of electrolysis was modally robust in that it *could happen* in more than one possible way and be identified and re-identified by different epistemic agents over time.

For example, independently of Grotthuss, Humphry Davy arrived at the same conclusion about water electrolysis, from a series of observations concerning

---

[29] In Faraday's own words, 'the chemical decomposing action of a current *is constant for a constant quantity of electricity*, notwithstanding the greatest variation in its sources, in its intensity, in the site of electrodes used, in the nature of conductors (or non-conductors . . .) through which it is passed, or in other circumstances' (Faraday 1833/2012, vol. I, p. 207, emphasis in original).

electrified water in gold cones, agate cups, tubes of wax, tubes of resin, and so forth.[30] Like any good architectural blueprint, Grotthuss's model with its chain of positive and negative charges was amenable to being scaled up or down. In this case, the 'scaling-up' metaphor translates into how the charged 'electrolytes' (as Faraday called them) became the charged 'idle wheels' in Maxwell's honey-comb model of the ether (1861–2/1890), which in turn served as an inferen-tial blueprint for a different phenomenon: that of electromagnetic induction. In Maxwell's model (see Bokulich 2015 and Massimi 2019c), ethereal vortices represented the magnetic field and its strength while idle wheels among vortices represented the electric displacement associated with the magnetic field. Such ethereal vortices accompanied by charged particles resurfaced with J.J. Thomson and his 'Faraday tubes', still described in the Silliman Lectures of 1906 as a model for electrostatic induction.

Grotthuss's 1805–6 chain model for water electrolysis and J.J. Thomson's Faraday tubes for electrostatic induction are perspectival models. They were representing electricity from two different vantage points: the electrochem-ical and the electromagnetic perspectives. After all, the two phenomena—electrolysis (P_1) and cathode rays bending (P_2)—are different in nature. The former takes place at the scale of the molecules of chemical electrolytes. The latter occurs in the interaction between magnetism and electricity.

The relevant data-to-phenomena inferences in each case were also perspec-tival. Consider, for example, the wildly diverging views that existed throughout the late eighteenth and early nineteenth century about the nature of electricity at work in electrochemistry: from Galvani's animal electricity to Volta's me-tallic electricity, which still informed Grotthuss's model, or the Victorian con-text of ether theory in late nineteenth-century Cambridge where Maxwell and Thomson developed their models (see, e.g., Siegel 1981), without mentioning the craftsmanship of glass-blowing and producing kelp-fluxed glass tubes (from Crookes to Thomson).

And yet, in spite of the perspectival inferences from the data, these two phenomena—electrolysis (P_1) and cathode rays bending (P_2)—proved to be modally robust in that each *could happen* in more than one possible way, and be re-identified over time. Moreover, as a distinctive type of model pluralism, perspectival modelling has a history of its own. The relevant models—from Grotthuss's to Thomson's—lie on a continuum, almost a genealogy. Modelling

[30] 'Water slowly distilled, being electrified either in gold cones or agate cups, did not evolve any fixed alkaline matter, though it exhibited signs of ammonia; but in tubes of wax, both soda and potash were evolved.... When water was electrified in vacuo scarcely any nitrous acid, and no volatile alkali, was formed.... Mr Davy ... thinks these electric energies are communicated from one particle to another of the same kind, so as to establish a conducting chain in the fluid, as acid matter is always found in the alkaline solutions through which it is transferred' (Davy 1807, pp. 247–250).

electricity required Grotthuss, no less than Davy, Faraday, Maxwell, and Thomson after him, to work multi-handed on a number of perspectival models qua inferential blueprints.

One of the distinctive challenges in this exploratory modelling exercise was to reconcile the continuous field-theoretic nature of phenomenon (P_2) with the discrete corpuscular nature of phenomenon (P_1). Thomson's Faraday tubes were meant to offer a solution. They were in their own way a remnant of Maxwell's ethereal vortices combined with discrete corpuscular opposite electric charges at each end—a distant cousin of Grotthuss's chain model,[31] as if Volta's pile with its opposite electric charges had been coupled with mechanical models of the ether for electromagnetism. Faraday tubes in turn enabled scientists at the turn of the twentieth century to make novel inferences about the relevant phenomena (P_1) and (P_2) and use the observed lawlike dependencies to ultimately infer what Thomson called the 'corpuscle'. But what should one say about the nature of the inferences here at play?

10.7. Chains of conditionals-supporting inferences

In Chapter 5, Section 5.7, I contended that the inferences supported by perspectival models can be expressed in terms of chains of indicative conditionals with suppositional antecedents. I also stressed that there is a clear division of modal labour between indicative and subjunctive conditionals in these inferences. Consider, for example, the difference between the following indicative conditional at play in this historical episode:

(E.1) If hydrogen and oxygen form a Grotthuss's chain, hydrogen is released at the negative electrode.

and the subjunctive conditional (denoting the subjunctive with A rather than E)

(A.1) Were electrodes to be immersed in water, hydrogen would be released at the negative electrode.

---

[31] 'We might, as we shall see, have taken the tubes of magnetic force as the quantity by which to express all the changes in the electric field; the reason I have chosen the tubes of electrostatic induction is that the intimate relation between electrical charges and atomic structure seems to point to the conclusion that it is the tubes of electrostatic induction which are most directly involved in the many cases in which electrical charges are accompanied by chemical ones' (Thomson 1891, p. 150).

The two conditionals conceal a crucial difference behind the syntactical difference between the present tense 'is released' and the subjunctive 'would be released'. Although the consequent is the same in both cases, the subjunctive mode in (A.1) conveys the objective possibility of hydrogen being released, were the antecedent condition to hold. But the indicative conditional (E.1) conveys instead an implicit (unpronounced) epistemic possibility concerning hydrogen being released, under the supposition of the antecedent.

In my philosophical lingo, the subjunctive mode (A.1) speaks to the stability of the event under the antecedent's holding—hydrogen's being released at the negative electrode whenever the electrodes are immersed in water—its objective (non-epistemic) possibility being grounded in the lawlike causal dependency between quantity of electricity and electrochemical decomposition observed by Faraday. Hydrogen would still be released if electrodes were immersed in water, regardless of the nature of the metal used for the electrodes, for example.

By contrast, the indicative mode speaks to our epistemic attitudes when we judge whether the phenomenon P_1 (water electrolysis) is likely to occur in the physically conceivable scenario described by Grotthuss's model in the antecedent. This is the realm of perspectival models and of how epistemic agents use these models to physically conceive the scenario captured by the antecedent. As per Chapter 5, Section 5.7, indicative conditionals such as (E.1) are epistemic conditionals with an implicit (unpronounced) modal. Along the lines of Angelika Kratzer (2012), I maintain that (E.1) can be regarded as a bare conditional which is implicitly modalized as follows:

(E.1*) If hydrogen and oxygen form a Grotthuss's chain, hydrogen *may* be released at the negative electrode.

The modal verb 'may' is again epistemic *not* in the sense of expressing the sheer belief of a particular epistemic agent or community, but in capturing instead possibilities concerning specific relations within the limits afforded by perspectivism. In this example, the implicit modal verb reflects the particular state of knowledge and perspectival model available to the epistemic community at the time to *think* and *talk* about what was objectively possible concerning the hydrogen, under the supposition that the water molecules formed an ionised chain as per Grotthuss's model. As explained in Chapter 5, I see indicative conditionals as key to the inferential reasoning supported by perspectival models. They tell us that 'Given the antecedent supposition, plus a number of auxiliary assumptions R, S, T, U, the consequent follows', where 'follows' can be understood in a variety of ways (inductively, deductively, abductively) on a case-by-case basis. For example, (E.1*) can be unpacked as follows:

Let us physically conceive of hydrogen and oxygen as forming a Grotthuss's chain; then—given auxiliary assumptions R, S, T, U—hydrogen may be released at the negative electrode.

Auxiliary assumptions R, S, T, U include water being a chemical compound rather than an element, electricity being able to decompose it. But also other claims that have now long been forgotten and abandoned, including the idea of an 'electropolar' system in nature (see Pauliukaite et al. 2017).

Indicative conditionals often enter into long chains of inferential reasoning spanning several phenomena indexed to different domains and evinced through perspectival data, methods, models, and techniques. Consider, for example, the following chain of indicative conditionals-supporting inferences:

(E.1) If hydrogen and oxygen form a Grotthuss's chain, hydrogen is released at the negative electrode.

(E.2) If ether vortices move as in Maxwell's honeycomb model, electric current is displaced.

(E.3) If a Faraday tube of electrostatic induction is stretched and broken, free atoms of electricity are produced (be it in metals or liquid electrolytes).

(E.4) If free atoms of electricity in metals are conceived along van't Hoff's kinetic theory of osmotic pressure (as Paul Drude did), dispersion of light and reflection of metal surfaces ensue.

(E.5) If carriers of metallic conductivity are conceived along the model of Drude's electron gases, the phenomenon of black-body radiation can be calculated.

(E.6) If the monovalent hydrogen ion is conceived along the lines of Planck's quantum hypothesis, the quantum of electricity (measured from the radiation constants) is equal to 4.67×10^{-10} in electrostatic units.

The inferential chain (E.1)–(E.6) allowed physicists around 1897–1906 to conclude that something was (electric charge) and what it was (a quantum of electricity with a well-defined measurable value). Electric charge as a fundamental unit of nature is not a Lockean nominal essence with necessary and sufficient conditions for membership. For it is not *just* an itemized list of phenomena $P_1, P_2, P_3, \ldots$ from electrolysis to electromagnetic induction, from metal conductivity to radiant heat. What is needed in addition is a *set of instructions* for epistemic communities—working across different scientific perspectives and willing to engage with one another—to reliably make informed decisions about how to proceed, what conclusions to draw, what tentative conclusions to discard, what further novel inferences to explore and probe about these phenomena and new ones too.

These instructions take the form of conditionals-supporting inferences like (E.1)–(E.6). Models are involved at different points in these inferences. Some of them are perspectival. But these are only a subset of a much larger family of scientific models routinely used to make these inferences, including phenomenological models such as Drude's electron gas, and theoretical models such as Planck's theory of black-body radiation.

How can a chain of conditionals-supporting inferences ever successfully deliver instructions as to how to proceed in the garden of forking paths? If these indicative conditionals (and their covert epistemic modals) are advanced by epistemic communities working within situated scientific perspectives at a particular time on the basis of limited evidence, how can they ever deliver any realist commitment on what there is?

One is reminded here of Marco Polo's answer to Kublai Khan in Italo Calvino's quote at the opening of this chapter. Situated communities recognize 'the little that is theirs, discovering the much they have not had and will never have'. Futures not achieved are indeed only branches of the past: but dead branches. Our walk in the inferential garden of forking paths is not some arbitrary meandering. At each step, and each branch point, the scientific paths taken must be explained and justified with our fellow travellers. The inferential game of giving and asking for reasons includes reasons for the futures achieved by our evolving kinds, and those for the dead branches we left behind as empty kinds.

In the example at stake, the instructions encoded by these conditionals-supporting inferences required comparing the phenomenon P_1, on the one hand, and its ratio E/M in the hydrogen emerging from Grotthuss's and Helmholtz's work on electrolysis (E.1), with the phenomenon P_2, on the other hand, and its e/m measured by Thomson's experiments on cathode rays underpinned by models such as Faraday tubes (E.3) building on Maxwell's ether model (E.2), and noticing a numerical discrepancy.

To resolve this, a new round of data-to-phenomena inferences was required that this time involved *forking subjunctive conditionals* at a key juncture (E.3) of the chain of indicative conditionals (recall I use 'A' for subjunctive conditionals):

(A.3.a) Were e bigger than E, e/m would be much bigger than E/M;

Or

(A.3.b) Were m much smaller than M, e/m would be much bigger than E/M.

As new data became available (using Wilson's technique of weighing water droplets that condense around negative charges, and Townsend's experiments on the charge of ions produced by X-rays) and more refined measurements made possible thanks to Thomson's cathode rays experiments and improved glass tubes, the choice could reliably be made in favour of (A.3.b). This led Thomson to conclude that his corpuscle had a mass much smaller than the hydrogen atom.

From there the step to the next further inference that there *is* a quantum of electricity was a short one. A further round of data-to-phenomena inferences was required, this time involving the comparison of Thomson's value for e (as per A.3.b) with Planck's value derived from his theory of black-body radiation (via E.4–E.6). And again, the discrepancy between the two opened up yet another inferential forking path at a key juncture (E.6) with the following subjunctive conditionals:

(A.6.a) Were e a semi-classical quantity, its value would be derived from the laws of classical electrodynamics;
Or
(A.6.b) Were e a quantum of electricity, its value would be derived from the laws of black-body radiation.

Further measurement obtained by Rutherford, Regener, Perrin, and Millikan, among others, settled the choice for Planck's (A.6.b) eventually. In Rorty's language, we all stand on Planck's 'grid' today in taking Planck's constant e as a minimal natural unit. Electricity got quantized alongside black-body radiation. Fast forward a century, and *die kleine h* has established itself as a new fundamental unit in the International System of Units (SI) replacing *le grand Kilo*.

This is no argument against fundamental physical constants, of course. If anything, this is an argument to the effect that the physical constant e and, more broadly, the natural kind *electron* are the outcome of conditionals-supporting inferences. These inferences were enabled by lawlike dependencies among relevant features at work in each and every one of the different phenomena that were historically identified and eventually grouped under the sortal concept *electron*. Ultimately, it is the lawlike dependencies in phenomena, the way they enter into forking subjunctive conditionals at key junctures, and how in doing so they inform communities about choosing which path to take that underpin the *truth-conductive nature* of conditionals-supporting inferences.

Historically, the identification of the relevant lawlike dependencies in the phenomenon of cathode rays bending constituted the main hurdle. Thomson's experiments in 1897 and his ability to reliably settle for (A.3.b) gained him the Nobel Prize, no matter how mistaken his beliefs about corpuscles in Faraday tubes. Truth-conducive conditionals-supporting inferences are reiterated and enabled by an ever-growing number of phenomena in the open-ended grouping. Over time, they reliably lead epistemic communities to agree *that* something is *and what* it is.

This is how real historical communities over time learn how to navigate the space of what is possible: that is, by comparing a plurality of modally robust phenomena so as to make more and more refined inferences on what might be the case at every twist and turn.

This procedure is entirely fallibilist, anti-foundationalist, and revisable. It does not start from metaphysically given building blocks. It takes seriously the situated nature of our scientific knowledge, our starting always *from somewhere*, in the form of model-based inferential reasoning with epistemic indicative conditionals. It is truth-conducive in giving and asking for reasons as to why some paths are taken and others are not along the way.

10.8. Coda: what remains of truth?

This brief foray into the history of the electric charge around 1897–1906 shows how a historically identified grouping of phenomena became over time the natural kind *electron*. To complete the picture of NKHF, this chapter has focused on the last condition (iii) in my definition. I have made three main points:

- Modally robust phenomena $P_1, P_2, \ldots$ display lawlike dependencies among features that are captured by subjunctive conditionals;
- These subjunctive conditionals enter at key forking junctions in long inferential chains of indicative conditionals
- Indicative conditionals are epistemic conditionals about the phenomena at stake, under antecedently held conceivable scenarios by the models.

In other words, the antecedents of these indicative conditionals invite us to physically conceive certain scenarios under particular models. The consequents express claims of knowledge under the supposition of the scenarios. Following Kratzer (2012), I suggested that the consequent of an indicative conditional hides a modal verb, as when the bare conditional (E.1) is rewritten as (E.1*). The fully fledged epistemic conditionals express modal knowledge claims that agents entertain when using a variety of scientific models to make inferences about phenomena. Conclusions about what natural kinds exist are reached by epistemic communities willing to engage with one another across a plurality of scientific perspectives. But—one might insist at this point—what makes their lengthy sequences of conditionals-supporting inferences *truth-conducive*? If scientific knowledge is genuinely perspectival in the way described, why even bother with 'truth'? If anything, is not there a lingering danger of scepticism about knowledge at play in perspectivism? The problem is well expressed by Barry Stroud (2020, pp. 147–148):

> Could it be that perspectivism perhaps expresses a certain sympathy with this tradition of doubt or suspicion about knowledge? I speculate here, but the idea of knowledge is so directly connected with the idea of truth, which is independent of human beings' holding the attitudes they do towards it, that perhaps perspectivism sees more promise in shifting the focus away from the idea of knowledge as such, and looking instead to other human attitudes or responses involved in explaining what we want to understand about the whole enterprise of what we call human knowledge. . . . And whatever the goal, can we really understand what we most want to understand about the enterprise of human knowledge by thinking of those who investigate the world as exercising only the concepts needed for the less-committal epistemic attitudes and responses that perspectivism concentrates on, not a concept of knowledge that implies truth and so apparently resists perspectival treatment?

The direction Stroud ultimately recommends to perspectivists shares features of a variety of no-knowledge-centred accounts of science: from Elgin's (2017) non-factive scientific understanding to Potochnik's (2017). However, the view I have developed in this book is indeed centred on knowledge, claims of knowledge, and claims that are modal in flavour too. So I ought to say something to justify my use of 'truth-conducive'.

Behind Stroud's remark lies a long-standing and deeply entrenched view of knowledge (and knowledge as implying truth) that sits uncomfortably with the perspectival realist narrative I have endeavoured to offer. According to this entrenched view, truth is the *aim* of science. It is what scientific inquiry should be about in the sense of *converging towards* some final true story about the way the natural world is. Those who share stronger metaphysical intuitions about the way the world is and how science tracks this (metaphysics-first) ontology—be it an ontology of properties or kinds or something else—will remain unmoved by my account. And this is of course as is to be expected. For my goal in this book has not been to offer winning arguments against a metaphysics-first approach to science. I do not have such arguments—nor can I see any against epistemology-first realist accounts either. The whole point of this book has been to show that *if* one accepts an epistemology-first stance on the realism debate, *then* there is a bottom-up story to be told (from data to phenomena to kinds) that can open up a different flavour of realism about science. But it should also be clear by now that perspectival realism—as I have presented it—is far from traditional convergent realist accounts.

There is no truth with a capital T at the end of the inquiry because there is *nothing* to converge to. No 'hidden goings on' of any kind, and no Humean mosaics. There is nonetheless an external world teeming with modally robust phenomena which scientists engage with by picking a way through the garden of inferential forking paths as it cuts across different scientific perspectives. The 'windows on reality' that perspectival$_2$ representations afford open up in this process.

As discussed in Chapter 5 (Section 5.7), following Kratzer (2012), truth conditions and assertability conditions easily come apart as one walks along the inferential garden of forking paths. For assertability conditions, speakers' evidence at the time and in their *situated perspective* is all that counts, but not so for truth conditions. For example, Thomson was justified to entertain the indicative conditional (E.3) on the basis of the evidence he had back in the 1890s despite the fact that the same evidence did not constitute a truth condition for it. Scientific perspectives do not ratify their own claims of knowledge.

Claims of knowledge must instead be assessable from the point of view of other scientific perspectives, as discussed in Chapter 5, Section 5.7. In this historical episode, for example, Planck's quantum perspective offered a standpoint from which the indicative conditionals-supporting inferences and associated claims by Helmholtz, Drude, Thomson, et al. could all be evaluated.

This cross-perspectival assessment is key to the notion of perspectival truth that I see at work in perspectival realism. It combines perspectival pluralism about models with a non-convergentist yet still realist account of truth *across perspectives*. Day to day, whenever truth conditions are vague, scientists typically rely on assertability conditions and specific pieces of contextually available evidence to advance knowledge claims.

But ultimately, the evolution of our NKHF and their projectibility and unanimity do not depend on the assertability conditions but on the cross-perspectival truth conditions for our knowledge claims. And this presupposes the willingness of epistemic agents occupying different scientific perspectives to engage in the inferential game of giving and asking for reasons as to why some knowledge claims are retained and others are withdrawn; why some paths continue and others (futures not achieved) become abandoned branches.

The wider implications of this view for how to think about the multicultural situatedness of scientific knowledge—and the epistemic injustices that arise when engagement with other epistemic communities go badly wrong—are the topic of my final Chapter 11.

11
Multiculturalism and cosmopolitanism in science

The atlas depicts cities which neither Marco nor the geographers know exist or where they are, though they cannot be missing among the forms of possible cities: a Cuzco on a radial and multipartite plan which reflects the perfect order of its trade, a verdant Mexico on the lake dominated by Montezuma's palace. . . .

'I think you recognize cities better on the atlas than when you visit them in person', the emperor says to Marco, snapping the volume shut.

And Polo answers, 'Traveling, you realize that differences are lost: each city takes to resembling all cities, places exchange their form, order, distances, a shapeless dust cloud invades the continents. Your atlas preserves the differences intact: that assortment of qualities which are like the letters in a name'.

Italo Calvino (1972/1997) *Invisible Cities*, pp. 124–125[1]

11.1. Introduction

The 'atlas preserves the differences intact', warns Calvino's Marco Polo; yet 'Traveling, you realize that differences are lost'. It is this counterpoint between the multiculturalism inherent in perspectival pluralism and the cosmopolitanism latent in realism that this chapter explores. Not as an afterthought to perspectival realism, but as a natural continuation of it (indeed, as a prolegomena to another possible book to write). Throughout this book, I have endeavoured to spell out the kind of realism in science that emerges from perspectival$_2$ representations about modelling. But here I want to return to my original question. How are

---

Perspectival Realism. Michela Massimi, Oxford University Press. © Oxford University Press 2022.
DOI: 10.1093/oso/9780197555620.003.0014

wonderfully diverse *human beings*—occupying a plurality of *historically and culturally situated perspectives*—able to form reliable knowledge of the natural world? Are not the limits of our knowledge also the limits of our historically and culturally situated perspectives?

As I suggested in Chapter 2, the importance of perspectival$_1$ representations lies in the unique epistemic resources that *situatedness* opens up. Perspectival$_1$ representations offer unique epistemic standpoints afforded by particular geographical, socioeconomic, political, historical, and cultural locations. As an example, consider the following perspectival$_1$ representation of the river Congo in a local artefact known as a *lukasa*, or 'memory board', used by the Luba community (Figure 11.1—on *lukasas*, see Roberts 1998; Roberts and Roberts 1996; Reefe 1977).

An oral culture, the Luba rely on court historians known as 'men of memory'—or Mbudye historians—to pass on the story of their origin, kingdom, and culture. They learn a formulaic repertoire, whose protagonist is the mythical hunter and Luba ancestor Mbidi Kiluwe fighting the cruel enemy Nkongolo Mwamba. The stories recount the fights, and the subsequent union of Mbidi with Nkongolo's sister, from which the first Luba king (Kalala Ilunga) was born. A *lukasa* is a visual mnemonic, a wooden tablet with encrusted beads of different sizes, colours, and configurations: a big blue bead represents Mbidi while a red bead represents Nkongolo, sometimes surrounded by a circle of beads (his allies). A gash in the wood stands for the river Congo; curved lines of beads for the reeds along the river, and white beads for white birds in the river (see Reefe 1977, p. 50).

Touching the beads on the *lukasa*, Mbudye historians tell the mythical story of the Luba people again and again. Each time the story may take a different twist. Each narrator might add details or omit others. But the bead configuration in the *lukasa* provides the boundaries within which the formulaic repertoire unfolds. *Lukasas* are perspectival representations of the history of Luba people and their kingdom. In my terminology (Chapter 2), they offer a perspectival$_1$ representation of the river Congo as seen *from the point of view* of the Luba living there. The uniqueness of the epistemic access that a *lukasa* gives to the history of the Luba comes from the situatedness of the representation—expressed in the choice of the materials, the colour codes of the beads, and the wood craftmanship. It is the situatedness of perspectival$_1$ representations in a kaleidoscope of cultures that I want to return to here in more detail.

Perspectival realism is irreducibly pluralist: multiple cultures had, have, and will have their own distinctive scientific perspectives. There is no default scientific perspective as a given 'view from nowhere'. Nor is there an ideal unity to which scientific perspectives can converge—either as a Peircean limit or a Kantian regulative idea. The vanishing points of perspectival$_2$ representations are, indeed, 'vanishing': imaginary standpoints that do not stand for anything (be it a final Theory of Everything or a Humean mosaic or any other metaphysical posit).[2]

Balancing the *distinctive situatedness* of each scientific perspective and their *collective ability* to contribute to scientific knowledge over time is like walking a tightrope. The authoritarian temptation towards homogenizing, universalizing, amalgamating, merging, or overlapping other cultures lingers still. Sandra Harding has captured this tension in what she calls 'no universal realism, no radical relativism'. She notes that

the very best science is always an assemblage of heterogeneous elements. It is precisely this pragmatic heterogeneity that gains the results of scientific practice at least some small but significant degree of independence from hegemony of the theory being testing. This heterogeneity insures that some elements of scientific practice are at least relatively autonomous from the scientist's theoretical commitment and thus can indeed do the kind of critical work for which the sciences are so valued. (Harding 2015, p. 124)

Harding's comments resonate with Alison Wylie's collaborative archaeological projects with Native American communities (see Wylie 2020), and her observation that

archeologists who take seriously the claims of Native Americans routinely argue that collaborative practice enriches their research practice in any number of ways, not only adding useful detail but generating new questions and forms of knowledge... [W]hen these projects succeed they powerfully illustrate the virtues of extending the cognitive-social norms of Longino's proceduralist account of objectivity—specifically her "tempered equality of intellectual authority"—beyond the confines of the scientific community (Wylie 2015, p. 192.)

The pluralism of scientific practices inherent in perspectival modelling aligns with Longino's proceduralist account of objectivity, Harding's warning against universal realism, and Wylie's plea for a 'plurality of pluralisms'. But the balancing act between the *particular situatedness* of each scientific perspective and their *collective ability* to contribute to scientific knowledge over time is only partly captured in the usual dichotomous terms—whether Harding's 'universal realism' and 'radical relativism'; Ronald Giere's (2006a, pp. 5 and 13) 'objectivist realism' and 'silly relativism'; Donna Haraway's (1988) 'totalization' and 'relativism';[3] or

---

[2] For a reading of Kant's regulative ideas as 'imaginary standpoints', see Massimi (2017a, 2018c).
[3] 'Relativism and totalization are both "god tricks" promising vision from everywhere and nowhere equally and fully, common myths in rhetorics surrounding Science. But it is precisely in the

Eduardo Viveiros de Castro's (1998a, 1998b) 'universalism' and 'relativism' in the context of what he calls Amerindian 'cosmological perspectivism' (see also Rossi 2015, p. 27, for an insightful discussion of perspectivism as 'a mode of historiographic writing').

These formulations hide another, more profound dichotomy: between the reality of multiculturalism and the aspiration of cosmopolitanism; the particular practices in historically and culturally situated contexts and the unanimity in what passes muster as 'scientific knowledge'. It is a tension that scholars in literary theory, anthropology, sociology, legal and political philosophy have long recognized. In the words of Walter Benn Michaels (2017, p. 60), it is a choice between 'the universalism of the cosmopolitan and the particularism of the multicultural, . . . between cosmopolitanism's relaxed view of cultural changes and mixes and multiculturalism's preference for cultures which are preserved and kept separate'.

Surprisingly, philosophers of science have not engaged with this dichotomy. Debates on realism in science have often revolved around the antitheses local vs universal, particular vs global, practices vs theories. But it is the tension between the *particular multicultural situatedness* of scientific perspectives and their *collective cosmopolitan ability* to contribute to scientific knowledge that underlies these antitheses. It is this tension that I explore in this final chapter. In Section 11.2, I tease out what is at stake in the idea of multiculturalism in science in terms of what I call 'interlacing' scientific perspectives and I offer some normative pointers for it. It is one thing to *historically describe* how many epistemic communities have met over time and collectively contributed to scientific knowledge. It is another thing to try to clarify the epistemic duties and rights that come with multicultural encounters. In Section 11.3 I distinguish between two possible ways of understanding the notion of *situatedness* and clarify how situated perspectives interlace in 'historical lineages'. I illustrate all this in Section 11.4, with a brief example from the early history of magnetism.

In Section 11.5, I warn against two varieties of epistemic injustice affecting scientific narratives: *epistemic severing* and *epistemic trademarking*. I discuss remedies required for these injustices in the form of 'reinstating' the severed communities as belonging to a 'scientific world citizenship'.

My argument goes from the value of multiculturalism for science to what I call a non-classist, non-elitist form of scientific cosmopolitanism. In the final Section 11.6, I make a plea for a kind of cosmopolitanism in science that I see as necessary to deliver on what the United Nations Declaration of Human Rights in its Article 27 (1) calls the right of 'Everyone . . . to share in scientific advancement

politics and epistemology of partial perspectives that the possibility of sustained, rational, objective inquiry rests' (Haraway 1988, p. 584).

and its benefits'. Ultimately, discussions about pluralism and realism about science ought to speak to the rights and obligations that come with scientific knowledge production; who benefits from it; who gets excluded from it; and what perspectival realism has to say about these epistemic injustices.

11.2. Multicultural science and interlacing scientific perspectives

Think again of J.J. Thomson, doing his experiments with cathode rays in 1897, within what in Chapter 10 I labelled the 'Faraday–Maxwell' scientific perspective. The Faraday–Maxwell scientific perspective is one among myriad other historically and culturally situated perspectives. Understanding what makes this perspective historically and culturally distinct would require a much longer historical narrative than I offered in Chapter 10.

Such a narrative would take in modelling practices (e.g. Faraday tubes and ether models) and the wider sociocultural context in which they developed in Victorian Britain (but not, for example, in the Germany of Max Planck). It would include details about the popularity of other theories at the time and the relation to varieties of 'spiritual' ethers invoked in spiritualistic practices fashionable among the Victorians. Historians have indeed produced such narratives (see, e.g., Kragh 2002; Noakes 2005; Wilson 1971).

Attention to the local, particular sociocultural context is crucial to understanding how a scientific perspective could have flourished, entrenched itself, and eventually become defunct. Being thus situated explains the identity, so to speak, of each scientific perspective: each of them is the unique product of its own time and culture. Cathode rays continue to be used in labs across the world. J. J. Thomson's experiments continue to be repeated in front of generations of physics students. But no physics teacher today would dream of mentioning Thomson's ether theory, or his belief in Faraday tubes (and the underlying Victorian beliefs in spiritual ethers at the time). How should one then understand the *philosophical* relevance of the notion of being *historically and culturally* situated? Is this just a matter for historians of science and scholars in cultural studies? What can philosophers of science contribute?

First, I think it would be hasty to conclude from the *situatedness* of scientific perspectives that they are no bigger than the sociocultural context where they first flourished. The Faraday–Maxwell perspective is not confined to Victorian Cambridge. Its modelling practices, experimental designs, and technological tools continued to inform Heinrich Hertz's experiments on electromagnetic waves in Germany and Guglielmo Marconi's research on radio waves. Its research outcomes continue to underwrite our contemporary uses of cathode rays in physics labs across the planet. Scientific perspectives do *span over time* and stretch *beyond specific geographical, sociocultural, and even national boundaries.*

Scientific knowledge claims and their underlying perspectival practices are exportable, and typically are exported. Recall, for example, how scientists working on palaeoproxy data for reconstructing the past climate may avail themselves of data from boreholes (Chapter 4.b); or how Washington's petrologist practice of classifying rocks in the Naples area ultimately informed Goldschmidt's cosmochemical practices in Norway (Chapter 4.a).

Second, it would be a mistake to think of the *situatedness* of any scientific perspective in terms of 'shared membership' of an epistemic community or belonging to a 'shared scientific homeland'. Being *situated* should not be equated with 'enrolling' in a particular epistemic community. Scientific knowledge by its very nature is cosmopolitan: it does not grow in a silo but through exchanges, trades, and cultural encounters. The typical tacit assumption that scientific history comes neatly divided into pre-packaged, historically and culturally *well-insulated* scientific perspectives is highly questionable.

For such an assumption is often a double-edged sword: it defines the identity of a perspective as much as it unjustly severs its links with others which might have been instrumental to its flourishing. 'Shared membership' is often classist and elitist. For example, one ought not to identify the Faraday–Maxwell perspective with some 'shared membership' of certain field-theoretical assumptions and modelling practices qua exclusive intellectual repository of the Cavendish Lab in Victorian Cambridge. For doing so would lose sight of the bigger picture in which the perspective became possible in the first instance and thrived. It would, for example, unjustly cut out swathes of the society of the time. Scottish kelp-makers but also local glassware artisans and glass-blowers, among others, would not be given their due credit for contributing to the techniques that were key to the Faraday–Maxwell perspective.

I believe the assumption of well-insulated scientific perspectives is a remnant of a deeply rooted view that I am going to call 'Kuhnian communitarianism'. By it I mean the view that scientific knowledge is *defined by* the specific historical-geographical-cultural *membership* of particular epistemic communities sharing what the early Kuhn called a 'paradigm'.

Kuhn advocated a communitarian view, where scientific knowledge is produced in fairly well-insulated scientific paradigms, often competing with one another, pitted against one another, and with a successor eventually supplanting the previous one. After a scientific revolution and a change in paradigm, 'normal science' is defined by canonical texts (be it the *Almagest* of Ptolemaic astronomy or the *Principia* of Newtonian mechanics). Scientific terms such as *mass* and *weight*, or inertial mass vs gravitational mass and associated nomic generalizations (e.g. Newton's second law), are said to be learned from canonical texts such as the *Principia*. This is how, according to Kuhnian communitarianism, scientific knowledge gets passed on from one generation to the next *belonging to the same scientific community* in periods of normal science—until the time comes when anomalies accumulate, trigger a crisis, and a new paradigm comes to the fore.

But a historically *insulated* and culturally *homogeneous* scientific paradigm is hard to find. Moreover, scientific progress quickly became mysterious under Kuhnian communitarianism. Hence, the debates about the incommensurability thesis and so-called Kuhn-loss (see Bird 2000; Wray 2011), the baffling succession of one paradigm after another with no commensurable methodologies or taxonomic concepts; and the even more perplexing analogy with Gestalt-switch to explain consensus-gathering around a new paradigm whose language—as Kuhn reminded us—required bilingualism instead of translation.

The mystery, of course, is only apparent. Indeed, it is an artefact of Kuhnian communitarianism and its associated view of how one paradigm is replaced by another one, like a piece on a chessboard being taken by an opponent's piece. This hardly ever happens. To start with, at any given historical point there is typically a pluralism of practices. In Europe at the turn of the nineteenth century there were at least three different perspectives on the nature of the electric charge. Similarly, in contemporary studies on dyslexia one can identify a number of different perspectives (neurobiological, cognitive, and educational).

The wonderfully diverse, pluralistic, and fluid scientific perspectives that are always at play in scientific inquiry do not lend themselves to being confined in historically insulated and culturally homogeneous silos. Historically-culturally situated scientific perspectives *are no more natural kinds* than 'Natural Kinds with a Human Face' (NKHF) themselves.

To be perfectly clear and avoid ambiguities here, there is no denying that policies of assimilation, homogenization, and exclusion have *historically* taken place. But what Kuhn the historian rightly identified as a description of power structures should not be reified into a philosophical view of how science works, or better *ought* to work. It is the latter, not the former, that I have a quarrel with. I think it is important here to keep distinct the historical descriptive component and the philosophical normative one, and not pattern the latter on the former.

Kuhnian communitarianism, as I see it, with its normal science inscribed in textbooks, curricula, and shared lexicons, tacitly and unwittingly buys into a kind of 'scientific homogenizing'. Those who oppose a scientific paradigm are epistemically disenfranchised and institutionally disempowered. Minority views are excluded from the dominant paradigm. And a condition for gaining 'scientific citizenship' or simply recognition of one's work is to adopt the main scientific paradigm, its language, its laws and conceptual taxonomy.

Leaving behind Kuhnian communitarianism means rejecting the philosophical assumption that science evolves via epistemic *membership of one historically-culturally sufficiently insulated scientific perspective.* This is something that historians of science have rediscovered with their kaleidoscopic approach to science and increasing emphasis on material cultures, rather than on canonical textbooks, systems of beliefs, or scientific theories.

As soon as attention shifts to material cultures, crafted objects, and tools— or, in my language, to the modelling techniques, experimental tools, and technological resources available to any epistemic community to *reliably* advance scientific knowledge claims—the pluralistic, diverse, and fluid nature of scientific perspectives becomes evident. *Pace* Kuhnian communitarianism, scientific knowledge travels across cultures and times and, I argue in the rest of this chapter, it is inherently cosmopolitan.

Scientific cosmopolitanism, in my idiolect, has nothing to do with scientific 'globalization',[4] the 'integrating' of scientific perspectives in the sense of 'melting', 'merging', 'overlapping', or 'hybridizing'[5] historically and culturally situated perspectives, with all the troublesome colonialistic attributes implicit in such expressions.

Neither does scientific cosmopolitanism imply subscribing to a *lingua franca*,[6] a vestige of the Western colonial-imperialistic past. Historically and culturally situated scientific perspectives have been able to travel, trade, and thrive, or—in my idiolect—'interlace' with one another, not only in the absence of but in fact *thanks to the absence* of a lingua franca.

Scientific perspectives 'interlace' in the sense that without ever losing their historical and cultural identity they nevertheless *collectively* feed into cosmopolitan

---

[4] See Sandra Harding (2015, Ch. 4) for a discussion of the problems she sees in well-meaning calls for 'integrating' indigenous cultures with modern Western science as articulated, for example, by Susantha Goonatilake (1998) as an attempt to salvage minority cultures that would otherwise go extinct.

[5] On the notion of 'hybridity', the historian of science Anna Winterbottom (2016, pp. 2–3), for example, writes: 'Hybrid is a word applied to animals, plants and people. As Steven Shapin argues, both scientific and social knowledge are in some sense inevitably hybrid, since what we know about natural phenomena is always mediated through our knowledge of the people who describe them.... The term "hybrid" has also been used previously to describe global encounters, especially in the Atlantic context; indeed, it has attracted some criticism for its ubiquity in this context. Nonetheless, the concept of hybridity has been useful in moving the global history of ideas forward from an earlier language that relied on the concept of centre and periphery and the assumption that both science and commerce were essentially European creations, exported and modified to a greater or lesser extent'.

[6] As Gobbo and Russo (2020, pp. 196–197) write: 'The expression "lingua franca" comes from Latin. . . . It was proposed originally by Hugo Schuchardt, . . . for him, the *lingua franca* was a *Vermittlungssprache*, a "mediation language", that emerged because of the trading in the Mediterranean Sea during the Middle Ages between speakers of Romance languages such as Castilian, Catalan, Provençal, Ligurian, Venetian, once in contact with Arabs and Turks. It was a sort of unstable pidgin for the domain-specific purpose of trading. . . . Clearly, the original *lingua franca*, and pidgins in general, do not respect this requisite. In fact, the club of the languages of science is very exclusive: according to *Ethnologue*, there are currently more than 7000 living languages in the world; however, if we check all the original scientific production—even in a large sense, including Western and Eastern antiquity—the languages of science in all the history of humankind are less than 20 (Gordin 2015, Ch. 1)'.

knowledge. Consider a piece of Scottish tweed and how the woollen fibres are interlaced so that it is almost impossible to discern one from the others. Similarly, it is difficult to disentangle an individual situated scientific perspective from others with which it has *historically interlaced*.

Here it is important not to conflate 'interlacing' with the already introduced notion of 'intersecting' scientific perspectives familiar from previous Chapters. 'Intersecting' refers to a methodological feature of scientific perspectives, 'interlacing' to a historical one. Scientific perspectives 'intersect' (sometimes synchronically, as with the case studies of Chapters 4.b and 4.c, other times diachronically, as with the case study of the nucleus in Chapter 4.a) whenever more than one perspective is required to refine the reliability of the claims of knowledge advanced. This is the case when we bring borehole data to bear on paleoproxy data in climate science; or when the educationalists's perspective is brought to bear on the cognitive psychologists's one, just to refer back to our case studies.

By contrast, 'interlacing' captures how historically a number of situated scientific perspectives have encountered and traded with one another some of their tools, instruments and techniques. As a result of these encounters and trades, some of these tools changed their *use*, so that tracking the particular history of any such tool via interlaced scientific perspectives becomes a way of tracking the evolution of knowledge concerning particular phenomena elicited by that tool in what I call a 'historical lineage' (more on it in the next Section). Tracking such evolution through historical lineages is an integral part of how historically identified and open-ended groupings of phenomena were sorted into the evolving kinds we know and love, as discussed in Chapters 9 and 10.

That scientific perspectives have historically interlaced is a fact. What ought to be said about it from a philosophical-normative point of view? The nature of the interlacing matters in each case. Some interlacing has clearly been the product of exploitative trade encounters, exclusionary projects of political assimilation and marginalization, and colonialist ambitions. Any such encounters where one scientific perspective imposes itself upon others for the purpose of intellectual dominance, political oppression, and socioeconomic exclusion result in endemic varieties of epistemic injustice.[7] In Section 11.5, I concentrate on two varieties of epistemic injustice originating from interlacing going badly wrong—what I call *epistemic severing* and *epistemic trademarking*—and I outline the nature of the epistemic remedies for them. But before I turn to those, one question still looms large: when is the interlacing philosophically virtuous? In other words, under

---

[7] On the notion of epistemic injustice, see Fricker (2007). For a philosophy of science discussion of these themes, see Fernández Pinto (2020a, 2020b).

what conditions does the historical interlacing *not* lead to epistemic injustices? Here I can only sketch some normative pointers in increasing order of strength:

1. The interlacing has to respect the historical situatedness and cultural identity of each perspective: interlacing should not come at the cost of homogenizing what is distinctive about the scientific perspectives involved.
2. The interlacing should be based on mutually and reciprocally agreed upon norms and methods for knowledge production sharing among the relevant epistemic communities. It goes without saying that acquiring and using other communities' knowledge, tools, and techniques without their informed consent and approval and without historically acknowledging them is unethical and unjust.
3. The interlacing should make transparent the mechanisms and pathways through which each situated epistemic community with its perspectival practices has contributed to knowledge production. There should be no merging or blending or obfuscating the specific contribution of each community without giving proper credit.
4. The interlacing should prevent exploitative systems of scientific knowledge appropriation resulting in the commodification of knowledge to the exclusive socioeconomic benefit of one community at the expense of others who have also contributed to knowledge production with their perspectival practices.

Epistemic severing—as I describe it in Section 11.5—is the violation of pointer 3. Epistemic trademarking violates 4 (building on the violation of 3). Both presuppose a violation of 1 and 2. Returning to 1, how can perspectival realism offer a normative antidote against the perennial risk of interlacing at the cost of homogenizing? How to think in philosophical terms about the historical and cultural situatedness of each perspective in the interlacing? I address these questions in the next Section.

11.3. Historical lineages and two notions of situatedness

I'd like to think of the *situatedness* of scientific perspectives not in terms of the self-contained sociocultural-national context in which they emerge, but as part of a 'historical lineage'—an open-ended, ever-growing, and irreducibly entwined body of scientific knowledge claims grounded in well-defined scientific practices and in their experimental, modelling, and technological resources. It is useful here to distinguish two senses in which a scientific perspective can be said to be historically and culturally situated: *situated in* and *situated for*.

A scientific perspective is always *situated in* the scientific practice of a given epistemic community at a given historical time. As my definition from the outset (Chapter 1) made clear, to have a scientific perspective is not just to endorse a body of scientific knowledge claims. Equally important, it means to have at one's disposal experimental, theoretical, and technological resources to *reliably* advance those scientific knowledge claims and methodological-epistemic principles that can *justify* the reliability of the scientific knowledge claims advanced. The material tools and instruments, the craftmanship of particular techniques— be it Joule's paddle-wheel, Lavoisier's ice calorimeter, or hydraulic engineering in Alhambra and Villa D'Este—are an integral element of the community's ability to make inferences from data to phenomena and to encounter modally robust phenomena.

Material cultures—like the paper and the blue dye in architectural blueprints, or the wooden carving and the beads in *lukasas*—are the medium that makes perspectival$_1$ representations possible.[8] Recognizing, acknowledging, and preserving what is historically and culturally distinctive in each and every material culture is vital for the situatedness of a scientific perspective (qua *situated in*). Epistemic severing begins with discarding, disparaging, or homogenizing the material cultures, techniques, and tools of particular epistemic communities. For example, it begins when one fails to acknowledge the labour of Scottish coastal communities in the kelp industry since the eighteenth century behind the thriving glassware production that proved instrumental for reliably advancing electrical researches at the time of Crookes and Thomson. It begins also when one fails to appreciate how the unreliability of an instrument like Lavoisier's ice calorimeter was in fact key to the eventual downgrading of caloric from an in-the-making kind to an empty kind at the turn of the nineteenth century.

Being *situated in* is the notion that is relevant to my epistemological analysis in this book and whose far-reaching implications for two varieties of epistemic injustice will become clear in the rest of this chapter. But there is another sense in which a scientific perspective can be said to be historically and culturally situated: *situated for* specific purposes and epistemic needs. For example, one can say that kelp-making was a situated practice of Scottish Hebridean communities since the eighteenth century *for the purpose* of the local economy at a time when the Napoleonic wars had made it difficult for the local glass manufacturers in Leith and Glasgow (but also in England, with Bristol and Liverpool being thriving centres) to import from Spain natural alkali (the ash called 'barilla' obtained by burning Mediterranean saltwort—see Clow and Clow 1947; Kennedy 2017).

---

[8] The importance of the medium of the representation and its materiality is key to Tarja Knuuttila's (2011) artefactual approach to model-based representation, with which I completely agree here.

Situated in and *situated for* are intertwined notions as much as scientific knowledge production is itself intertwined with historical and socioeconomic factors of the society in which these practices took place. I cannot hope or claim to have done justice to the notion of 'situated for' in this book as I am no historian or sociologist of science.[9] But I do hope that the distinction helps clarify the broader philosophical issue that matters for my analysis here. Scientific perspectives are situated in historically and culturally well-defined practices, including their experimental tools and wider material cultures.

The philosophically virtuous 'interlacing' is one where situated perspectives, each in its own epistemic right, are located, so to speak, alongside others in an open-ended and ever-growing historical lineage *without violating the aforementioned normative pointers*. It is in this specific philosophical sense that—*pace* Kuhnian communitarianism—historical lineages span and ramify beyond geographical, national, and sociocultural boundaries. They have a history, evolve, and branch out rather than statically demarcate well-defined territories, scientific homelands or shared memberships.

Any encounter and re-encounter with the natural world is therefore one where one entangles one's own situated scientific perspectives again and again in 'foreign grafts' (to use Waldron's [1992] apt phrase). Such cultural and historical encounters deserve more attention among philosophers of science. On the one hand, they are the key junctures of complex historical lineages where the historical 'interlacing' becomes visible so that it is possible to track the evolution of knowledge concerning particular phenomena. On the other hand, they are also the critical points of any historical lineage where interlacing risks going badly wrong and epistemic injustices might creep in. I illustrate the first of these two aspects in the next section and I turn to the second in Section 11.5.

Consider as an example the history of magnetism. The Anglophone canon of historical resources often maintains that the first systematic study of magnetism began with William Gilbert's *De Magnete* in 1600. Gilbert in the first pages of his famous treatise gives his own account of the origins of knowledge about the magnetic properties of the lodestone. He mentions Lucretius, Aquinas, Plato, and Marsilio Ficino (the latter credited with the mystical view that the magnetic direction of the lodestone was directly caused by the constellation of Ursa Major). Paracelsus is there too, and the sailors of the Italian town of Amalfi, to whom Gilbert attributed the invention of the mariner's compass—a claim that apparently led to the six hundredth anniversary celebration of this alleged invention in Amalfi in 1901 (see Mitchell 1932, pp. 123–124; Smith 1992).

---

[9] I refer the reader to the work of Schaffer (1989, 1997), Steinle (2005/2016), and Werrett (2019), among many others, for some illustrative examples.

344   THE WORLD AS WE PERSPECTIVALLY MODEL IT

And in this swirl of names Gilbert also cited the Venetian Marco Polo, who—he claims—'about the year 1260 learned the art of the compass in China' (Gilbert 1600/1958, p. 7). This too has been disputed by a number of scholars, who trace it to a mistake in translations of Polo's *Il Milione* (see Smith 1992, p. 29), where apparently the word 'compass' did not originally appear. But why do these details about *De Magnete* matter? They matter because if my analysis so far is on the right path, by zooming out on the story of Chapter 10, we should be able to see that J. J. Thomson's experiments with cathode rays bending in a magnetic field in 1897 belonged to the same historical lineage from which—over six hundred years earlier—the mariner's compass and associated astronomical-nautical practices originated. This historical lineage is worth a closer philosophical look with an eye to tracking the evolution of knowledge concerning particular phenomena about the Earth's magnetic field. I attend to it in the next section.

11.4. Han geomancers, the monk from St Albans, and Norse sailors

Historical lineages come with responsibilities and accountability. The history of magnetism, like any other, turns out to be a lot less linear, a lot less Western-centric ('Italo-centric' in this case) and more zig-zagging than the historical canon suggests. It did not start with the sailors of Amalfi and Marco Polo to reach Gilbert and eventually Newton's 'magnetic effluvia' in the *Opticks*, all the way up to the Victorian Cambridge of James Clerk Maxwell and J. J. Thomson. Instead, it seems to have started around the time of the Chinese Han dynasty. According to Joseph Needham, the Han geomancers, including Wang Chung's *Discourses Weighed in the Balance* (83 CE), were among the first to refer to a 'south-controlling spoon', believed to be a spoon made of lodestone which, placed on a diviner's board, could be used for divination.[10] In the following centuries, during the Sung dynasty, geomancers developed a 'wet' compass with the lodestone inside a wooden fish floating in water. And from there it developed into a 'dry' compass where a magnetized needle (sometimes suspended from a silk fibre) replaced the original lodestone (see Needham 1970).

Needham (1969/1972, p. 72, fn. 1) argued that such 'south-pointing' early compasses originated from Chinese symbolism of the emperor as the pole star facing south to his realm, and that from their original geomantic use, around the eleventh century CE or possibly even earlier, they were used at sea. Whether

---

[10] As Needham (1969/1972, pp. 72–73) clarifies: 'It is true that this device is a reconstruction from a text, and that an actual spoon made of lodestone has not so far been found in any tomb. But during the following thousand years there are constant literary references to a "south pointer" which can only be explained if something of this kind existed'.

the compass reached the West directly overland from China or was independently developed in the West remains uncertain.[11] But by the time of the Sung dynasty, the Chinese Tsêng San-I (1174 CE *Mutual Discussions*) was in command of a theory of magnetic declination a few centuries before ca. 1440, when the 'German makers of portable sun-dials embodying compasses by which to set the noon line begin to make special marks on their dials which showed empirical knowledge of declination' (Needham 1970, p. 244).

'The fact that these theories were not of the modern type'—Needham (1969/1972, p. 75) remarks—'does not entitle us to ignore them. The whole discovery had arisen from a divination procedure or cosmical magic, but what carried it forward was the Chinese attachment to a doctrine of action at a distance, or wave-motion through a continuum, rather than direct mechanical impulsion of particles; atomism being foreign to them, this it was which led them on to see nothing impossible in the pole-pointing property of a stone or of iron which had touched it'. And it is hard not to see analogies with the continuum physics that one can find in Gilbert's studies of magnetism, Newton's 'magnetic effluvia' in Query 22 of the *Opticks*, all the way to Faraday's 'magnetic lines of force' in the early nineteenth-century.

Most importantly, we see here intertwined the two aforementioned notions of situatedness. For the Han–Sung craftmanship of wet and later dry compasses was *situated in* a scientific practice which delivered knowledge of important phenomena about the Earth's magnetic field and that sits in a historical lineage alongside Gilbert's later research in *De Magnete* all the way up to Faraday and Thomson in the nineteenth century. The Han–Sung craftmanship of compasses was also *situated for* a particular purpose: divination in this case. It was part of a complex network of cosmological beliefs with a clear sociopolitical undertone (concerning the absolute power of the Chinese emperor over his realm). The two senses of 'being situated' are clearly intertwined in this example.

It is through these crafted objects and associated practices that Han–Sung geomancers first encountered the modally robust phenomenon of the Earth's

---

[11] Smith (1992, p. 24) argues, for example, that 'The overland route remains possible, but again, decisive evidence is lacking. Consequently, until the link is established, I will adopt the provisional hypothesis that the two compasses of China and Europe were invented and evolved independently'. Needham (1970, p. 247) suggests that, 'Since the crucial couple of centuries before Alexander Neckham (+1190) has so far afforded no trace or clue from intermediary regions such as the Arabic-Persian culture-area or the literatures of the Indian sub-continent, the possibility arises that transmission from China occurred not in the maritime context at all, but by some overland route through the hands of surveyors and astronomers who were primarily interested in establishing the "meridian" for the purpose of cartography and sun-dial clocks'. Needham adds: 'It is certainly a striking fact that as late as the +17th century the needles used in the compasses of surveyors and astronomers all pointed to the south, in contradistinction to the north-pointing sailors' needles—exactly as the Chinese needles had done for perhaps as much as a millennium previously' (p. 247). Other scholars have pointed out the presence of lodestone artefacts among the Mexican Olmec well before the Chinese ones (see Carlson 1975).

magnetic field. The situatedness of their perspectival₁ representation of the cosmos (ruled by the south-pointing spoon/emperor on a diviner's board) resides in the *use* they made of these tools *for* divination. *From their point of view*, these very early compasses were divination tools. Yet it is through these tools *situated in* these practices that they were able to acquire knowledge about a real robust phenomenon concerning the Earth's magnetic field, a phenomenon about which it was possible to tease out a network of inferences (e.g. concerning its directionality and magnetic declination) as later scholars of the Ming dynasty did.[12]

Some of the very first mentions of a 'dry' compass made of a magnetized needle appeared in a couple of treatises between 1175 and 1204 by the Benedictine monk Alexander Neckham from St Albans Abbey, near London (see Smith 1992, pp. 33ff). Such references are in the context of sailors having to find their way with a needle 'pointing north' through overcast skies (Neckham *De Naturis Rerum* quoted in Smith 1992, p. 37). They proliferated through a number of medieval authors—from the French poet Guyot de Provins, who worked at the court of Frederick Barbarossa, to the Icelandic poet Snorri Sturluson, who allegedly in 1213 received a mariner's compass as a gift; from Michael Scot, the Scottish astrologer to the court of the Holy Roman Emperor Frederick II, to Peter Peregrinus de Maricourt, working at the Sicilian court of Charles I of Anjou and whose *Epistola de magnete* (1269) is widely regarded as among the first explicit statements that directionality of the lodestone is affected not by stellar constellations but by the 'poles of the world' (see Smith 1992 for an extensive analysis of all these authors).

Similar knowledge of the magnetic compass used at sea has been recorded in Persian sources as early as 1232–1233 CE, and the Rasulid sultan of Yemen Al-Ashraf Umar ibn Yūsuf (ca. 1294–1297 CE) wrote one of the early treatises on the magnetic compass (see Schmidl 1996, pp. 84 and 88–89).

It was through both astronomical/surveyor practices necessary to establish the magnetic declination so important for cartography, sun-dial clocks, *and* nautical techniques among Chinese, Italian, and Norse sailors that knowledge of terrestrial magnetism eventually entered modern science. Where does one situated scientific perspective begin? And where does it end? Should one say that the Han geomancers with their 'south-controlling spoon' made of lodestone occupy a different scientific perspective from those of the much later Sung dynasty with their 'dry' compass made of a magnetized needle? How does the latter relate to the

---

[12] The geomantic use of the Chinese compass seems to have lasted for centuries. Still in ca. 1619 in the so-called Selden map rediscovered in 2008 at the Bodleian Library in Oxford, the compass rose on this Chinese map seemed to have been associated to a geomantic-cosmological compass that John Selden obtained with the map (see Batchelor 2013, p. 47). I am very grateful to Rebekah Higgitt and Simon Schaffer for helpful pointers on this topic.

scientific perspective of Petrus Peregrinus working at the Sicilian court of the King of Anjou? Or to the nautical perspectives of Italian, Icelandic, and Norse sailors?

A few hundred years separate the Chinese 'floating fish' from Peregrinus's description of a mariner's compass, in a completely different geographical-political-sociocultural context. And yet, echoing Calvino's Marco Polo, travelling, one realizes that differences are often lost; that tools and material artefacts—like a magnetized 'floating fish'—resemble the magnetized needles used in deep-sea sailing by Norse sailors. By 'interlacing' the nautical perspective with the geomantic/astronomical one, common experimental tools and techniques were traded: their *use* changed, and sometimes dramatically, as in this example. An instrument *for* divination became an instrument *for* navigation.[13] And an instrument *for* navigation became eventually an instrument *for* electromagnetic research—from the use of a magnetic needle in Ørsted's and Faraday's experiments on electromagnetic induction all the way to Thomson's exhausted glass tubes for cathode rays in a magnetic field.

While 'intersecting' perspectives matters *methodologically* for refining the *reliability* of knowledge claims over time (recall the case studies in Chapters 4.a, 4.b, and 4.c), their 'interlacing' in a historical lineage matters for tracking the evolution of knowledge concerning some phenomena. Without ever losing their historical and cultural situatedness (qua situated *in*), scientific perspectives nevertheless interlaced with others. Some of their tools and techniques changed use and function (situated *for*) as a result of these encounters. The notion of 'interlacing scientific perspectives' gives us then an important glimpse into the exploratory nature of objects, tools, artefacts, and techniques, and how their *situatedness for* is part of complex and highly non-linear historical lineages. Multicultural science is not some kind of fragmentism of scientific perspectives, which siloes the cultural identity of each perspective in a way that is insular, and, worse, pits one against another.

At the same time, the unanimity of evolving kinds discussed in Chapter 10 is not the outcome of any successful universalizing process, or convergence towards some final endpoint. For it does not merge, integrate, or overlap the distinct scientific perspectives that—each in its own way—contributed to knowledge of phenomena and natural kinds. There is no lurking globalism in the zig-zagging ways in which scientific knowledge evolves over time, by taking some inferential paths rather than others.

---

[13] One of the questions left open by Needham was whether these early prototypes of a compass were effectively used for nautical purposes rather than cosmogonic/geomantic ones. Some authors have conjectured that because Chinese trading routes were of relatively short distances and the trading was seasonal (dictated by monsoons), the Chinese compass might not have been primarily used for nautical purposes (see Davies 2018).

348   THE WORLD AS WE PERSPECTIVALLY MODEL IT

In the longest historical vistas, this unanimity can only arise from thousands of scientific perspectives that historically interlaced with one another. We would not have had J. J. Thomson's achievements without the mariner's compass and the ability of earlier historical perspectives to use their tools and techniques to identify modally robust phenomena and acquire more and more reliable knowledge of magnetism—from the cosmogonic/astronomical practices of Han–Sung geomancers to the nautical ones of Mediterranean and Norse sailors. By the same token, we would not have had Thomson's discoveries without Ebenezer Everett mastering the glass-blowing techniques necessary to produce Thomson's bespoke cathode rays, and without concerted efforts to produce high-quality glass from vegetable ashes (be it Spanish barilla or Scottish kelp).

Multiculturalism starts from here: from the 'recognition'[14] of the role that these myriad historical perspectives have played in shaping scientific knowledge as we have it today. Some philosophers of science may find such recognition difficult—Popper's demarcation criterion still looms large. Who among the philosophers would want to entangle the history of magnetism with geomantics, chemistry with alchemy, or astronomy with astrology? Yet, as I said already, the historical interlacing of perspectives is a fact such that trying to disentangle the Newton of the *Principia* from the Newton of the *Opticks*, or Dalton's chemistry from imponderable fluids, or the mariner's compass from the magnetic field studied by Faraday does violence to history.

Recognizing this *historical multiculturalism* poses no threat to science or the reliability of scientific knowledge. Perspectival pluralism, as I have characterized it, offers fluid, ever-shifting inferential networks within which scientific knowledge forms, grows, and evolves, and realist commitments—from data to phenomena to kinds—coalesce. It captures the interconnectedness of scientific perspectives—their methodological intersecting and historical interlacing—to explain how they collectively contribute to scientific knowledge that *can be shared*.

This is the beginning of an argument for reconciling multiculturalism with cosmopolitanism in science: the distinctive historical-cultural situatedness of countless scientific perspectives with the cosmopolitan nature of scientific knowledge that—via the Chinese Silk Road, Mediterranean or Viking sea routes, Bedouin caravans along North African routes and myriad others—travelled across cultures and times.

---

[14] I use 'recognition' in quotes as it has become a word of art in social and political philosophy since the work of Fraser (1995), Honneth (1996), Taylor (1992), and Young (1997). Fraser and Young are reprinted in Willett (1998).

11.5. Two varieties of epistemic injustice in science

How can one move from multiculturalism in science so understood to scientific cosmopolitanism? And what about the tension between multiculturalism and cosmopolitanism well-known from the history of social, ethical, and political philosophy? How does that tension translate in the domain of scientific knowledge?

Historically, multiculturalism has stressed the role of pluralism and the recognition of minority groups and the cultural identity of often historically marginalized, exploited, and endangered communities (see, e.g., Fraser 2003; Kymlicka 2007; Taylor 1992). Multiculturalism has traditionally been advocated in political theory to reject assimilation or exclusion of such non-dominant groups; to call out injustices done by policies of assimilation or exclusion; and to offer remedies for those injustices (see Kymlicka 1996, 2001).

Cosmopolitanism, by contrast, has traditionally been invoked as a philosophical-normative position that seeks what is common among epistemic communities whose historical and cultural identity might be far apart (see Appiah 2006; Robbins and Horta 2017; Waldron 1992; for a recent discussion of cosmopolitanism in the philosophy of medicine, see Broadbent 2019, Ch. 7).[15] It emphasizes fluidity in commercial trades and historical-cultural encounters as a way of countering what Jeremy Waldron (1992, p. 781) has called 'the staple claim of modern nationalism . . . that we always have belonged to specific, defined, and culturally homogenous people'.

When it comes to multiculturalism and cosmopolitanism *in science*, and their respective value when thinking about *who* science really ought to be *for* and how one *ought to think* about scientific advances, a good starting point is to chart two main varieties of epistemic injustice often to be found in narratives about scientific knowledge and that arise from 'interlacing' gone badly wrong, as outlined in Section 11.2. I am going to call them *epistemic severing* and *epistemic trademarking*.

Epistemic severing affects narratives about scientific knowledge production that tend to surgically excise the contributions of particular communities either within the same scientific perspective or across culturally diverse perspectives. Severing is an act of informational injustice[16] in how scientific

---

[15] The history of both multiculturalism and cosmopolitanism is huge, nuanced, and complex, and it goes well beyond my remit and scope in this chapter to even attempt at summarizing either of them. As such I will not go here into the details of varieties of cosmopolitanism that have been put forward especially in the context of cultural anthropology, diasporic studies, and sociology of science (see, e.g., Beck 2004; Derrida 2001; Gilroy 2010; Latour 2004; Stengers 2010/11; Watson 2011, for some examples).

[16] By 'informational justice' I mean the broad framework that focuses on 'equitable inclusion of people, groups, and communities as they themselves are sources of information, and they actively contribute to, seek, process, and analyse information' (Atkins and Mahmud 2021, p. 1). Much of the

knowledge production *gets narrated* in scientific textbooks and canons. It should not be confused with the (epistemically more innocuous) omission or epistemic 'blinkering', as one might call it, that inevitably accompanies any scientific narrative, where the narrator can of course always choose to foreground some pieces of information and background others depending on their relevance to the narrative and on basis of important ethical considerations too.[17]

Epistemic severing is the act of cutting off specific historically and culturally situated communities to historically remove or blur their contributions to 'historical lineages' in the scientific knowledge production. This can sometimes be done on various socioeconomic-ethnic grounds, as when it is epistemically and socially inconvenient for the ruling class to admit the contribution to scientific achievements resulting from manual labour of the working class. This is what happened in the example of the kelp labourers among the Scottish coastal communities.[18] Electromagnetic research from Crookes to Thomson could flourish in the second half of the nineteenth century thanks to long-standing situated practices of producing lead-free glass from natural alkali made possible by kelp labourers in Scotland and before them barilla labourers in Spain.

Yet these communities are hardly ever even mentioned when writing narratives about scientific knowledge production, needless to say about realism in science. And it is not just that the story would be too long, too complicated or too meandering to tell. It would be epistemically insufferable for the scientific canon. This is an example of an epistemic injustice in narratives about scientific knowledge. And the problem only gets amplified if one considers communities across culturally diverse perspectives.

A story of developments in telecommunications that fails to mention local indigenous knowledge about gutta-percha as key in the production of underwater

current work in this field concerns information and communication technology (ICT), with a focus on notions such as 'information poverty' and 'information inequities' in the way in which particular data are collected, analysed, and used by various social groups (see, e.g., Eubanks 2011). However, there is an additional and no less important aspect of informational justice that concerns not so much *access to* data and information already available, but instead the very *production of scientific knowledge qua information*—in the forms of data, inferred phenomena, modelling practices to elicit them, experimental tools, and so forth—and how that information gets passed on, shared, and traded from one epistemic community to another in the seamless process of knowledge production.

[17] I am very grateful to Adrian Currie, Catherine Kendig, S. Andrew Schroeder, and John Turney for comments on this point.
[18] For example Gray (1951, p. 200) notes that 'it was more profitable for the landlords to become active entrepreneurs, organising the surplus labour of their estate to produce, at fixed rates, from the raw material which they kept in their own hands. Thus, the landlord could force the labourers to work on a product that would be entirely at his disposal. With the weed that lay on land rented by the small tenants—and most of the land in Highland properties was coming into this category—this was easy to accomplish. Such tenants were usually without leases and their terms of tenure could be arbitrarily changed from year to year'.

telegraph cables commits epistemic severing towards the indigenous Kadayans and Murut communities of Borneo.[19] Epistemic severing is also committed by a narrative on the invention of Hindu-Arabic numerals that fails to mention the role played by the North African nomadic culture of Bedouins[20] in the transmission of such knowledge to the Mediterranean communities.

The outcome in each case is a distorted historical lineage, an excised scientific canon that is the mirror image of the ruling class. A crucial scientific achievement gets credited to some kind of 'passing the baton' among the main civilizations (the Indian, the Arabic, and the European) as if they were operating in an epistemic and social vacuum. Severing is the opposite of interlacing. Or, better, it is interlacing gone badly wrong from a philosophical-normative point of view: interlacing that violates normative pointers 1–3 as outlined in Section 11.2. As these examples show, at the receiving end of epistemic severing are typically what get referred to as *minority communities*, where minority is here understood as synonymous with under-represented communities that on various grounds (e.g. class, ethnicity, gender) are not the dominant, ruling, scientific-canon-writing ones.

Epistemic severing can happen in various ways. It can happen as a wilful act on behalf of the narrator to exclude the contribution to scientific knowledge production of under-represented communities. But it can equally happen in a non-wilful—yet still culpable—way as a result of socioeconomic structures and epistemic norms that place an emphasis on particular modalities of scientific knowledge production over others (e.g. textual rather than oral, codified in educational curricula rather than artisanal, universal rather than local knowledge). In the latter case, the narrator who fails to acknowledge such contributions, while often unaware of them, would still commit an epistemic injustice—one resulting

---

[19] Gutta-percha is a rubbery substance obtained from the sap of some trees natives to Malaysia but also South America. English settlers from the Asian colonies were responsible for bringing samples back to Kew Gardens, London, and soon the manufacturing of gutta-percha for underwater telegraph cables but also the production of golf balls took off in Victorian Britain (see Burbidge 1879; Williams 1964).

[20] Consider as an example the Hindu-Arabic system of numerals. As a scientific perspective, it did not remain confined to the Baghdad of the Abbāsid culture where it flourished. It travelled well beyond it. The merchants exchanging beads, spices, leather, or beeswax along the coasts of North Africa in medieval times were using different currencies and units of measure and spoke different languages. Trading goods made it necessary for them to develop knowledge of accounting techniques, valuation practices, number systems, and interest rates rules that could facilitate trading encounters. Such knowledge travelled with the merchants. The early Hindu mathematical treatises of Āryabhata in the sixth century CE and Bhāskara in the seventh century reached ninth-century Abbāsid Baghdad via a flurry of Arabic translations (see Al-Khalili 2010, Ch. 5; Sarton 1927). And from Abbāsid Baghdad where Al Khwārizmī wrote his landmark *Kitab al-Jebr* (*Algebra*), practice-based knowledge of commercial arithmetic travelled along the trading routes of Syria and Egypt and eventually reached the Pisan merchants in the Algerian town of Bugia. Among them was Leonardo Fibonacci, whose *Liber Abaci* in 1202 marks the beginning of what scholars have seen as a 'financial revolution' in thirteenth-century Tuscany (see Goetzmann 2005, Ch. 7; Rashed 1994).

from structural inequities in the way in which scientific knowledge production is narrated and portrayed within particular societies.

Epistemic severing therefore is not simply failing to 'recognize' the contribution of these communities. It slashes through the very fabric of scientific knowledge production; it tears apart the historical interlacing of situated scientific perspectives that is ultimately responsible for the growth and evolution of scientific knowledge. By severing the historical lineages at the junctures where minority communities feature, epistemic severing damages and jeopardizes the very possibility of understanding the processes through which scientific knowledge become possible over time. Failing to do justice to the historical fact that scientific knowledge is indeed the outcome of a seamless multicultural web makes it hard to acknowledge ways in which today it continues to benefit from interlacing perspectives of minority communities.[21]

The remedy to epistemic severing cannot therefore just be a 'recognition remedy' (to use again here Nancy Fraser's [1995] and Charles Taylor's [1992] expression). Recognition remedies are of course very important: for example, the five asteroids named after Aboriginal people from Mer Island as a tribute to their ancient astronomical knowledge for navigation and agriculture.[22] But the contributions to scientific history of these communities do not just deserve to be recognized. Epistemic severing can only be undone by 'reinstating' these communities in epistemological narratives about scientific knowledge production.

'Reinstatement remedies' call for restoring the role and position from which these communities had been unjustly removed in scientific narratives. Reinstatement remedies take different forms. In what follows, I focus on three of them. They are inter-related and might be labelled as follows: reinstatement of authorship in the co-creation of knowledge, reinstatement of historical role in the production of knowledge, and reinstatement of intellectual ownership for scientific knowledge. For example, to start with the first one, when it comes to indigenous knowledge, there is the imperative to collaborate with indigenous communities so that their epistemic agency is fully 'reinstated' in authorship of scientific narratives (see Leonard et al. 2020; TallBear 2014, 2016; Wylie 2020). In the words of Kim TallBear (2014, emphases added):

If what we want is democratic knowledge production that serves not only those who inquire and their institutions, but also those who are inquired upon (and appeals to 'knowledge for the good of all' do not cut it), we must soften that boundary erected long ago *between those who know* versus *those from whom the*

---

[21] I thank Jon Turney for helpful comments on this point.
[22] See https://www.abc.net.au/news/2020-08-15/asteroids-named-in-honour-of-indigenous-australians/12557778.

raw materials of knowledge production are extracted. Part of doing this is broadening the conceptual field—thinking more expansively about what counts as risk (ontological harms?) and rightful benefit (institution building and community development?) in the course of building knowledge. It is also helpful to think creatively about the research process as a relationship-building process, as a professional networking process with colleagues (not 'subjects'), as an *opportunity for conversation and sharing of knowledge*, not simply data gathering.

Thus, the first kind of reinstatement remedies consists in fostering practices of co-creation of knowledge, via *'conversation and sharing of knowledge',* in TallBear's words.[23]

A second example of reinstatement remedy concerns more directly historical-philosophical narratives about scientific knowledge production. A significant volume of work in the history of science[24] and social studies of science (see, e.g., Bhambra 2007; Santos and Meneses 2020) in recent times has gone in this direction. In the history and philosophy of science, important work has been done to reassess practices of the past and their role in understanding key historical moments (e.g. think of Chang 2012a on the Chemical Revolution; or work in ethnobiology, e.g. Kendig 2015; Ludwig 2018b; Ludwig and Weiskopf 2019). But debates on realism in science have failed to catch up with these trends.

Philosophical discussions about realism and scientific knowledge have continued to be run—by and large—from a view from nowhere. Perspectival realism can offer an alternative. For it shows how the very process of scientific knowledge production requires a kind of multiculturalism that I have so far described in terms of 'interlacing' perspectives. The main realist take-home message of this book is that the reliability of scientific knowledge—what we most treasure about science—demands and requires several scientific perspectives whose methodological intersecting and historical interlacing is responsible for our ability to move from data to phenomena and from phenomena to kinds.

As a project in the epistemology of science, perspectival realism does not cut a motorway through the inferential garden of forking paths. It follows instead the zig-zagging paths as they run across multiple historically and culturally situated

---

[23] Increasing attention to the methodological need to collaborate with local indigenous communities and rectify possible 'equivocations' that may arise in the use of different languages and ontological categories is also stressed in recent work by Furlan et al. (2020, p. 7): 'The problem with multiple worlds passing by unnoticed in texts and for researchers is that they reproduce the power structure of occidental science. The invisibility of those worlds homogenizes people, and then flattens out the possibility of dialogue between worlds.' Along similar lines, Weiskopf (2020) has argued for the abandonment of any presumption behind what he calls the 'integration project' and stressed the need for 'knowledge coordination' instead.

[24] The postcolonial studies literature is too vast for me to be able to do justice to it here, but see, among others: Cañizares-Esguerra (2001), Ophir and Shapin (1991), and Shapin (1998). On the role of technicians and artisans in science, see Schaffer (1989, 1997, 2004) and Werrett (2019).

scientific perspectives. Perspectival realism starts with the historical fact about the seamless multicultural web and aims to provide an epistemological framework and associated realist commitments.

The second kind of reinstatement remedies call for abolishing the dichotomy 'minor' vs 'major' communities, patterned as it is on power structures and socioeconomic inequalities. The goal is to question and challenge the very epistemic framework (and its underlying power structures and inequalities) around which the labelling of 'minor' vs 'major' arose and upon which realism—qua monolithic view from nowhere—could flourish in the first instance.

But there is also a third kind of reinstatement remedies, which speaks more directly to another variety of epistemic injustice—closely connected with epistemic severing—, i.e. *epistemic trademarking*. Epistemic severing is a precondition for epistemic trademarking. Having severed the very historical lineages of knowledge production, the next step typically involves the appropriation[25] and branding of entire bodies of knowledge claims, with associated practices, as a 'trademark' of one particular epistemic community. Epistemic trademarking manifests itself in the fencing, and ultimately often merchandising of portions of scientific knowledge that are the products of myriad interlaced scientific perspectives.

STS scholars[26] and social anthropologists such as Marilyn Strathern, for example, have well documented how patenting truncates, fragments, and segments long and complex social networks. In so doing, knowledge gets commodified and rights of ownership applied.[27] This is what happened when the seedlings of trees imported from Malay to Kew Gardens and the techniques learned from the Kadayans and Murut of Borneo to collect the latex became effectively a trademark in the flourishing industry of telegraphic cables (see Noakes 2014, p. 124).

But epistemic trademarking does not refer so much to the privatization and patenting of scientific knowledge. By contrast with existing critiques of patenting systems, I see epistemic trademarking as yet another consequence of interlacing gone badly wrong on normative grounds. Epistemic trademarking builds on epistemic severing to further exploit merchandising rights arising from scientific

---

[25] In this respect, the origin of epistemic trademarking can be identified in broader and well-studied mechanisms of epistemic exclusion affecting marginalized groups in what Kristie Dotson (2014) has called 'epistemic oppression' and Emmalon Davis (2018) has characterized as 'epistemic appropriation'. The peculiar way in which these mechanisms play out in the case of scientific knowledge—via what I call *epistemic trademarking*—has not received enough attention to date in the philosophy of science literature.

[26] Mario Biagioli, for example, has done important work on the costs and benefits of intellectual property and patenting (see Biagioli 2019; Biagioli et al. 2011). See also Hayden (2011).

[27] 'In another case, forty names to a scientific article became six names to a patent application; the rest did not join in. The long network of scientists that was formerly such an aid to knowledge becomes hastily cut. Ownership thereby curtails relations between persons; owners exclude those who do not belong. . . . [T]he prospect of ownership cut into the network' (Strathern 1996, p. 524).

knowledge to the exclusive socioeconomic benefit of one epistemic community at the expense of others (in violation of the aforementioned normative pointer 4. in Section 11.2).

Traditionally, in economic theory and law, trademark protection has two main functions: 'the prevention of consumer confusion' and the protection against 'dilution by blurring' (Sunder 2018, pp. 217 and 219). The first is an argument that a market where consumers are not confused about the source and quality of trademarked products is more competitive, with more choice, higher-quality goods, and lower prices. The second refers to preventing the dilution of the originality and uniqueness of the trademarked products through unauthorized reproductions and unmonitored uses.

By analogy, I see epistemic trademarking operating along these two dimensions. Applying an 'epistemic trademark' to a piece of scientific knowledge or technological innovation is indeed a way of preventing consumer confusion. But it is equally a way of laying claim to merchandising rights. And it is also a powerful way of ringfencing the uniqueness and originality of a piece of knowledge or innovation as if it were the exclusive product of either one agent alone, or one particular epistemic community over any other that might have contributed to the process of scientific knowledge production, and that—once severed from the historical lineage—loses any rights to it. In either case, epistemic trademarking is the expression of merchandising concerns about scientific knowledge qua 'epistemic good' that can be commodified and traded to a target consumer audience.

By its very nature, epistemic trademarking harms forms of scientific knowledge that do not easily lend themselves to commodification and merchandising; for example, when the knowledge is collective rather than individual, oral instead of written, or passed on from one generation to the next rather than codified in scientific canons and epistemic norms.

Epistemic trademarking—like epistemic severing—can occur either within the same historically situated scientific perspective or across culturally diverse ones. Let me unpack this difference. Consider how even in common parlance scientific outputs and achievements are marked with someone's name: Newton's laws, Lavoisier's oxygen, Maxwell's equations, and the Boltzmann constant are just a few examples from the history of physics. Attaching names to a scientific result (be it equations, laws, constants, models, particles, chemical elements, etc.) is a way of rightly recognizing authorship and tracing back the original idea to its legitimate owner. This is of course a common and uncontroversial practice, and key to copyright laws and patent rights (where applicable).

As with epistemic severing (vs epistemic blinkering), here too one should not confuse epistemic trademarking with the (epistemically innocuous) epistemic 'labelling', so to speak, which is ubiquitous in science and does not necessarily presuppose that *severing* has occurred. In other words, it is possible to epistemically 'label' without severing. But I do not take it to be possible for epistemic trademarking to occur without severing. Then the question arises: when does a label or mark become an 'epistemic trademark'? The transition from an (epistemically innocent) label or mark to an 'epistemic trademark' occurs when there is an *epistemic overstretch* of the former to include not just a particular scientific output or achievement of someone, who is legitimately recognized as its *intellectual owner*, but also swaths of knowledge claims that for various historically contingent reasons become associated with that specific label or mark.

Consider the following examples: the passage from Newton's laws to 'Newtonian mechanics'; or from Maxwell's equations to 'Maxwellian electromagnetic theory'; or from Lavoisier's oxygen to 'Lavoisierian chemistry'. The former are marks. The latter are epistemic trademarks that designate entire bodies of scientific knowledge claims under the aegis of Newton or Maxwell or Lavoisier. Within those bodies of knowledge claims lie particular contributions and scientific achievements that predate Newton or Maxwell or Lavoisier, respectively.

For example, under 'Maxwellian electromagnetic theory' one would typically count not just Maxwell's equations for the electromagnetic field but an entire corpus of knowledge about electromagnetic phenomena that begins with Ørsted's experiments in 1820s Denmark and continues with Faraday's experiments in 1830s England, without counting the whole tradition of electrical researches with exhausted glass tubes that developed in the mid-to-late 19th century and was made possible by the manufacture of lead-free glass, produced by professionally trained glass-blowers, using alkali sources obtained from ashes of burnt seaweed (kelp) and, later, synthetic soda. The specific contribution of each epistemic community—including glass-blowers and kelp-makers—is lost in the name 'Maxwellian electromagnetic theory'. The emphasis is placed on the theory rather than on the tools, technological and experimental resources, or better the wider scientific perspective (qua historically and culturally situated scientific practice) in which the theory was embedded and became possible.

One might be tempted to brush this observation under a rug as epistemically moot. A label is after all just an expedient shorthand to refer to something. The historians know these historical lineages, it could be argued, and they can choose to zoom in and out of them depending on the focus of their narrative. Scientists and the public, however, do not need to be constantly reminded of them in daily discussions and common parlance.

Yet, I argue, what lies in the label is important. A label or mark in scientific discourse (e.g. 'Maxwell's equations') becomes an *epistemic trademark* (e.g. 'Maxwellian electromagnetic theory') when it ends up *concealing* the complex historical lineages and *blurring* the epistemic contributions of various communities. To be clear, in these examples there is no culpability on the part of individual epistemic agents (be it Maxwell, or similar) in the process of transforming a mark into an epistemic trademark, for there is no reasonable assumption of a wilful act of severing in the agent's intentions. Often the relevant epistemic agents are long dead well before an epistemic trademark associated with their name is even coined (as in this example). Thus, *epistemic trademarking* is first and foremost a structural phenomenon of how scientific narratives (or a particular kind thereof) get off the ground and tacitly enter public discourse as a result of specific epistemic norms that codify scientific knowledge production in particular societies.

It is the perceived need to protect epistemic goods (like the body of knowledge behind 'Newtonian mechanics' or 'Maxwellian electromagnetic theory' or 'Lavoisierian chemistry') under a trademark so as to avoid 'consumer confusion' with rival products, so to speak. In this case, the rival products included Aristotelian physics, which was still lingering in the medieval impetus theories of Oresme and Abu Al-Barakāt; the 'electromagnetic worldview' associated with ether theories; and Joseph Priestley's 'dephlogisticated air' and associated chemistry, respectively. The epistemic trademark shows its efficacy by branding bodies of knowledge as epistemic goods that can be easily recognized and commodified for the use of a particular consumer audience. 'Consumer confusion' was removed, and acceptance secured, by trademarking the electromagnetic theory—a hodgepodge of ether models and electrical fluid views at the time of Maxwell—under the epistemic trademark associated with Maxwell's equations.

Historical examples of epistemic trademarking abound. However, it would be hasty to conclude that this is just a historical phenomenon, or maybe one concerning narratives about physics in particular, whose past historiographical tendency to portray science as the product of a 'lone genius' has done much damage in cutting out entire communities from scientific narratives. Epistemic trademarking is very much an ongoing epistemic injustice in science affecting in equal measure the biomedical sciences, as the next example shows, going back to culturally diverse scientific perspectives.

In the biomedical sciences, the commodification of scientific knowledge for merchandising purposes relies more than ever on the practice of epistemic severing and epistemic trademarking. At the receiving end, there are ethnic minorities and local communities whose local knowledge—often orally transmitted from one generation to the next—is particularly vulnerable to these types of epistemic injustices. Indeed, epistemic trademarking is the very epistemic mechanism that structurally underpins the phenomenon known as 'biopiracy.'

Biopiracy is the 'unauthorized and uncompensated taking and use of biological resources for valuable research purposes' (Mattix 1999, p. 529; see also Mgbeoji 2006 and Shiva 2016). It is a derivative of bioprospecting, the search by pharmaceutical and biomedical companies for genetic and biochemical materials in nature that can be exploited commercially. In this case, the epistemic trademark finds its tangible incarnation in commercial trademarks that often overstretch swaths of knowledge claims contributed by a number of different communities.

One of the oft-quoted examples of biopiracy concerns the use of local knowledge about the rosy periwinkle plant.[28] The plant known in botany as *Catharantus roseus*—and with local names such as for example *tonga trongatsy*—has long been known in ethnobotany for its medicinal properties. Tea made with its leaves was believed in the Philippines to help with diabetes, among other conditions. The story goes that Canadian-trained surgeon C.D. Johnston in Jamaica became interested in collecting the leaves, drying them, and sending them to Robert Laing Noble, who was the Associate Director of the Collip Medical Research Laboratory at the University of Western Ontario (see Duffin 2000, on which I draw here). His brother Clark Noble played an important role in the discovery of insulin.

While the hypothesized anti-diabetic properties of the plant had already been refuted in the late 1920s, Robert Noble and his team was able to identify other unexpected medical properties. His collaborator Charles Beer was able to isolate vinca alkaloids with powerful anti-cancer effects. The discovery of vinblastine was announced in 1958 by Noble and Beer and the first clinical trial for the anti-cancer drug started in 1959 run by the pharmaceutical company Eli Lilly & Co.[29] This marks the beginning of the production on a global scale of vincristine and vinblastine, two compounds used in anti-cancer drugs.

Rosy periwinkle continues to be harvested, dried, and collected today in southern regions of Madagascar, where the highest-quality vinca alkaloids can be found. Local rural Malagasy communities continue to provide labour for large international corporations. Recent studies (e.g. Neimark 2012) have remarked how despite international protocols such as the Convention on Biological Diversity, very few socioeconomic benefits trickle down to the local Malagasy community. By and large, the community continues to rely on traditional methods for harvesting, drying, and transporting (often for long distances and on foot) the dried plants to central facilities where subsidiaries for multinational companies collect them for international transport. Often such tasks fall on older women in

---

[28] See for example https://www.europarl.europa.eu/news/en/headlines/world/20121203STO04309/biopiracy-protecting-genetic-resources-in-developing-countries

[29] 'During 1958, Lilly adopted Beer's method and quickly began trials on animal tumours with VLB (called Lilly 29060-LE). Johnson [*Irving S. Johnson, director of research at Eli Lilly*, M.M.] reported the anti-tumour effects at two meetings in the following year. . . . The question of who discovered the anti-cancer properties of the Vinca alkaloids seems less clear now. . . . Two teams were involved in the process researching the issue from two different angles, although they relied on the same well-established methods in biochemistry. Behind the teams were a number of lesser known (and less *knowable*) individuals, in Jamaica, Australia and the Philippines, who had investigated the healing properties of these plants' (Duffin 2000, pp. 165 and 167).

the Malagasy community, who can pick periwinkle in the wild and carry 'up to 5 to 10kg per trip. . . Some buyers estimate that close to half of all root bundles are brought to market by older women' (Neimark 2012, p. 436):

> Better quality does not always translate into better prices for the peasant producers because per kilo returns on periwinkle has remained low for years. These meagre prices have resulted in especially low margins for the peasant producers as compared to others in the chain. (p. 433)

Like the Hebridean kelp-makers of the eighteenth/nineteenth century, contemporary Malagasy periwinkle producers continue to be at the receiving end of the commodification of scientific knowledge that I have called epistemic trademarking.

This case, among others (including neem trees in India for toiletries and insecticides or the African katempfe plant for a calorie-free sweetener—see Ostergard et al. 2001), has raised questions about how indigenous knowledge of biodiversity, for example, can be subject to intellectual property laws. Some have remarked how the current WTO system of Trade-Related aspects of Intellectual Property Rights (TRIPS) ends up protecting the interest of multinational corporations at the expense of communities who, despite providing the knowledge, often 'are among the least likely to benefit from the resulting drugs, much less even hear about them or reap any monetary benefits at all' (Jiang 2008, pp. 30–31).

Legal scholars have been considering alternative options such as benefit-sharing agreements that multinational companies must sign when using, for example, the National Cancer Institute's Natural Product Collection, whose biodiverse materials mostly come from the Global South (see Jiang 2008, p. 32). In other cases, compensation takes the form of shares in royalties for the development of new drugs using indigenous plant and animal extracts while retaining patent rights for the corporation (see Stone 1992).

Tools for legally protecting indigenous knowledge—such as the UN Convention on Biological Diversity (CBD) adopted at the Earth Summit in Rio de Janeiro in 1992 and the Nagoya Protocol in 2010—have so far aimed to facilitate 'access and benefit sharing (ABS) mechanisms' and biopiracy is now recognized as a 'serious violation of indigenous peoples' rights' (see UN 2014, p. 2).[30]

---

[30] The full UN text states: 'Indigenous peoples have greatly contributed to developing and preserving unique knowledge on ecosystems, but unfortunately, there is still inadequate regulation of the use of biological resources. Some firms take advantage of lack of legal regulation in order to take indigenous peoples' knowledge and thus obtain patents. In doing so, they deny prior traditional knowledge and are able to keep all the profits resulting from the use of genetic resources for

Yet the normative-legal force of the CBD and Nagoya Protocol remains at the mercy of individual nation-states, whose sole sovereignty on the public land, its biodiverse materials and resources, and local knowledge means that local people's rights are ultimately still dependent upon nation-states' decisions.

On 21 November 2019, the UN General Assembly in its 'Implementation of the Convention on Biological Diversity and Its Contribution to Sustainable Development' (UN 2019b) reiterated the 2050 Vision for Biodiversity in engaging indigenous people and local communities—IPLC, as defined in Chapter 8—(art. 8(d)); lamented 'the limited progress made by its parties in the implementation of the Nagoya Protocol' (art. 25); and stressed 'the importance of the engagement of the private sector and relevant stakeholders, as well as indigenous people and local communities, women and youth', in the implementation of the Convention (art. 40).

Behind the battle for recognizing intellectual property rights to varieties of local knowledge there are huge economic interests. But this debate is also eye-opening for philosophical discussions about realism in science and the nature of scientific knowledge: *who produces* scientific knowledge, and *who gets to benefit* from it. Epistemic severing cuts off the interlacing of scientific perspectives upon which scientific knowledge historically grows and evolves. Epistemic trademarking applies the trademark to well-defined portions or fragments of this vast and open-ended historical and cultural interlacing in order to advance merchandising rights.

Epistemic trademarking as an epistemic injustice associated with scientific knowledge production calls for more than ABS as a legal tool for sharing benefits with the source countries and communities, whose local knowledge about the methods and practices for harvesting biodiverse materials proves commercially lucrative. Epistemic trademarking can only be remedied by *reinstating intellectual ownership* of the relevant knowledge to local communities. In other words, it is important to distinguish between the harm that epistemic trademarking does from that caused by the commercial practice of appropriating and merchandising local botanical knowledge, for example.

Even in the best case scenarios where fair ABS mechanisms are in place, credit to the origin of the biodiverse material is given, and dividends are paid back to local communities, there would still be epistemic injustice—I contend—if the story that gets narrated is the story of progress in biomedical innovations where local knowledge is marginalized and downgraded to a mere repository of traditional wisdom. Legal scholar Madhavi Sunder has crisply summarized the

themselves. This illegitimate misappropriation of genetic resources and associated traditional knowledge, without prior informed consent nor any sharing of resulting benefits, commonly known as biopiracy, is a serious violation of indigenous peoples' rights' (2014, p. 2).

grounds on which intellectual property rights should be recognized for traditional knowledge:

> Reifying the public domain may have the unintended effect of congealing traditional knowledge as 'the opposite of property', presenting poor people's knowledge as the raw material of innovation—ancient, static, and *natural*—rather than as intellectual property—modern, dynamic, *scientific, and cultural invention.* Under this view, traditional knowledge holders may receive remuneration for conserving biodiversity and contributing the raw materials of innovation, but they are not recognized as intellectual property holders in their own right. (Sunder 2007, pp. 100–101, emphases in original)

Understanding the epistemic structural mechanism behind biopiracy implies grasping the deeper source of widespread epistemic injustices in science. Biopiracy is only one manifestation, glaring as is with its merchandising implications, of a wider and more subtle variety of epistemic injustice affecting scientific knowledge production and narratives thereof. Redistribution remedies by themselves (in the form of ABS or similar) are not sufficient. For the problem is not so much or only about 'giving back' scientific knowledge, or not restricting access to it. The bigger substantive problem is how to reclaim as one's own portions of knowledge that have been appropriated, re-used, and eventually trademarked by others. Hence the third kind of reinstatement remedies is about the *intellectual ownership* of scientific knowledge.

And while legal scholars continue to the debate on IP rights, philosophers of science too have a responsibility to analyze the mechanisms behind this variety of epistemic injustice. The dynamic, fluid, open-ended, inferential nature of scientific knowledge as I have described it—a spectacular product of myriad interlacing perspectives—does not lend itself to the logic of severing and trademarking. It undercuts precisely the dichotomy that Sunder highlights as a barrier to the legal protection of local communities' knowledge.

The historical naturalism at play in the inferentialist view of natural kinds (Chapter 8) should remind us of the role of local knowledge in historically identifying *relevant groupings* of phenomena (biodiverse phenomena included) and fine-graining or coarse-graining their description to respond to specific epistemic needs. Ringfencing portions of the inferential network and labelling them with a trademark diminishes the contributions of other communities. Their role in this seamless process of scientific knowledge production is reduced to that of wardens of the raw materials, in Sunder's words (2007, pp. 100–101).

But there is no such thing as 'raw materials'. There are modally robust phenomena—such as the coagulation of gutta-percha's latex known to the Kadayans before telegraphic cables, or suspected medicinal effects from the rosy periwinkle known to local communities (in the Philippines as well as in the Caribbean and Madagascar) well before anti-cancer alkaloids were extracted, patented, and industrially produced on a global scale. A phenomena-first ontology like the one I advocate undercuts any argument designed to ghettoize traditional knowledge as sheer repository of raw materials. There is more. The perspectival pluralism that I see as compatible with realism makes trademarking an epistemic injustice before it even qualifies as a socioeconomic injustice in the dress of biopiracy.

Epistemic trademarking is first and foremost an epistemic injustice in (i) privileging some (commerce-apt) phenomena over others; (ii) appropriating local knowledge necessary to infer those phenomena (e.g. knowledge about specific practices, harvesting techniques, sites); (iii) severing the contribution of communities that are the local repository of such knowledge; and (iv) commodifying such knowledge for the benefit of a few. Epistemic trademarking is an epistemic injustice in that it undercuts the situatedness of scientific knowledge as much as epistemic severing breaks down historical lineages.

Perspectival realism offers a philosophical framework (surely not the only possible one) for understanding scientific knowledge and how it grows over time that is stridently at odds with epistemic narratives friendly to epistemic severing and trademarking. For it does not treat scientific knowledge as a decontextualized, deracinated process. It shows how the realist question ultimately boils down to the question of reliable scientific knowledge production. And the latter can in turn only be addressed by considering how a plurality of situated scientific perspectives have methodologically intersected and historically interlaced (within normative boundaries). Any attempt at severing and trademarking jeopardizes the very fabric of reliable scientific knowledge production and what makes it possible at all.

One consequence of this move is that epistemic communities—be they Malagasy women, Hebridean kelp-makers, or Yucatán beekeepers—who are often exploited and marginalized in the socioeconomic dynamics of knowledge production, can legitimately reclaim as their own rights to scientific knowledge production. That knowledge belongs to them as much as it belongs to anyone else in the relevant historical lineage.

To conclude, a clear path then emerges between multiculturalism and cosmopolitanism, which traditionally have seemed to be in tension with one another. Endorsing multiculturalism in science does not amount to insulating each scientific perspective and treating it as an island. Nor is embracing cosmopolitanism in science tantamount to defending globalization with its epistemic pitfalls (of postcolonial memory) about 'merging' or 'integrating'.

What is needed is a multiculturally tempered cosmopolitanism in science that does not epistemically trademark scientific knowledge and guarantees instead equal rights to partake in scientific advancements. This is the argument from multiculturalism in science to what I call non-classist scientific cosmopolitanism.

11.6. Non-classist scientific cosmopolitanism

Books about scientific realism often start or end with discussions about the *aims of science*, or the *goals of science*. Does science aim to tell us a true story about nature? Or does it aim to be empirically adequate? Is problem-solving or scientific understanding the goal of science? Typically, answers to these questions determine the flavour of realism or anti-realism that each author espouses. But I have carefully avoided any discussion of *aims* and *goals* in this book.

I do not know whether specific scientific communities have particular aims or goals. Possibly they do. If so, I would imagine they do not necessarily match those around which the realism and anti-realism debate has often revolved. The passion with which metaphysical discussions on the many-worlds interpretation of quantum mechanics are joined among philosophers of physics contrasts with the daily pragmatism (mixed with scepticism) I have often observed among particle physicists. Whatever the aims of science might be, they are likely to be localized and discipline-specific.

However, this book would be incomplete without a discussion of something equally important that is often absent in the philosophy of science literature on realism: namely, 'duties' and 'rights' attaching to scientific knowledge. Let us start with duties.

When thinking about duties and obligations in science, familiar ethical questions come to mind: integrity in the production and replication of data; accuracy in measurements; non-biased algorithms; eco-friendly innovations; and countless other examples.

But my invoking duties and obligations here does not point to these daily concerns of bioethics committees, ethics of AI groups, food standards agencies, and similar. It is an invitation instead to shift attention away from the aims of scientific knowledge towards the equally important matter of what duties and obligations scientific knowledge carries with it.

The discussion in Section 11.5 pointed out two varieties of epistemic injustice. I suggested that three kinds of reinstatement remedies (authorship, historical role, and intellectual ownership) are required to counteract epistemic severing and epistemic trademarking. But on *whose shoulders* should the responsibility of carrying out such remedies fall?

It would be a bad outcome if these responsibilities were somehow delegated entirely to discussions among historians and philosophers of science. For these remedies fall on all of us: they fall on national governments responsible among other matters for public health care; on pharmaceutical companies bioprospecting, designing, and distributing drugs; on multinational corporations who exploit local knowledge and local labour for meagre returns. Reinstatement remedies are duties that fall on everyone, for they are what I am going to call 'cosmopolitan obligations towards science'.

Cosmopolitan obligations are—in legal lingo—obligations pertaining to 'world citizenship'. The term 'cosmopolitan' means 'citizen of the world' in a long-standing tradition in social, political, and ethical philosophy. The general philosophical idea is that 'all human beings, regardless of their political affiliation, are (or can and should be) citizens in a single community' (Kleingeld and Brown 2019). One of the questions that this thousand-year-old tradition has, however, left untouched concerns how best to understand the role of science. *What (if anything) makes scientific knowledge 'cosmopolitan'? More pressingly, how ought one to understand the nature of cosmopolitanism in science?*

Traditionally, three main strands of cosmopolitanism can be identified: cosmopolitanism as 'an attitude of enlightened morality'; cosmopolitanism identified with 'hybridity, fluidity, and recognizing the fractured and internally riven character of human selves and citizens'; and cosmopolitanism as a 'normative philosophy for carrying the universalistic norms of discourse ethics beyond the confines of the nation-state' (Benhabib 2006, pp. 17–18).

In the third strand are philosophers like Karl Jaspers, Hannah Arendt (Benhabib 2006, pp. 13–16), and Seyla Benhabib herself, whose 'cosmopolitan norms of justice' can be traced back to the notion of a *Weltbürgerrecht* (cosmopolitan right).[31] Benhabib, rightly in my view, draws attention to the UN Declaration of Human Rights (UNDHR)[32] in 1948 as a watershed in the transition from international to cosmopolitan norms of justice.

And since there can be no duties if there are no rights (see O'Neill 2013, 2018), a good starting point in answering the questions I laid out earlier is indeed the UNDHR, which in its Article 27 (1) says something important about science and technology:

---

[31] Kant (1795/2006) introduced the notion of a 'cosmopolitan right' (*Weltbürgerrecht*). He understood it as a 'right to hospitality' that foreigners can reclaim as their own, and as a right that went beyond the domestic and international right. As Benhabib writes, '*[H]ospitality* is not to be understood as a virtue of sociability, as the kindness and generosity one may show to strangers who come to one's land or who become dependent on one's act of kindness through circumstances of nature or history; hospitality is a right that belongs to all human beings insofar as we view them as potential participants in a world republic. Following Kant, Arendt likewise argues that "crimes against humanity" are not violations of moral norms alone, but violations of the rights of humanity in our person' (Benhabib 2006, p. 22, emphasis in original). See on this point also Kleingeld (2011, 2016) and Waldron (2006).
[32] See https://www.un.org/en/about-us/universal-declaration-of-human-rights.

27 (1) Everyone has the right freely to participate in the cultural life of the community, to enjoy the arts and to *share in scientific advancement and its benefits*. (Emphasis mine.)

The history of this article can be glimpsed from the *travaux préparatoires* of the UNDHR. The Mexican and the Cuban members of the drafting committee insisted on having the 'rights of the individual as an intellectual worker, scientist, or writer' recognized (UN 1950, p. 71). The request was modelled on Article 13 of the American Declaration on the Rights and Duties of Man (1948), which referred to the right for every person 'to participate in the benefits that results from intellectual progress, especially scientific discoveries', alongside the 'right to the protection' of moral and material interests resulting from 'his inventions, or any literary, scientific or artistic works of which he is the author'.[33] Unsurprisingly, the UNDHR accompanies 27 (1) with article 27 (2), which recognizes the 'right to the protection of the moral and material interests resulting from any scientific, literary and artistic production of which he is the author'. Similar provisions are to be found in the 1966 International Covenant on Economic, Social and Cultural Rights (ICESCR), Article 15 (b) 'to enjoy the benefits of scientific progress and its applications' and more recently in the so-called *Venice Statement* organised by UNESCO in 2009.

In all these cases, intellectual property rights were packaged as part of an overarching right for 'everyone' to partake in culture, arts, and science, or to 'share in' what the UNDHR calls 'scientific advancement and its benefit' (rather than 'scientific discoveries'). This human rights approach to scientific advancement signals an important shift: *from* scientific discoveries as individualistic achievements *towards* science as a human activity without which, in the words of the Mexican drafting members, 'no social progress would be possible' (UN 1950, p. 71). Treating 'sharing in scientific advancement and its benefits' as a human right implies democratizing and de-commodifying scientific knowledge. It makes it possible to think of science as carrying cosmopolitan obligations. For where there are rights, there are also obligations to respect those rights.

An overarching right 'to share in scientific advancement and its benefits' ramifies into a number of others: for example, the right to health, the right to education, and the right to food, among many others. The emphasis in the UNDHR's carefully chosen language is on '*Everyone* has the right *freely* . . .', as if there were indeed a genuine, cosmopolitan world citizenship in which everyone partakes and by virtue of which everyone enjoys these rights, including the right of 'sharing in scientific advancement and its benefits'.

---

[33] See http://hrlibrary.umn.edu/oasinstr/zoas2dec.htm.

But *who* is 'everyone'? The vague formulation of Article 27 (1) left unspecified what the terms 'the cultural life of the community' and 'scientific advancement and its benefits' refer to. The risk remains of assuming some dominant form of community qua repository of cultural life and scientific advancement (see Donders 2008, p. 3).

Unsurprisingly, scholars in critical theory, postcolonial studies, sociology, and anthropology point out that 'everyone' is often a disguised proxy for the Western-centric 'we' (what the Romans called *nos majestatis*). They have called for varieties of cosmopolitanism that do not advance universalistic globalizing Western-centric claims disguised under cosmopolitan lingo.

Among them, Homi Bhabha's *vernacular cosmopolitanism* involves 'the commitment to a "right to difference in equality" . . . a political process that works toward the shared goals of a democratic rule . . . rather than simply acknowledging already constituted "marginal" political identities or entities' (Bhabha 2017, pp. 145–146). Boaventura de Sousa Santos has defended what he and others refer to as 'subaltern cosmopolitanism or cosmopolitanism of the oppressed' as a 'counter-hegemonic globalization . . . focussed on the struggle against social exclusion' (Santos 2002, pp. 458–459). And drawing on African ubuntu ethics, Michael Onyebuchi Eze observes that:

> The quest to eradicate boundaries presupposes social mobility. But if social mobility is a privilege for people from certain regions of the world or economic caliber, does not cosmopolitanism become on the one hand intuitively exclusive and on the other hand tendentiously imperialist? . . . The elitism of the cosmopolitan ideal thus enunciated replicates the Western citizen of the ordinary traveler or non-Western citizen with advanced social capital. At the opposite end of the spectrum are the immigrants—the unseen cosmopolitans. . . . Cosmopolitanism as a product of culture emerges at the intersection of cultural experiences, a confluence of narratives (note: not hybridity). The cosmopolitan identity is not a hybrid identity; it is rather a confluent identity. The idea of cosmopolitanism as a hybrid evokes subjective possession, what may be termed *colonization of subjectivity*. The other is grafted only as an offshoot of another culture. This other is dominated and although expressing sprouts of individuality still depend on dominant culture for lifeline. Our new cosmopolitanism is an advocacy for cultural difference in a dialectic relation; a meadow of creative interactionism; of human communities in confluence relation with one another. (Eze 2017, pp. 95, 96, 106, emphasis in original)

When it comes to scientific knowledge, the aforementioned reinstatement remedies must be part of a non-classist, non-elitist scientific cosmopolitanism.

Ultimately, recognizing the multiculturalism inherent in science is a stepping-stone towards a non-classist scientific cosmopolitanism. Affirming the normatively virtuous historical interlacing of scientific perspectives is to resist the temptation of constructing them as scientific 'silos'. But it is also a way of starting a conversation about *who* scientific knowledge is really *for, who produces* scientific knowledge over time, and *who* ought to be benefiting from it. Without that, epistemological narratives about science, nature, and realism risk being inward-looking academic exercises. I hope these scattered reflections are at least an invitation to a more sustained conversation on this topic, which deserves a book treatment in its own right.

This book has come to an end. It has been a journey through the perspectival nature of realism I see at play in science. One where the correspondence theory of truth still applies, but the correspondence is not understood as between theoretical statements and metaphysical posits; between, for example, what the best theory in mature science about the electron says and presumptive metaphysical posits about the electron. The correspondence is instead between claims of knowledge that have been retained over time and across perspectives, on the one hand, and modally robust phenomena reliably inferred from data, on the other.

Perspectival pluralism is not just a desirable feature of scientific methodology, useful for explanation and modelling. It is the very engine of how scientific knowledge claims get refined and revised over time, and of how reliability accrues (via intersecting perspectives). Such pluralism tells also a different story about how to sort phenomena into natural kinds, and about the malleability and fluidity of those kinds. Most of all, the pluralism of scientific perspectives I have advocated in this book is a plea for a variety of realism in science that is not historically exclusionary, that recognizes the fluidity, 'confluence' in Eze's words, or historical interlacing of scientific perspectives (in my words) without attempting to unify, merge, overlap, or hybridize them.

Seen through the lenses of perspectival realism, science does not fulfil any particular aim or goal—be it truth, empirical adequacy, or scientific understanding. It instead serves the needs of a multicultural 'scientific world citizenship' that can legitimately reclaim as its own the cosmopolitan 'right' of 'sharing in scientific advancement and its benefits'. But that carries also cosmopolitan obligations in ensuring no one is deprived of such a right and remedies are in place to address stark inequalities in the historical access to scientific knowledge and its benefits.

Philosophers of science need not tell scientists what they ought to do (why would we even want to?). But our work must go beyond merely describing scientific practice. I see our job as opening up a critical space to talk about scientific knowledge, and to offer 'journeys of exploration' through which it is possible to see scientific knowledge in a transformative way. This is in part what I call the 'social function of philosophy of science' (building on J. D. Bernal—see Massimi 2019b).

Like Marco Polo's journeys of exploration, the final reward is not necessarily to stumble into cities which—once conceived and thought to be possible—might well turn out to be non-existent. The whole purpose of the journey is instead to transform the epistemological landscape and to discover that the differences preserved by the atlas are in fact much more fluid and interlaced as the journey goes on. And with it also the rediscovery of rights and obligations that we owe to our fellow travellers (past, present, and future) in sharing scientific knowledge, its advancements, and its benefits.

References

Ade, P.A.R., N. Aghanim, M. Arnaud, et al. Planck Collaboration (2015) 'Planck 2015 results: XIII. Cosmological parameters', *Astronomy & Astrophysics* 594, A13.

Aghanim, N., M. Ashdown, J. Aumont, et al., Planck Collaboration (2016) 'Planck intermediate results: XLVI. Reduction of large-scale systematic effects in HFI polarization maps and estimation of the reionization optical depth', *Astronomy & Astrophysics* 596, A107.

Akerib, D.S., S. Alsum, H.M. Araújo, et al., LUX Collaboration (2017) 'Results from a search for dark matter in the complete LUX exposure', *Physical Review Letters* 118, 021303(8).

Al-Khalili, J. (2010) *Pathfinders: The Golden Age of Arabic Science.* London: Penguin Books.

Alexandrova, A. (2010) 'Adequacy-for-purpose: the best deal a model can get', *The Modern Schoolman* 86, 295–300.

Alfaro Bates, R.G., J.A. González Acereto, J.J. Ortiz Díaz, et al. (2010) *Caracterización palinológica de las mieles de la península de Yucatán.* Universidad Autónoma de Yucatán, Comisión Nacional para el Conocimiento y Uso de la Biodiversidad, Mérida, Yucatán, México.

Alpher, R.A., H. Bethe, and G. Gamow (1948) 'The origin of chemical elements', *Physical Review* 73, 803–804.

Amaldi, E., O. D'Agostino, E. Fermi, B. Pontecorvo, F. Rasetti, and E. Segrè (1935) 'Artificial radioactivity produced by neutron bombardment. II', *Proceedings of the Royal Society of London A* 149, 522–558.

Ankeny, R., and S. Leonelli (2016) 'Repertoires: a post-Kuhnian perspective on scientific change and collaborative research', *Studies in History and Philosophy of Science A* 60, 18–28.

Ankeny, R.A., and S. Leonelli (2020) *Model Organisms.* Cambridge: Cambridge University Press.

Appiah, K.A. (2006) *Cosmopolitanism: Ethics in a World of Strangers.* New York: W.W. Norton.

Arabatzis, T. (2006) *Representing Electrons: A Biographical Approach to Theoretical Entities.* Chicago: University of Chicago Press.

Arlo-Costa, H. (2019) 'The Logic of Conditionals', *Stanford Encyclopedia of Philosophy* (Summer 2019 Edition), Edward N. Zalta (ed.). https://plato.stanford.edu/archives/sum2019/entries/logic-conditionals/.

Armstrong, D. (1985) *What Is a Law of Nature?* Cambridge: Cambridge University Press.

Armstrong, D. (1989) *A Combinatorial Theory of Possibility.* Cambridge: Cambridge University Press.

Armstrong, D. (1993) 'A world of states of affairs', *Philosophical Perspectives* 7, 429–440.

Armstrong, D. (1997) *A World of States of Affairs.* Cambridge: Cambridge University Press.

Armstrong, D. (2004) *Truth and Truth-Makers.* Cambridge: Cambridge University Press.

Aston, F. (1922/1966) 'Mass spectra and isotopes, Nobel Lecture, December 12, 1922', in *Nobel Lectures, Chemistry 1922–1941*. Amsterdam: Elsevier Publishing Company, pp. 7–20. https://www.nobelprize.org/prizes/chemistry/1922/aston/lecture/.

Aston, F. (1924) 'Atomic species and their abundance on the Earth', *Nature* 113 (2837), 393–395.

Atkins, L.C., and A. Mahmud (2021) 'Informational justice: equity of access, implementation, and interaction', in W. Leal Filho, A.M. Azul, L. Brandli, A.L. Salvia, P.G. Özyvar, and T. Wall (eds), *Peace, Justice and Strong Institutions: Encyclopedia of the UN Sustainable Development Goals*, 1–12. https://doi.org/10.1007/978-3-319-71066-2_80-1.

ATLAS Collaboration (2012) 'Observation of a new particle in the search for the Standard Model Higgs boson with the ATLAS detector at the LHC', *Physics Letters B* 716, 1–29.

ATLAS Collaboration (2015) 'Summary of the ATLAS Experiment's sensitivity to supersymmetry after LHC Run 1—interpreted in the phenomenological MSSM', *Journal of High Energy Physics* 134, 1–74.

Austin, J.L. (1961) 'Unfair to facts', in *Philosophical Papers* (eds J.O. Urmson and G.J. Warnock). Oxford: Clarendon Press, pp. 102–122.

Baggott, J. (2000) *Quantum Reality*. Oxford: Oxford University Press.

Bailer-Jones, D. (2003) 'When scientific models represent', *International Studies in Philosophy of Science* 17, 59–74.

Balcerak Jackson, M. (2016) 'On the epistemic value of imagining, supposing, and conceiving', in A. Kind and P. Kung (eds), *Knowledge through Imagination*. Oxford: Oxford University Press, pp. 42–60.

Ball, P. (2018) *Beyond Weird*. London: Penguin Books.

Bartlett, J.H., Jr (1932) 'Structure of atomic nuclei', *Physical Review* 41, 370–371.

Batchelor, R. (2013) 'The Selden Map rediscovered: a Chinese map of East Asian shipping routes c. 1619', *Imago Mundi* 65, 37–63.

Beck, U. (2004) 'The truth of others: a cosmopolitan approach', *Common Knowledge* 10, 430–449.

Beebee, H. (2013) 'How to carve across the joints in nature without abandoning Kripke-Putnam semantics', in S. Mumford and M. Tugby (eds), *Metaphysics and Science*. Oxford: Oxford University Press, pp. 141–163.

Beebee, H., and N. Sabbarton-Leary (eds) (2010) *The Semantics and Metaphysics of Natural Kinds*. London: Routledge.

Belkin, H.E., and T. Gidwiz (2020) 'The contributions and influence of two Americans, Henry S. Washington and Frank A. Perret, to the study of Italian volcanism with emphasis on volcanoes in the Naples area', in B. De Vivo, H.E. Belkin, and G. Rolandi (eds), *Vesuvius, Campi Flegrei, and Campanian Volcanism*. Amsterdam: Elsevier, pp. 9–32.

Benhabib, S. (2006) 'The philosophical foundations of cosmopolitan norms', in S. Benhabib, *Another Cosmopolitanism: with commentaries by Jeremy Waldron, Bonnie Honig, Will Kymlicka* (ed. R. Post). Oxford: Oxford University Press, pp. 13–44.

Bensaude-Vincent, B. (1992) 'The balance: between chemistry and politics', *The Eighteenth Century* 33, 217–237.

Berger, R.W. (1974) 'Garden cascades in Italy and France, 1565–1665', *Journal of the Society of Architectural Historians* 33, 304–322.

Berto, F., and T. Schoonen (2018) 'Conceivability and possibility: some dilemmas for Humeans', *Synthese* 195, 2697–2715.

Bethe, H.A. (1937) 'Nuclear physics: B. Nuclear dynamics, theoretical', *Review of Modern Physics* 9, 69–244.

Bethe, H.A., and R.F. Backer (1936) 'Nuclear physics: A. Stationary states of nuclei', *Review of Modern Physics* 8, 82–229.

Betz, G. (2009) 'Underdetermination, model-ensembles and surprises: on the epistemology of scenario-analysis in climatology', *Journal for General Philosophy of Science* 40, 3–21.

Betz, G. (2015) 'Are climate models credible worlds? Prospects and limitations of possibilistic climate prediction', *European Journal for Philosophy of Science* 5, 191–215.

Bhabha, H. (2017) 'Spectral sovereignty, vernacular cosmopolitans, and cosmopolitan memories', in B. Robbins and P.L. Horta (eds), *Cosmopolitanisms*. New York: New York University Press, pp. 141–152.

Bhambra, G.K. (2007) *Rethinking Modernity: Postcolonialism and the Sociological Imagination*. Basingstoke: Palgrave.

Biagioli, M. (2019) 'Weighing intellectual property: can we balance the costs and benefits of patenting?', *History of Science* 57, 140–163.

Biagioli, M., M. Woodmansee, and P. Jaszi (eds) (2011) *Making and Unmaking Intellectual Property: Creative Production in Legal and Cultural Perspective*. Chicago: University of Chicago Press.

Bird, A. (2000) *Thomas Kuhn*. Chesham: Acumen and Princeton: Princeton University Press.

Bird, A. (2007) *Nature's Metaphysics: Laws and Properties*. Oxford: Oxford University Press.

Bird, A. (2018) 'The metaphysics of natural kinds', *Synthese* 195, 1397–1426.

Boerhaave, H. (1732/1735) *Elements of Chemistry*. London: J. Pemberton et al.

Bogen, J., and J. Woodward (1988) 'Saving the phenomena', *Philosophical Review* 97, 303–352.

Bohr, A., and B.R. Mottelson (1953) 'Collective and individual-particle aspects of nuclear structure', *Matematisk-fysiske Meddelelser* 27 (16), 1–174.

Bohr, A., and B.R. Mottelson (1969) *Nuclear Structure*, Vol. I. Reading, MA: W.A. Benjamin.

Bohr, A., and B.R. Mottelson (1975) *Nuclear Structure*, Vol. II. Reading, MA: W.A. Benjamin.

Bohr, N. (1936) 'Neutron capture and nuclear constitution', *Nature* 137, 344–348.

Bohr, N., and J. Wheeler (1939) 'The mechanism of nuclear fission', *Physical Review* 56, 426–450.

Bokulich, A. (2011) 'How scientific models can explain', *Synthese* 180, 33–45.

Bokulich, A. (2014) 'How the tiger bush got its stripes: "how possibly" vs "how actually" model explanations', *The Monist* 97, 321–338.

Bokulich, A. (2015) 'Maxwell, Helmholtz, and the unreasonable effectiveness of the method of physical analogy', *Studies in History and Philosophy of Science* 50, 28–37.

Bokulich, A. (2017) 'Models and explanation', in L. Magnani and T. Bertolotti (eds), *Handbook of Model-Based Science*. Dordrecht: Springer, pp. 103–118.

Bokulich, A. (2018a) 'Searching for non-causal explanations in a sea of causes', in A. Reutlinger and J. Saatsi (eds), *Explanation beyond Causation: Philosophical Perspectives on Non-Causal Explanations*. Oxford: Oxford University Press, pp. 141–163.

Bokulich, A. (2018b) 'Using models to correct data: paleodiversity and the fossil record', *Synthese*. https://doi.org/10.1007/s11229-018-1820-x.

Bokulich, A., and W. Parker (2021) 'Data models, representation, and adequacy for purpose', *European Journal for Philosophy of Science* 11 (31), 1–26.

Bonan, G. (2015) 'Surface energy fluxes', in *Ecological Climatology: Concepts and Applications* (Third Edition). Cambridge: Cambridge University Press, pp. 193–208.

Boon, M. (2020) 'The role of disciplinary perspectives in an epistemology of scientific models', *European Journal for Philosophy of Science* 10, 31.

Borges, J.L. (1941/2000) 'The garden of forking paths', reprinted in *Fictions* (trans. A. Hurley). London: Penguin Books, pp. 75–86.

Boyd, R. (1990) 'Realism, conventionality, and "realism about"', in G. Boolos (ed.), *Meaning and Method: Essays in Honor of Hilary Putnam*. Cambridge: Cambridge University Press, pp. 171–196.

Boyd, R. (1991) 'Realism, anti-foundationalism and the enthusiasm for natural kinds', *Philosophical Studies* 61, 127–148.

Boyd, R. (1992) 'Constructivism, realism, and philosophical method', in J. Earman (ed.), *Inference, Explanation, and Other Frustrations*. Berkeley: University of California Press, pp. 131–198.

Boyd, R. (1999a) 'Homeostasis, species, and higher taxa', in R. Wilson (ed.), *Species: New Interdisciplinary Essays*. Cambridge, MA: MIT Press, pp. 141–186.

Boyd, R. (1999b) 'Kinds as the "workmanship of men": realism, constructivism, and natural kinds', in J. Nida-Rümelin (ed.), *Rationalität, Realismus, Revision: Proceedings of the Third International Congress, Gesellschaft für Analytische Philosophie*. Berlin: de Gruyter, pp. 52–89.

Boyd, R. (2010) 'Realism, natural kinds, and philosophical method', in H. Beebee and N. Sabbarton-Leary (eds), *The Semantics and Metaphysics of Natural Kinds*. London: Routledge, pp. 212–234.

Brading, K. (2012) 'Newton's law-constitutive approach to bodies: a response to Descartes', in A. Janiak and E. Schliesser (eds), *Interpreting Newton: Critical Essays*. Cambridge: Cambridge University Press, pp. 13–32.

Bradley, L.D., R.J. Bouwens, H.C. Ford, et al. (2008) 'Discovery of a very bright strongly-lensed galaxy candidate at z ≈ 7.6', *The Astrophysical Journal* 678, 647–658.

Brandom, R.B. (1998) 'Insights and blindspots of reliabilism', *The Monist* 81, 371–392.

Brandon, R.N. (1990) *Adaptation and Environment*. Princeton: Princeton University Press.

Bratman, E.Z. (2020) 'Saving the other bees', *Conservation & Society* 18, 387–398.

Breit, G., and E.P. Wigner (1936) 'Capture of slow neutrons', *Physical Review* 49, 519–531.

Briffa, K.R., and T.J. Osborn (1999) 'Seeing the wood from the trees', *Science* 284, 926–927.

Brigandt, I., and P.E. Griffiths (2007) 'The importance of homology for biology and philosophy', *Biology & Philosophy* 22, 633–641.

Broadbent, A. (2019) *Philosophy of Medicine*. Oxford: Oxford University Press.

Brown, N. (2016) *In Silico Medicinal Chemistry: Computational Methods to Support Drug Design*. RSC Theoretical and Computational Chemistry Series No. 8. Cambridge: Royal Society of Chemistry.

Bueno, O., and S.A. Shalkowski (2014) 'Modalism and theoretical virtues: towards an epistemology of modality', *Philosophical Studies* 172, 671–689.

Burbidge, E.M., G.R. Burbidge, W.A. Fowler, and F. Hoyle (1957) 'Synthesis of the elements in stars', *Review of Modern Physics* 29, 547–650.

Burbidge, F.W. (1879) 'Notes on gutta percha and caoutchouc-yielding trees', *Journal of the Straits Branch of the Royal Asiatic Society* 3, 52–61.

Burian, R. (1997) 'Exploratory experimentation and the role of histochemical techniques in the work of Jean Brachet 1938–52', *History and Philosophy of the Life Sciences* 19, 27–45.

Burles, S., K.M. Nollett, and M.S. Turner (2001) 'Big bang nucleosynthesis predictions for precision cosmology', *The Astrophysical Journal* 552, L1–L5.

Bursten, J.R.S. (2014) 'Microstructure without essentialism: a new perspective on chemical classification', *Philosophy of Science* 81, 633–653.

Bursten, J.R.S. (2016) 'Conceptual analysis for nanoscience', *The Journal of Physical Chemistry Letters* 7, 1917–1918.

Bursten, J.R.S. (2018) 'Smaller than a breadbox: scale and natural kinds', *British Journal for the Philosophy of Science* 69, 1–23.

Bursten, J.R.S. (ed.) (2020a) *Perspectives on Classification in Synthetic Sciences: Unnatural Kinds.* London: Routledge.

Bursten, J.R.S. (2020b) 'Computer simulations', in C. Mody and J. Martin (eds), *Between Making and Knowing: Tools in the History of Materials Research.* Singapore: World Scientific, pp. 195–206.

Callanan M., A. Repp, M. McCarthy, and M. Latzke (1994) 'Children's hypotheses about word meanings: is there a basic level constraint?', *Journal of Experimental Child Psychology* 57, 108–138.

Calvino, I. (1972/1997) *Invisible Cities* (trans. W. Weaver). London: Vintage, The Random House Group Limited.

Canagarajah, S. (1993) 'Up the garden path: second language writing approaches, local knowledge, and pluralism', *TESOL Quarterly* 27, 301–306.

Canagarajah, S. (2002) 'Reconstructing local knowledge', *Journal of Language, Identity, and Education* 1, 243–259.

Cañizares-Esguerra, J. (2001) *How to Write the History of the New World: Histories, Epistemologies, and Identities in the Eighteenth-Century Atlantic World.* Stanford: Stanford University Press.

Carlson, J.B. (1975) 'Lodestone compass: Chinese or Olmec Primacy? Multidisciplinary analysis of an Olmec hematite artifact from San Lorenzo, Veracruz, Mexico', *Science* 5 (185), 753–760.

Carroll, J.M., J. Solity, and L.R. Shapiro (2016) 'Predicting dyslexia using prereading skills: the role of sensorimotor and cognitive abilities', *Journal of Child Psychology and Psychiatry* 57, 750–758.

Carter, J.A. (2020) 'Virtue perspectivism, externalism, and epistemic circularity', in A. Crețu and M. Massimi (eds), *Knowledge from a Human Point of View.* Dordrecht: Springer, pp. 123–140.

Cartwright, N. (1997) 'Models: the blueprints for laws', *Philosophy of Science* 64, S292–S303.

Cartwright, N. (1999) *The Dappled World.* Cambridge: Cambridge University Press.

Cartwright, N. (2019) *Nature, the Artful Modeler,* The Paul Carus Lectures 23, Open Court.

Cartwright, N., J. Cat, L. Fleck, and T.E. Uebel (1996) *Otto Neurath: Philosophy between Science and Politics.* Cambridge: Cambridge University Press.

Casati, R., and A. Varzi (2020) 'Events', *Stanford Encyclopedia of Philosophy* (Summer 2020 Edition), E.N. Zalta (ed.). https://plato.stanford.edu/archives/sum2020/entries/events/.

Caurier, E., G. Martínez-Pinedo, F. Nowacki, A. Poves and A.P. Zuker (2005) 'The shell model as a unified view of nuclear structure', *Reviews of Modern Physics* 77, 427–488.

Cavendish, H. (1784/2010) 'Experiments on airs', *Philosophical Transactions* 74, pp. 119–153. Reprinted in *The Scientific Papers of the Honourable Henry Cavendish*, ed. Edward Thorpe, Cambridge: Cambridge University Press, pp. 161–181.

Chakravartty, A. (2007) *A Metaphysics for Scientific Realism*. Cambridge: Cambridge University Press.

Chakravartty, A. (2010) 'Perspectivism, inconsistent models, and contrastive explanation', *Studies in History and Philosophy of Science* 41, 405–412.

Chakravartty, A. (2011) 'Scientific realism and ontological relativity', *The Monist* 94, 157–180.

Chakravartty, A. (2017) *Scientific Ontology: Integrating Naturalized Metaphysics and Voluntarist Epistemology*. Oxford: Oxford University Press.

Chalmers, D.J. (2002) 'Does conceivability entail possibility?', in T.S. Gendler and J. Hawthorne (eds), *Conceivability and Possibility*. Oxford: Oxford University Press, pp. 145–200.

Chang, H. (2004) *Inventing Temperature: Measurement and Scientific Progress*. Oxford: Oxford University Press.

Chang, H. (2012a) *Is Water H_2O? Evidence, Realism and Pluralism*. Dordrecht: Springer.

Chang, H. (2012b) 'Joseph Priestley (1733–1804)', in A.I. Woody, R.F Hendry, and P. Needham (eds), *Handbook of the Philosophy of Science. Vol. 6: Philosophy of Chemistry*. Amsterdam: Elsevier, pp. 55–62.

Chang, H. (2016) 'The rising of chemical natural kinds through epistemic iteration', in C. Kendig (ed.), *Natural Kinds and Classification in Scientific Practice*. London: Routledge, pp. 33–46.

Chang, H. (2022) *Realism for Realistic People: A New Pragmatist Philosophy of Science*. Cambridge: Cambridge University Press.

Chirimuuta, M. (2015) *Outside Color*. Cambridge, MA: MIT Press.

Chirimuuta, M. (2019) 'Charting the Heraclitean brain', in M. Massimi and C.D. McCoy (eds), *Understanding Perspectivism: Scientific Challenges and Methodological Prospects*. New York: Routledge, pp. 141–159.

Clay, M.M. (1987) 'Learning to be learning disabled', *New Zealand Journal of Educational Studies* 22, 155–172.

Clow, A., and N.L. Clow (1947) 'The natural and economic history of kelp', *Annals of Science* 5, 297–316.

Coh-Martínez, M.E., W. Cetzal-Ix, J.F. Martínez-Puc, et al. (2019) 'Perceptions of the local beekeepers on the diversity and flowering phenology of the melliferous flora in the community of Xmabén, Hopelchén, Campeche, Mexico', *Journal of Ethnobiology and Ethnomedicine* 15 (16), 1–16.

Collingwood, R.G. (1936/1976) *Human Nature and Human History*. London: Ardent Media.

Collins, H. (1985/1992) *Changing Order* (Second Edition). Chicago: University of Chicago Press.

Comte, A. (1830–42/1853) *The Positive Philosophy of August Comte* (ed. and trans. H. Martineau). London: J. Chapman.

Connor, C.M. (2010) 'Child characteristics–instruction interactions: implications for students' literacy skills development in the early grades', in S.B. Neuman and D.K. Dickinson (eds), *Handbook of Early Literacy Research*, Vol. 3. New York: Guilford Publications, pp. 256–275.

Connor, C.M., F.J. Morrison, and E.L. Katch (2004a) 'Beyond the reading wars: the effect of classroom instruction by child interactions on early reading', *Scientific Studies of Reading* 8, 305–336.

Connor, C.M., F.J. Morrison, and J.N. Petrella (2004b) 'Effective reading comprehension instruction: examining child by instruction interactions', *Journal of Educational Psychology* 96, 682–698.

Contessa, G. (2007) 'Scientific representation, interpretation, and surrogative reasoning', *Philosophy of Science* 74, 48–68.

Crane, E. (1999) *The World History of Beekeeping and Honey Hunting.* New York: Routledge.

Crețu, A.-M. (forthcoming) 'Perspectival instruments', *Philosophy of Science.*

Crețu, A.-M., and M. Massimi (eds) (2020) *Knowledge from a Human Point of View.* Dordrecht: Springer.

Crisfield, J. (1996) *The Dyslexia Handbook.* London: British Dyslexia Association.

Cross, W., J.P. Iddings, L.V. Pirsson, and H.S. Washington (1902) 'A quantitative chemico-mineralogical classification and nomenclature of igneous rocks', *Journal of Geology* 10, 555–690.

Crowther, J.G. (1974) *The Cavendish Laboratory 1874–1974.* Dordrecht: Springer.

Currie, A. (2014) 'Venomous dinosaurs and rear-fanged snakes: homology and homoplasy characterised', *Erkenntnis* 79, 701–727.

Currie, A. (2019) *Scientific Knowledge and the Deep Past: History Matters.* Cambridge: Cambridge University Press.

Currie, A. (2020) 'Bottled understanding: the role of lab work in ecology', *British Journal for the Philosophy of Science* 71, 905–932.

da Costa, A., and S. French (2003) *Science and Partial Truth: A Unitary Approach to Models and Scientific Reasoning.* Oxford: Oxford University Press.

Darcy, H. (1856) *Le fountains publique de la ville de Dijon.* Paris: Victor Dalmont.

Darden, L. (1987) 'Viewing the history of science as compiled hindsight', *AI Magazine* 8 (Summer), 33–41.

Daston, L. (1992) 'Objectivity and the escape from perspective', *Social Studies of Science* 22, 597–618.

Davies, S. (2018) 'Routes, rutters, navigational techniques and the development of navigational aids in traditional Chinese seagoing: the case of the compass', *Artefact* 8, 141–182.

Davis, E. (2018) 'On epistemic appropriation', *Ethics* 128, 702–727.

Davy, H. (1807) 'The Bakerian Lecture, on some chemical agencies of electricity', *Abstracts of the Papers Printed in the Philosophical Transactions of the Royal Society of London* 1 (1800–1814), 247–253.

Davy, H. (1812) *Elements of Chemical Philosophy.* Philadelphia: Bradford and Inskeep.

de Swart, J., G. Bertone, and J. van Dongen (2017) 'How dark matter came to matter', *Nature Astronomy* 1, 59.

Derrida, J. (2001) *On Cosmopolitanism and Forgiveness* (trans. M. Dooley and M. Hughes). London: Routledge.

Desan, C. (2014) *Making Money: Coin, Currency and the Coming of Capitalism.* Oxford: Oxford University Press.

Devitt, M. (2008) 'Resurrecting biological essentialism', *Philosophy of Science* 75, 344–382.

Devitt, M. (2010) 'Species have (partly) intrinsic essences', *Philosophy of Science* 77, 648–661.

Dieks, D. (2019) 'Quantum mechanics and perspectivalism', in O. Lombardi, S. Fortin, C. López, and F. Holik (eds), *Quantum World*. Cambridge: Cambridge University Press, pp. 51–70.

DiFrisco, J. (2019) 'Kinds of biological individuals: sortals, projectibility and selection', *British Journal for the Philosophy of Science* 70, 845–875.

Di Vecchia, P., G. Rossi, G. Veneziano, et al. (2017) 'Spontaneous *CP* breaking in QCD and the axion potential: an effective Lagrangian approach', *Journal of High Energy Physics*, 104.

Dodelson, S. (2011) 'The real problem with MOND', *International Journal of Modern Physics D* 20, 2749–2753.

Donders, Y. (2008) 'Cultural life in the context of human rights', UN Committee on Economic, Social and Cultural Rights, 9 May, E/C.12/40/13.

Dotson, K. (2014) 'Conceptualizing epistemic oppression', *Social Epistemology* 28, 115–138.

Douma, C.A., C. Agodi, H. Akimune, et al. (2020) 'Gamow–Teller strength distribution of ^{116}Sb and ^{122}Sb using the (^{3}He, t) charge–exchange reaction', *European Physical Journal* 56, 51.

Drude, P. (1900) 'Zur Elektronentheorie der Metalle', *Annelen der Physik* 1, 566–613.

Dudas, E., T. Gherghetta, Y. Mambrini, and K.A. Olive (2017) 'Inflation and high-scale supersymmetry with an EeV gravitino', *Physical Review D* 96, 115032.

Duffin, J. (2000) 'Poisoning the spindle: serendipity and discovery of the anti-tumor properties of the vinka alkaloids', *Canadian Bulletin of Medical History* 17, 155–192.

Duhem, P. (1908/1969) *To Save the Phenomena: An Essay on the Idea of Physical Theory from Plato to Galileo* (trans. E. Dolan and C. Maschler). Chicago: University of Chicago Press.

Dupré, J. (1981) 'Natural kinds and biological taxa', *Philosophical Review* 90, 66–90.

Dupré, J. (1993) *The Disorder of Things: Metaphysical Foundations of the Disunity of Science*. Cambridge, MA: Harvard University Press.

Dyslexia Action (2012) 'Dyslexia still matters: dyslexia in our schools today: progress, challenges and solutions'. Dyslexia Action, York, June. https://lemosandcrane.co.uk/resources/Dyslexia%20Action%20Policy%20Document_SINGLE%20PAGES.PDF.

Edgington, D. (2020) 'Indicative conditionals', *Stanford Encyclopedia of Philosophy* (Fall 2020 Edition), E.N. Zalta (ed.). https://plato.stanford.edu/archives/fall2020/entries/conditionals/.

Elgin, C.Z. (2017) *True Enough*. Oxford: Oxford University Press.

Elliott, J.G., and E.L. Grigorenko (2014) *The Dyslexia Debate*. Cambridge: Cambridge University Press.

Elliott, K. (2007) 'Varieties of exploratory experimentation in nanotoxicology', *History and Philosophy of the Life Sciences* 29, 313–336.

Ellis, B. (2001) *Scientific Essentialism*. Cambridge: Cambridge University Press.

Elsasser, W.M. (1933) 'Sur le principe de Pauli dans les noyaux', *Journal of Physics* 4, 549–556.

Elsasser, W.M. (1934) 'Sur le principe de Pauli dans les noyaux. II', *Journal of Physics* 5, 389–397.

Ereshefsky, M. (2004) 'Bridging the gap between human kinds and biological kinds', *Philosophy of Science* 71, 912–921.

Ereshefsky, M. (2010) 'Darwin's solution to the species problem', *Synthese* 175, 405–425.

Ereshefsky, M. (2012) 'Homology thinking', *Biology & Philosophy* 27, 381–400.

Ereshefsky, M. (2018) 'Natural kinds, mind independence, and defeasibility', *Philosophy of Science* 85, 845–856.

Ereshefsky, M., and T. Reydon (2015) 'Scientific kinds', *Philosophical Studies* 172, 969–986.

Eronen, M.I. (2013) 'No levels, no problems: downward causation in neuroscience', *Philosophy of Science* 80, 1042–1052.

Etter, M.C. (1990) 'Encoding and decoding hydrogen-bond patterns of organic compounds', *Accounts of Chemical Research* 23, 120–126.

Eubanks, V. (2011) *Digital Dead End: Fighting for Social Justice in the Information Age.* Cambridge, MA: MIT Press.

Eze, M.O. (2017) 'I am because you are: cosmopolitanism in the age of xenophobia', *Philosophical Papers* 46, 85–109.

Falconer, I. (1987) 'Corpuscles, electrons and cathode rays: J.J. Thomson and the "discovery of the electron"', *British Journal for the History of Science* 20, 241–276.

Falconer, I. (2001) 'Corpuscles to electrons', in J. Buchwald and A. Warwick (eds), *Histories of the Electron.* Cambridge, MA: MIT Press, pp. 77–100.

Fara, D.G. (2008) 'Relative-sameness counterpart theory', *Review of Symbolic Logic* 1, 167–189.

Fara, D.G. (2012) 'Possibility relative to a sortal', in K. Bennett and D. Zimmermann (eds), *Oxford Studies in Metaphysics*, Vol. 7. Oxford: Oxford University Press, pp. 3–40.

Faraday, M. (1830) 'On the manufacture of glass for optical purposes', *Philosophical Transactions of the Royal Society* 120, 1–57.

Faraday, M. (1833/2012) *Experimental Researches in Electricity* (3 vols). Cambridge: Cambridge University Press.

Faye, J., M. Urchs, and U. Scheffler (eds) (2001) *Things, Facts and Events.* Amsterdam: Rodopi.

Feest, U., and F. Steinle (eds) (2012) *Scientific Concepts and Investigative Practice.* Berlin: de Gruyter.

Fehr, C. (2001) 'The evolution of sex: domains and explanatory pluralism', *Biology and Philosophy* 16, 145–170.

Fehr, C. (2006) 'Explanations of the evolution of sex: a plurality of local mechanisms', in S.H. Kellert, H.E. Longino, and C.K. Waters (eds), *Scientific Pluralism.* Minneapolis: University of Minnesota Press, pp. 167–189.

Fernández Pinto, M. (2020a) 'Ignorance, science, and feminism', in S. Crasnow and K. Intemann (eds), *The Routledge Handbook of Feminist Philosophy of Science.* London: Routledge, pp. 225–235.

Fernández Pinto, M. (2020b) 'Commercial interests and the erosion of trust in science', *Philosophy of Science* 87, 1004–1013.

Fine, A (1984/1991) 'The natural ontological attitude', reprinted in R. Boyd, P. Gasper, and J.D. Trout (eds), *Philosophy of Science.* Cambridge, MA: MIT Press, pp. 261–278.

Fine, A. (2018) 'Motives for research', *Spontaneous Generations: A Journal for the History and Philosophy of Science* 9, 42–45.

Fischer, B., and F. Leon (eds) (2017) *Modal Epistemology after Rationalism.* Dordrecht: Springer.

Fischer, M.A. (2018) 'Towards an excursion flora for Austria and all the Eastern Alps' *Botanica Serbica* 42, 5–33.

Fisher, G., A. Gelfert, and F. Steinle (2021) 'Exploratory models and exploratory modeling in science: introduction', *Perspectives on Science* 29, 355–358.

Fisher, S.E., and C. Francks (2006) 'Genes, cognition and dyslexia: learning to read the genome', *Trends in Cognitive Science* 10, 250–257.

Foucault, M. (1966/2001) *The Order of Things: An Archaeology of the Human Sciences*. London: Routledge.

Franklin-Hall, L.R. (2015) 'Natural kinds as categorical bottlenecks', *Philosophical Studies* 172, 925–948.

Fraser, N. (1995) 'From redistribution to recognition? Dilemmas of justice in a "post-socialist" age', *New Left Review* 212, 68–93.

Fraser, N. (2003) 'Social justice in the age of identity politics: redistribution, recognition, and participation', in N. Fraser and A. Honneth, *Redistribution or Recognition? A Political-Philosophical Exchange*. London: Verso, pp. 7–109.

Freedman, W.L., B.F. Madore, D. Hatt, et al. (2019) 'The Carnegie–Chicago Hubble Program: VIII. An independent determination of the Hubble constant based on the tip of the red giant branch', *The Astrophysical Journal* 882, 1–29.

French, S. (2002) 'A phenomenological approach to the measurement problem: Husserl and the foundations of quantum mechanics', *Studies in History and Philosophy of Modern Physics* 22, 467–491.

French, S. (2014) *The Structure of the World*. Oxford: Oxford University Press.

French, S. (2020) 'From a lost history to a new future: is a phenomenological approach to quantum mechanics viable?', in H. Wiltsche and P. Berghofer (eds), *Phenomenological Approaches to Physics*. Dordrecht: Springer, pp. 205–226.

Fricker, M. (2007) *Epistemic Injustice: Power and the Ethics of Knowing*. Oxford: Oxford University Press.

Friedman, M. (2020) 'Newtonian methodological abstraction', *Studies in History and Philosophy of Modern Physics* 72, 162–178.

Frigg, R. (2010a) 'Fiction in science', in J. Woods (ed.), *Fictions and Models: New Essays*. Munich: Philosophia Verlag, pp. 247–287.

Frigg, R. (2010b) 'Fiction and scientific representation', in R. Frigg and M.C. Hunter (eds), *Beyond Mimesis and Convention: Representation in Art and Science*. Dordrecht: Springer, pp. 97–138.

Frigg, R., and J. Nguyen (2016) 'The fiction view of models reloaded', *The Monist* 99, 225–242.

Frigg, R., and J. Nguyen (2020) 'Scientific representation', *Stanford Encyclopedia of Philosophy* (Spring 2020 Edition), E.N. Zalta (ed.). https://plato.stanford.edu/archives/spr2020/entries/scientific-representation/.

Frigg, R., E. Thompson, and C. Werndl (2015a) 'Philosophy of climate science part I: observing climate change', *Philosophy Compass* 10, 953–964.

Frigg, R., E. Thompson, and C, Werndl (2015b) 'Philosophy of climate science part II: modelling climate change', *Philosophy Compass* 10, 965–977.

Frith, U. (1986) 'A developmental framework for developmental dyslexia', *Annals of Dyslexia* 36, 69–81.

Frith, U. (2002a) 'Culture, brain and dyslexia', in E. Hjelmquist and C. von Euler (eds), *Literacy in the New Millennium: A Festschrift in Honour of Ingvar Lundberg at the Occasion of His 65th Birthday*. London: Whurr Publishers, pp. 179–191.

Frith, U. (2002b) 'Resolving the paradox of dyslexia', in G. Reid and J. Wearmouth (eds), *Dyslexia and Literacy: Theory and Practice*. Chichester: John Wiley & Sons, pp. 45–68.

Furlan, V., N.D. Jiménez-Escobar, F. Zamudio, and C. Medrano (2020) '"Ethnobiological equivocation" and other misunderstandings in the interpretation of natures', *Studies in History and Philosophy of Science C* 84, 101333.

García-Pulido, L.J. (2016) 'The mastery in hydraulic techniques for water supply at the Alhambra', *Journal of Islamic Studies* 27, 355–382.

Gärdenfors, P. (1986) 'Belief revisions and the Ramsey Test for conditionals', *Philosophical Review* 95, 81–93.

Gay, H. (1996) 'Invisible resource: William Crookes and his circle of support, 1871–81', *British Journal for the History of Science* 29, 311–336.

Gelfert, A. (2016) *How to Do Science with Models*. Dordrecht: Springer.

Gelman, S.A. (2003) *The Essential Child: Origins of Essentialism in Everyday Thought*. Oxford: Oxford University Press.

Gelman, S.A., and E.M. Markman (1987) 'Young children's inductions from natural kinds: the role of categories and appearances', *Child Development* 58, 1532–1541.

Gelman, S.A., and H.M. Wellman (1991) 'Insides and essences: early understanding of the non-obvious', *Cognition* 38, 213–244.

Gendler, T.S., and J. Hawthorne (eds) (2002) *Conceivability and Possibility*. Oxford: Oxford University Press.

Ghiselin, M. (1974) 'A radical solution to the species problem', *Systematic Zoology* 23, 536–544.

Giere, R. (2006a) *Scientific Perspectivism*. Chicago: University of Chicago Press.

Giere, R. (2006b) 'Perspectival pluralism', in S.H. Kellert, H.E. Longino, and C.K. Waters (eds), *Scientific Pluralism*. Minneapolis: University of Minnesota Press, pp. 26–41.

Giere, R. (2010) 'An agent-based conception of models and scientific representation', *Synthese* 172, 269–281.

Giere, R. (2013) 'Kuhn as perspectival realist', *Topoi* 32, 53–57.

Gilbert, W. (1600/1958) *De Magnete* (trans. P.F. Mottelay). New York: Dover Publications.

Gilroy, P. (2010) 'Planetarity and cosmopolitics', *British Journal of Sociology* 61, 620–626.

Gobbo, F., and F. Russo (2020) 'Epistemic diversity and the question of lingua franca in science and philosophy', *Foundations of Science* 25, 185–207.

Godfrey-Smith, P. (2006) 'The strategy of model-based science', *Biology and Philosophy* 21, 725–740.

Godfrey-Smith, P. (2020) 'Models, fictions, and conditionals', in A. Levy and P. Godfrey-Smith (eds), *The Scientific Imagination: Philosophical and Psychological Perspectives*. Oxford: Oxford University Press, pp. 154–177.

Godman, M. (2015) 'The special science dilemma and how culture solves it', *Australasia Journal of Philosophy* 93, 491–508.

Godman, M., A. Mallozzi, and D. Papineau (2020) 'Essential properties are super-explanatory: taming metaphysical modality', *Journal of the American Philosophical Association* 3, 316–334.

Goeppert Mayer, M. (1948) 'On closed shells in nuclei', *Physical Review* 74, 235–239.

Goeppert Mayer, M. (1949) 'On closed shells in nuclei. II', *Physical Review* 75, 1969–1970.

Goeppert Mayer, M. (1963) 'The shell model', *Nobel Lecture*, 12 December, pp. 20–37. https://www.nobelprize.org/uploads/2018/06/mayer-lecture.pdf.

Goeppert Mayer, M. (1964) 'The shell model', *Science* 145, 999–1006.

Goetzmann, W.N. (2005) 'Fibonacci and the financial revolution', in W.N. Goetzmann and K.G. Rouwenhorst (eds), *The Origins of Value: The Financial Innovations That Created Modern Capital Markets*. Oxford: Oxford University Press, pp. 123–144.

Goldman, A. (1986) *Epistemology and Cognition*. Cambridge, MA: Harvard University Press.

Goldman, A. (2004) 'Sosa on reflective knowledge and virtue perspectivism', in J. Greco (ed.), *Ernest Sosa and His Critics*. Oxford: Blackwell, pp. 86–95.

Goodman, N. (1947) 'The problem of counterfactual conditionals', *Journal of Philosophy* 44, 113–128.

Goodman, N. (1978) *Ways of Worldmaking*. Indianapolis: Hackett.

Goonatilake, S. (1998) *Towards a Global Science*. Bloomington: Indiana University Press.

Gordin, M. (2015) *Scientific Babel*. Chicago: University of Chicago Press.

Gordin, M. (2018) *A Well-Ordered Thing: Dmitri Mendeleev and the Shadow of the Periodic Table*. Princeton: Princeton University Press.

Gray, A., and J.J. Dobbie (1898) 'On the connection between electrical properties and the chemical composition of different kinds of glass', *Proceedings of the Royal Society of London* 63, 38–44.

Gray, M. (1951) 'The kelp industry in the Highlands and islands', *Economic History Review* 4, 197–209.

Greco, J. (ed.) (2004) *Ernest Sosa and His Critics*. Oxford: Blackwell.

Griffith, P. (1999) 'Squaring the circle: natural kinds with historical essences', in R. Wilson (ed.), *Species: New Interdisciplinary Studies*. Cambridge, MA: MIT Press, pp. 209–228.

Grimm, S.R. (2012) 'The value of understanding', *Philosophy Compass* 7, 103–117.

Grüne-Yanoff, T. (2009) 'Learning from minimal economic models', *Erkenntnis* 70, 81–99.

Grüne-Yanoff, T., and C. Marchionni (2018) 'Modeling model selection in model pluralism', *Journal of Economic Methodology* 25, 265–275.

Guggenheimer, K. (1934) 'Remarques sur la constitution des noyaux atomiques. I', *Journal of Physics* 5, 253–256.

Haack, S. (1993) *Evidence and Inquiry: Towards Reconstruction in Epistemology*. Oxford: Blackwell.

Hacking, I. (1982) 'Experimentation and scientific realism', *Philosophical Topics* 13, 154–172.

Hacking, I. (1983) *Representing and Intervening: Introductory Topics in the Philosophy of Natural Science*. Cambridge: Cambridge University Press.

Hacking, I. (1991) 'A tradition of natural kinds', *Philosophical Studies* 61, 109–126.

Hacking, I. (1999) *The Social Construction of What?* Cambridge, MA: Harvard University Press.

Hacking, I. (2002) *Historical Ontology*. Cambridge, MA: Harvard University Press.

Hacking, I. (2007a) 'Natural kinds: rosy dawn, scholastic twilight', *Royal Institute of Philosophy Supplement* 61, 203–239.

Hacking, I. (2007b) 'The contingencies of ambiguity', *Analysis* 67, 269–277.

Hager, K.M., and W. Gu (2014) 'Understanding the non-canonical pathways involved in p53-mediated tumor suppression', *Carcinogenesis* 35, 740–746.

Häggqvist, S., and Å. Wikforss (2018) 'Natural kinds and natural kind terms: myth and reality', *British Journal for the Philosophy of Science* 69, 911–933.

Hall, E. (1994) *The Arnolfini Betrothal: Medieval Marriage and the Enigma of van Eyck's Double Portrait*. Berkeley: University of California Press.

Hall, N. (2015) 'Humean reductionism about laws of nature', in B. Loewer and J. Schaffer (eds), *A Companion to David Lewis*. Oxford: Blackwell, pp. 262–277.

Haraway, D. (1988) 'Situated knowledges: the science question in feminism and the privilege of partial perspective', *Feminist Studies* 14, 575–599.

Harbison, C. (1991) *Jan van Eyck: The Play of Realism*. London: Reaktion Books.

Harding, S. (1986) *The Science Question in Feminism*. Ithaca, NY: Cornell University Press.

Harding, S. (1991) *Whose Science? Whose Knowledge? Thinking from Women's Lives*. Ithaca NY: Cornell University Press.

Harding, S. (1995) 'Strong objectivity: a response to the new objectivity question', *Synthese* 104, 331–349.

Harding, S. (2015) *Objectivity and Diversity: Another Logic of Scientific Research.* Chicago: University of Chicago Press.

Harkins, W.D. (1921) 'Isotopes: their numbers and classifications', *Nature* 107, 202–203.

Hartmann, S. (1999) 'Models and stories in hadron physics', in M.S. Morgan and M. Morrison (eds), *Models as Mediators: Perspectives on Natural and Social Science.* Cambridge: Cambridge University Press, pp. 326–346.

Hartsock, N.C.M. (1997) 'Comments on Hekman's "Truth and method": truth or justice?', *Signs* 22, 367–374.

Hartsock, N.C.M. (1998) 'The feminist standpoint revisited', in *The Feminist Standpoint Revisited and Other Essays.* Boulder, CO: Westview Press, pp. 227–248.

Haslam R.H., J.T. Dalby, R.T. Johns, et al. (1981) 'Cerebral asymmetry in developmental dyslexia', *Archive of Neurology* 38, 679–682.

Hausfather, Z., and G.P. Peters (2020a) 'Emissions—the "business as usual" story is misleading', *Nature* 577, 618–620.

Hausfather, A., and G.P. Peters (2020b) 'RCP8.5 is a problematic scenario for near-term emissions', *Proceedings of the National Academy of Science USA* 117, 27791–27792.

Havstad, J.C. (2018) 'Messy chemical kinds', *British Journal for the Philosophy of Science* 69, 719–743.

Hawke, P. (2011) 'Van Inwagen's modal scepticism', *Philosophical Studies* 153, 351–364.

Hawley, K., and A. Bird (2011) 'What are natural kinds?', *Philosophical Perspectives* 25, 205–221.

Hayden, C. (2011) 'No patent, no generic: pharmaceutical access and the politics of the copy', in M. Biagioli, P. Jaszi, and M. Woodmansee (eds), *Making and Unmaking Intellectual Property: Creative Production in Legal and Cultural Perspective.* Chicago: University of Chicago Press, pp. 285–304.

Healey, R. (2017) *The Quantum Revolution in Philosophy.* Oxford: Oxford University Press.

Heilbron, J. (1993) 'Weighing imponderables and other quantitative science around 1800', *Historical Studies in the Physical and Biological Sciences* 24, 1–337.

Heisenberg, W. (1934) 'General theoretical considerations on the structure of the nucleus', in Institut International de Physique Solvay (ed.), *Structure et propriétés des noyaux atomiques: rapports et discussions du Septième Conseil de Physique tenu à Bruxelles du 22 au 29 Octobre 1933.* Paris: Gauthier-Villars, pp. 289–323.

Hekman, S. (1997) 'Truth and method: feminist standpoint theory revisited', *Signs* 22, 341–365.

Hendry, R.F. (2006) 'Elements, compounds and other chemical kinds', *Philosophy of Science* 73, 864–875.

Hendry, R.F. (2008) 'Microstructuralism: problems and prospects', in K. Ruthenberg and J. van Brakel (eds), *Stuff: The Nature of Chemical Substances.* Würzburg: Königshauen und Neumann, pp. 107–122.

Hendry, R.F. (2019) 'How (not) to argue for microstructural essentialism'. Unpublished manuscript.

Herschel, J.F.W. (1842) 'On the action of the rays of the solar spectrum on vegetable colours', *Abstracts of the Papers Printed in the Philosophical Transactions of the Royal Society of London*, Vol. 4 (1837–1843), 181–214.

Hesse, M. (1966) *Models and Analogies in Science.* Notre Dame, IN: University of Notre Dame Press.

Hodgson, C. (chair) (2019) 'The human cost of dyslexia'. Report from the All-Party Parliamentary Group for Dyslexia and other SpLDs, April. https://cdn.bdadyslexia. org.uk/documents/Final-APPG-for-Human-cost-of-dyslexia-appg-report.pdf.

Hoefer, C., and G. Martí (2019) 'Water has a microstructural essence after all', *European Journal for Philosophy of Science* 9, 12.

Hoefer, C., and G. Martí (2020) 'Realism, reference and perspective', *European Journal for the Philosophy of Science* 10, 38.

Holton, G. (1973) *Thematic Origins of Scientific Thought*. Cambridge, MA: Harvard University Press.

Honneth, A. (1996) *The Struggle for Recognition: The Moral Grammar of Social Conflicts*. Cambridge, MA: MIT Press.

Hoshika, S., N.A. Leal, M.-J. Kim, et al. (2019) 'Hachimoji DNA and RNA: a genetic system with eight building blocks', *Science* 363, 884–887.

Huang, S.P., H.N. Pollack, and P.Y. Shen (2000) 'Temperature trends over the past five centuries reconstructed from borehole temperatures', *Nature* 403, 756–758.

Hubbert, M.K. (1940) 'The theory of ground-water motion', *Journal of Geology* 48, 785–944.

Hughes, R.I.G. (1997) 'Models and representation', *Philosophy of Science* 64, S325–S336.

Hull, D. (1978) 'A matter of individuality', *Philosophy of Science* 45, 335–360.

Hynd G.W., and M. Semrud-Clikeman (1989) 'Dyslexia and brain morphology', *Psychological Bulletin* 106, 447–482.

Ippoliti, E., F. Sterpetti, and T. Nickles (eds) (2016) *Models and Inferences in Science*. Dordrecht: Springer.

Ippolito, M. (2013) *Subjunctive Conditionals: A Linguistic Analysis*. Cambridge, MA: MIT Press.

Ismael, J. (2016) 'How do causes depend on us? The many faces of perspectivalism', *Synthese* 193, 245–267.

Ivanova, M., and S. French (eds) (2020) *Aesthetics of Science: Beauty, Imagination and Understanding*. New York: Routledge.

Izadi, S., R. Anandakrishnan, and A.V. Onufriev (2014) 'Building water molecules: a different approach', *Journal of Physical Chemistry Letters* 5, 3863–3871.

Jackson, F., and P. Pettit (1992) 'In defense of explanatory ecumenism', *Economics and Philosophy* 8, 1–21.

Jacoby, F. (2021) 'Acids and rust: a new perspective on the Chemical Revolution', *Perspectives on Science* 29, 215–236.

Jeffrey, P.D., A.A. Russo, K. Polyak, et al. (1995) 'Mechanism of CDK activation revealed by the structure of a cyclinA–CDK2 complex', *Nature* 376, 313–320.

Jensen, J.H.D. (1965) 'The history of the theory of structure of the atomic nucleus', *Science* 147, 1419–1423.

Jiang, F. (2008) 'The problem with patents: traditional knowledge and international IP law', *Harvard International Review* 30, 30–33.

Johnson, K.E. (1992) 'Independent-particle models of the nucleus in the 1930s', *American Journal of Physics* 60, 164–172.

Johnson, K.E. (2004) 'From natural history to the nuclear shell model: chemical thinking in the work of Mayer, Haxel, Jensen, and Suess', *Physics in Perspective* 6, 295–309.

Johnson, S., and J. Boswell (1775/2020) *A Journey to the Western Islands of Scotland and The Journal of a Tour to the Hebrides* (eds J. Lynch and C. Barnes). Oxford: Oxford University Press.

Joule, J.P. (1850) 'On the mechanical equivalent of heat', *Philosophical Transactions of the Royal Society of London* 140, 61–82.

Juel, C., and C. Minden-Cupp, C. (2000) 'Learning to read words: linguistic units and instructional strategies', *Reading Research Quarterly* 35, 458–492.

Kaiser, W. (2001) 'Electron gas theory of metals: free electrons in bulk matter', in J. Buchwald and A. Warwick (eds), *Histories of the Electron*. Cambridge, MA: MIT Press, pp. 255–304.

Kant, I. (1755/1986) 'De igne' (AK 1: 369–385), English trans. in L.W. Beck, M.J. Gregor, J.A. Reuscher, and R. Meerbote (eds), *Kant's Latin Writings: Translations, Commentaries and Notes*. New York: Peter Lang.

Kant, I. (1795/2006) *Towards Perpetual Peace and Other Writings in Politics, Peace and History* (ed. P. Kleingeld). New Haven: Yale University Press.

Katzav, J. (2014) 'The epistemology of climate models and some of its implications for climate science and the philosophy of science', *Studies in History and Philosophy of Modern Physics* 46, 228–238.

Kellert, S.H., H.E. Longino, and C.K. Waters (2006) 'Introduction: the pluralist stance', in S.H. Kellert, H.E. Longino, and C.K. Waters (eds), *Scientific Pluralism*. Minneapolis: University of Minnesota Press, pp. vii–xxix.

Kendig, C. (2015) 'Homologizing as kinding', in C. Kendig (ed.), *Natural Kinds and Classification in Scientific Practice*. Abingdon: Routledge, pp. 106–125.

Kendig, C. (ed.) (2016a) *Natural Kinds and Classification in Scientific Practice*. Abingdon: Routledge.

Kendig, C. (2016b) 'What is proof of concept research and how does it generate epistemic and ethical categories for future scientific practice?', *Science and Engineering Ethics* 22, 735–753.

Kendig, C., and B.A. Bartley (2019) 'Synthetic kinds: kind-making in synthetic biology', in J.R.S. Bursten (ed.), *Perspectives on Classification in Synthetic Sciences: Unnatural Kinds*. London: Taylor & Francis, pp. 78–96.

Kendig, C., and J. Grey (2021) 'Can the epistemic value of natural kinds be explained independently of their metaphysics?', *British Journal for the Philosophy of Science* 72, 359–376.

Kennedy, C.J. (2017) 'Development of the skills and materials used in the Scottish glass industry from 1750', *Scottish Business and Industrial History* 31 (2), 18–43.

Kennedy, C.J., T. Addyman, K.R. Murdoch, and M.E. Young (2018) '18th- and 19th-century Scottish laboratory glass: assessment of chemical composition in relation to form and function', *Journal of Glass Studies* 60, 253–268.

Khalidi, M.A. (2013) *Natural Categories and Human Kinds: Classification in the Natural and Social Sciences*. Cambridge: Cambridge University Press.

Kirby, P. (2018) 'A brief history of dyslexia', *The Psychologist* 31, 56–59.

Kitcher, P. (1984) 'Species', *Philosophy of Science* 51, 308–333.

Kitcher, P. (1992) 'The naturalists return', *Philosophical Review* 101, 53–114.

Kitcher, P. (1993) *The Advancement of Science*. Oxford: Oxford University Press.

Kitcher, P. (2001) *Science, Truth, and Democracy*. Oxford: Oxford University Press.

Kiyotaki, N., and R. Wright (1989) 'On money as a medium of exchange', *Journal of Political Economy* 97, 927–954.

Kleingeld, P. (2011) *Kant and Cosmopolitanism*. Cambridge: Cambridge University Press.

Kleingeld, P. (2016) 'Kant's moral and political cosmopolitanism', *Philosophy Compass* 11, 14–23.

Kleingeld, P., and E. Brown (2019) 'Cosmopolitanism', *Stanford Encyclopedia of Philosophy* (Winter 2019 Edition), E.N. Zalta (ed.). https://plato.stanford.edu/archives/win2019/entries/cosmopolitanism/.

Knight, C. (2017) 'What is dyslexia? An exploration of the relationship between teachers' understandings of dyslexia and their training experiences', *Dyslexia* 24, 207–219.

Knobe, J., S. Prasada, and G.E. Newman (2013) 'Dual character concepts and the normative dimension of conceptual representation', *Cognition* 127, 242–257.

Knüsel, B., and C. Baumberger (2020) 'Understanding climate phenomena with data-driven models', *Studies in History and Philosophy of Science* 84, 46–56.

Knutti, R. (2010) 'The end of model democracy?', *Climatic Change* 10 (3–4), 2, 395–404.

Knutti, R., D. Masson, and A. Gettelman (2013) 'Climate model genealogy: Generation CMIP5 and how we got there', *Geophysical Research Letters* 40, 1194–1199.

Knutti, R., C. Baumberger, and G. Hirsch Hadorn (2019) 'Uncertainty quantification using multiple models—prospects and challenges', in C. Beisbart and N.J. Saam (eds), *Computer Simulation Validation*. Dordrecht: Springer, pp. 835–855.

Knuuttila, T. (2011) 'Modeling and representing: an artefactual approach to model-based representation', *Studies in History and Philosophy of Science* 42, 262–271.

Knuuttila, T. (2017) 'Imagination extended and embedded: artifactual and fictional accounts of models'. *Synthese*. https://doi.org/10.1007/s11229-017-1545-2.

Knuuttila, T., and M. Boon (2011) 'How do models give us knowledge? The case of Carnot's ideal heat engine', *European Journal for Philosophy of Science* 1, 309–334.

Knuuttila, T., and A. Loettgers (2013) 'Synthetic modelling and mechanistic account: material recombination and beyond', *Philosophy of Science* 80, 874–885.

Knuuttila, T., and A. Loettgers (2017) 'Modelling as indirect representation? The Lotka–Volterra model revisited', *British Journal for the Philosophy of Science* 68, 1007–1036.

Knuuttila, T., and A. Loettgers (2021) 'Biological control variously materialized: modeling, experimentation and exploration in multiple media', *Perspectives on Science* 29, 468–492.

Kobe, B., and B.E. Kemp (2010) 'Principles of kinase regulation', in R.A. Bradshaw and E.A. Dennis (eds), *Handbook of Cell Signaling*, Vol. 2 (Second Edition). Amsterdam: Elsevier, pp. 559–563.

Kolb, E.W., and A.J. Long (2017) 'Superheavy dark matter through Higgs portal operators', *Physical Review D* 96, 103540.

Kourany, J. (2002) *The Gender of Science*. Englewood Cliffs, NJ: Prentice Hall.

Kragh, H. (2000) 'An unlikely connection: geochemistry and nuclear structure', *Physics in Perspective* 2, 381–397.

Kragh H. (2002) 'The vortex atom: a Victorian theory of everything', *Centaurus* 44, 32–114.

Kratzer, A. (2012) *Modals and Conditionals: New and Revised Perspectives*. Oxford: Oxford University Press.

Kripke, S. (1980) *Naming and Necessity*. Oxford: Basil Blackwell.

Kuhn, T.S. (1957) *The Copernican Revolution: Planetary Astronomy in the Development of Western Thought*. Cambridge, MA: Harvard University Press.

Kuhn, T.S. (1962/1996) *The Structure of Scientific Revolution* (Third Edition). Chicago: University of Chicago Press.

Kuhn, T.S. (1990) 'Dubbing and re-dubbing: the vulnerability of rigid designation', in C.W. Savage (ed.), *Scientific Theories*. Minneapolis: University of Minnesota Press, pp. 298–318.

Kuhn, T.S. (2000) *The Road since Structure: Philosophical Essays 1970-1993.* Chicago: University of Chicago Press.

Kung, P. (2010) 'Imagining as a guide to possibility', *Philosophy and Phenomenological Research* 81, 620–663.

Kung, P. (2016) 'Imagination and modal knowledge', in A. Kind and P. Kung (eds), *Knowledge through Imagination.* New York: Oxford University Press, pp. 437–450.

Kymlicka, W. (1996) *Multicultural Citizenship: A Liberal Theory of Minority Rights.* Oxford: Oxford University Press.

Kymlicka, W. (2001) *Politics in the Vernacular: Nationalism, Multiculturalism, and Citizenship.* Oxford: Oxford University Press.

Kymlicka, W. (2007) *Multicultural Odysseys: Navigating the New International Politics of Diversity.* Oxford: Oxford University Press.

Ladyman, J. (2011) 'Structural realism versus standard scientific realism: the case of phlogiston and dephlogisticated air', *Synthese* 180, 87–101.

Ladyman, J., and D. Ross (2007) *Everything Must Go: Metaphysics Naturalized.* Oxford: Oxford University Press.

Lange, M. (2017) *Because without Cause: Non-Causal Explanations in Science.* Oxford: Oxford University Press.

LaPorte, J. (2004) *Natural Kinds and Conceptual Change.* Cambridge: Cambridge University Press.

Latour, B. (2004) 'Whose cosmos, whose cosmopolitics? Comments on the peace terms of Ulrich Beck', *Common Knowledge* 10, 450–492.

Laudan, L. (1981) 'A confutation of convergent realism', *Philosophy of Science* 48, 19–49.

Lavoisier, A.-L. (1799) *Elements of Chemistry* (trans. R. Kerr). London: G.G and J. Robinson, and T. Kay.

Le Bihan, S. (2017) 'Enlightening falsehoods: a modal view of scientific understanding', in S. Grimm, C. Baumberger, and S. Ammon (eds), *Explaining Understanding: New Perspectives from Epistemology and Philosophy of Science.* Abingdon: Routledge, pp. 111–135.

Lederberg, J. (1964) 'DENDRAL-64: a system for computer construction, enumeration & notation of organic molecules as tree structures and cyclic graphs. Part I. Notational algorithm for tree structures', NASA CR-57029. STAR No. N65-13158.

Lederberg, J. (1965) 'DENDRAL-64. Part II. Topology of cyclic graphs', NASA CR-68898. STAR No. N66-14074.

Lederberg, J. (1987) 'How DENDRAL was conceived and born', ACM symposium on the History of Medical Informatics, National Library of Medicine.

Lenhard, J., and E. Winsberg (2010) 'Holism, entrenchment and the future of climate model pluralism', *Studies in History and Philosophy of Modern Physics* 41, 253–262.

Leonard, K., J.D. Aldern, A. Christianson, et al. (2020) 'Indigenous conservation practices are not a monolith: Western cultural biases and a lack of engagement with indigenous experts undermine the study of land stewardship'. EcoEvoRxiv. 24 July. https://doi.org/10.32942/osf.io/jmvqy.

Leonelli, S. (2016) *Data-Centric Biology: A Philosophical Study.* Chicago: University of Chicago Press.

Leonelli, S. (2019) 'What distinguishes data from models?', *European Journal for Philosophy of Science* 9, 22.

Levins, R. (1966) 'The strategy of model building in population biology', *American Scientist* 54, 423–425.

Levy, A., and P. Godfrey-Smith (eds) (2020) *The Scientific Imagination*. Oxford: Oxford University Press.

Lewis, D. (1973) *Counterfactuals*. Oxford: Blackwell.

Lewis, D. (1976) 'Probabilities of conditionals and conditional probabilities', *Philosophical Review* 85, 297–315.

Lewis, D. (1986) *On the Plurality of Worlds*. Oxford: Blackwell.

Lewis, D. (1994) 'Humean supervenience debugged', *Mind* 103, 473–490.

Lindgren, S.D., E. De Renzi, and L.C. Richman (1985) 'Cross-national comparisons of developmental dyslexia in Italy and the United States', *Child Development* 56, 1404–1417.

Lindsay, R.K., B.G. Buchanan, E.A. Fegenbaum, and J. Lederberg (1980) *Applications of Artificial Intelligence for Organic Chemistry: The DENDRAL Project*. New York: McGraw-Hill.

Lipton, P. (2007) 'Review of Ron Giere's *Scientific Perspectivism*', *Science*, 316 (5826), 834.

Lloyd, E.A. (2010) 'Confirmation and robustness of climate models', *Philosophy of Science* 77, 971–984.

Lloyd, E.A. (2015) 'Model robustness as a confirmatory virtue: the case of climate science', *Studies in History and Philosophy of Science* 49, 58–68.

Lobier, M., and S. Valdois (2015) 'Visual attention deficits in developmental dyslexia cannot be ascribed solely to poor reading experience', *Nature Reviews Neuroscience* 16, 225.

Locke, J. (1689/1975) *An Essay Concerning Human Understanding* (ed. P.H. Nidditch). Oxford: Clarendon Press.

Longino, H. (1990) *Science as Social Knowledge: Values and Objectivity in Scientific Inquiry*. Princeton: Princeton University Press.

Longino, H. (1994) 'In search of feminist epistemology', *The Monist* 77, 472–485.

Longino, H. (2001) 'What do we measure when we measure aggression?', *Studies in History and Philosophy of Science* 32, 685–704.

Longino, H. (2002) *The Fate of Knowledge*. Princeton: Princeton University Press.

Longino, H. (2006) 'Theoretical pluralism and the scientific study of behaviour', in S.H. Kellert, H.E. Longino, and C.K. Waters (eds), *Scientific Pluralism*. Minnesota Studies in the Philosophy of Science XIX. Minneapolis: University of Minnesota Press, pp. 102–131.

Longino, H. (2013) *Studying Human Behavior: How Scientists Investigate Aggression and Sexuality*. Chicago: University of Chicago Press.

Lowe, E.J. (2009) *More Kinds of Being: A Further Study of Individuation, Identity, and the Logic of Sortal Terms*. Chichester: Wiley-Blackwell.

Ludwig, D. (2017) 'Indigenous and scientific kinds', *British Journal for the Philosophy of Science* 68, 187–212.

Ludwig, D. (2018a) 'Letting go of "natural kind": toward a multidimensional framework of nonarbitrary classification', *Philosophy of Science* 85, 31–52.

Ludwig, D. (2018b) 'Revamping the metaphysics of ethnobiological classification', *Current Anthropology* 59, 415–438.

Ludwig, D., and D. Weiskopf (2019) 'Ethnoontology: ways of world-building across cultures', *Philosophy Compass* 14 (9), e12621.

Lyons, T.D. (2002) 'Scientific realism and the pessimistic meta-modus tollens', in S. Clarke and T.D. Lyons (eds), *Recent Themes in the Philosophy of Science: Scientific Realism and Commonsense*. Dordrecht: Springer, pp. 63–90.

Lyons, T.D. (2006) 'Scientific realism and the *stratagema de divide et impera*', *British Journal for the Philosophy of Science* 57, 537–560.

Mach, E. (1897/1914) *The Analysis of the Sensations* (trans. C.M. Williams). Chicago: Open Court.

Mackintosh, R. (1977) 'The shape of nuclei', *Reports on Progress in Physics* 40, 731–789.

MacLeod, M., and T. Reydon (2013) 'Natural kinds in philosophy and in the life sciences: scholastic twilight or new dawn?', *Biological Theory* 7, 89–99.

Magnani, L., and N.J. Nersessian (eds) (2002) *Model-Based Reasoning: Science, Technology, Values*. Boston: Springer US.

Magnus, P.D. (2012) *Scientific Enquiry and Natural Kinds: From Planets to Mallards*. Basingstoke: Palgrave Macmillan.

Magnus, P.D. (2014) 'NK ≠ HPC', *Philosophical Quarterly* 64, 471–477.

Maisog, J.M., E.R. Einbinder, D.L. Flowers, et al. (2008) 'A meta-analysis of functional neuroimagining studies of dyslexia', in G.F. Eden and D.L. Flowers (eds), *Learning, Skill Acquisition, Reading and Dyslexia*. New York: Wiley-Blackwell, pp. 237–259.

Mäki, U. 2011. 'Models and the locus of their truth', *Synthese* 180, 47–63.

Mann, M.E., R.S. Bradley, and M.K. Hughes (1998) 'Global-scale temperature patterns and climate forcing over the past six centuries', *Nature* 392, 779–787.

Mann, M.E., R.S. Bradley, and M.K. Hughes (1999) 'Northern hemisphere temperatures during the past millennium: inferences, uncertainties, and limitations', *Geophysical Research Letters* 26, 759–762.

Markman, E. (1989) *Categorization and Naming in Children: Problems of Induction*. Cambridge, MA: MIT Press.

Markman E., and J. Hutchinson (1984) 'Children's sensitivity to constraints on word meaning: taxonomic vs thematic relations', *Cognitive Psychology* 16, 1–27.

Markman, E., and G.F. Wachtel (1988) 'Children's use of mutual exclusivity to constrain the meaning of words', *Cognitive Psychology* 20, 121–157.

Markman, E., J.L. Wasow, and M.B. Hansen (2003) 'Use of the mutual exclusivity assumption by young word learners', *Cognitive Psychology* 47, 241–275.

Massimi, M. (2005) *Pauli's Exclusion Principle: The Origin and Validation of a Scientific Principle*. Cambridge: Cambridge University Press.

Massimi, M. (2007) 'Saving unobservable phenomena', *British Journal for the Philosophy of Science* 58, 235–262.

Massimi, M. (2008) 'Why there are no ready-made phenomena: what philosophers of science should learn from Kant', in M. Massimi (ed.), *Kant and Philosophy of Science Today, Royal Institute of Philosophy Supplement* 63. Cambridge: Cambridge University Press, pp. 1–35.

Massimi, M. (2009) 'Review of Bas van Fraassen, *Scientific Representation: Paradoxes of Perspective*', *International Studies in Philosophy of Science* 23, 323–337.

Massimi, M. (2010) 'Galileo's mathematization of nature at the crossroad between the empiricist tradition and the Kantian one', *Perspectives on Science* 18, 152–188.

Massimi, M. (2011a) 'From data to phenomena: a Kantian stance', *Synthese* 182, 101–116.

Massimi, M. (2011b) 'Kant's dynamical theory of matter in 1755, and its debt to speculative Newtonian experimentalism', *Studies in History and Philosophy of Science* 42, 525–543.

Massimi, M. (2012a) 'Scientific perspectivism and its foes', *Philosophica* 84, 25–52.

Massimi, M. (2012b) '*Dwatery* ocean', *Philosophy* 87, 531–555.

Massimi, M. (2012c) 'Natural kinds, conceptual change, and the duck-billed platypus: LaPorte on incommensurability', in D. Dieks, W.J. Gonzalez, S. Hartmann,

M. Stoeltzner, and M. Weber (eds), *Probabilities, Laws, and Structures*. Dordrecht: Springer, pp. 201–216.

Massimi, M. (2014) 'Natural kinds and naturalised Kantianism', *Noûs* 48, 416–449.

Massimi, M. (2015) 'Walking the line: Kuhn between realism and relativism', in A. Bokulich and W. Devlin (eds), *Kuhn's Structure of Scientific Revolutions: 50 Years On*. Dordrecht: Springer, pp. 135–152.

Massimi, M. (2016) 'Bringing real realism back home: a perspectival slant', in M. Couch and J. Pfeifer (eds), *The Philosophy of Philip Kitcher*. Oxford: Oxford University Press, pp. 98–120.

Massimi, M. (2017a) 'What is this thing called "scientific knowledge"? Kant on imaginary standpoints and the regulative role of reason', *Kant Yearbook* 9, 63–83.

Massimi, M. (2017b) 'Review of S. Ruphy, *Scientific Pluralism Reconsidered*. University of Pittsburgh Press'. *BJPS Review of Books*. http://www.thebsps.org/reviewofbooks/stepha nie-ruphy-scientific-pluralism-reconsidered/.

Massimi, M. (2017c) 'Grounds, modality, and nomic necessity in the Critical Kant', in M. Massimi and A. Breitenbach (eds), *Kant and the Laws of Nature*. Cambridge: Cambridge University Press, pp. 150–170.

Massimi, M. (2017d) 'Laws of nature, natural properties, and the *robustly* best system', *The Monist* 100, 406–421.

Massimi, M. (2018a) 'Perspectivism', in J. Saatsi (ed.), *The Routledge Handbook of Scientific Realism*. London: Routledge, pp. 164–175.

Massimi, M. (2018b) 'Perspectival modelling', *Philosophy of Science* 85, 335–359.

Massimi, M. (2018c) 'Points of view: Kant on perspectival knowledge', *Synthese* Special Issue on *The Current Relevance of Kant's Method in Philosophy*, G. Gava (ed.). https:// link.springer.com/article/10.1007/s11229-018-1876-7.

Massimi, M. (2018d) 'Three problems about multiscale modelling in cosmology', *Studies in History and Philosophy of Modern Physics* 64, 26–38.

Massimi, M. (2018e) 'Four kinds of perspectival truth', *Philosophy and Phenomenological Research* 96, 342–359.

Massimi, M. (2018f) 'A *perspectivalist* better best system account of lawhood', in W. Ott and L. Patton (eds), *Laws of Nature*. Oxford: Oxford University Press, pp. 139–157.

Massimi, M (2019a) 'Two kinds of exploratory models', *Philosophy of Science* 86, 869–881.

Massimi, M. (2019b) '2017 Wilkins–Bernal–Medawar lecture: why philosophy of science matters to science', *Notes and Records: The Royal Society Journal for the History of Science* 73, 353–367.

Massimi, M. (2019c) 'Realism, perspectivism and disagreement in science', *Synthese*. https://doi.org/10.1007/s11229-019-02500-6.

Massimi, M. (2021) 'Cosmic Bayes: datasets and priors in the hunt for dark energy', *European Journal for the Philosophy of Science* 11, 29. https://doi.org/10.1007/ s13194-020-00338-1.

Massimi, M., and W. Bhimji (2015) 'Computer simulations and experiments: the case of the Higgs boson', *Studies in History and Philosophy of Modern Physics* 51, 71–81.

Massimi, M., and C.D. McCoy (eds) (2019) *Understanding Perspectivism: Scientific Challenges and Methodological Prospects*. New York: Routledge.

Mattix, C. (1999) 'The debate over bioprospecting on the public lands', *Natural Resources and Environment* 13, 528–532.

Maxwell, J.C. (1861–2/1890) 'On physical lines of force', reprinted in W.D. Niven (ed.), *The Scientific Papers of James Clerk Maxwell*. New York: Dover, pp. 451–513.

McEarlean, T.C. (2007) 'Archaeology of the Strangford Lough kelp industry in the 18th and early 19th century', *Historical Archaelogy* 41, 76–93.

McLeish, C. (2005) 'Scientific realism bit by bit. Part I: Kitcher on reference', *Studies in History and Philosophy of Science* 36, 667–685.

McLeish, T. (2019) *The Poetry and Music of Science: Comparing Creativity in Science and Art*. Oxford: Oxford University Press.

McTaggart, J.M.E. (1908) 'The unreality of time', *Mind* 17, 457–474.

Mgbeoji, I. (2006) *Global Biopiracy: Patents, Plants and Indigenous Knowledge*. Vancouver: UBC Press.

Michaels, W.B. (2017) 'Cosmopolitanism goes to class', in B. Robbins, P.L. Horta (eds), *Cosmopolitanisms*. Oxford: Oxford University Press, pp. 59–64.

Mignolo, W.D. (2000) *Local Histories/Global Designs: Coloniality, Subaltern Knowledges, and Border Thinking*. Princeton: Princeton University Press.

Millikan, R. (1998) 'Language conventions made simple', *Journal of Philosophy* 95, 161–180.

Millikan, R. (1999) 'Historical kinds and the "special sciences"', *Philosophical Studies* 95, 45–65.

Mitchell, A.C. (1932) 'Chapters in the history of terrestrial magnetism', *Terrestrial Magnetism and Atmospheric Electricity* 37, 123–124.

Mitchell, S. (2003) *Biological Complexity and Integrative Pluralism*. Cambridge: Cambridge University Press.

Mitchell, S. (2009) *Unsimple Truths*. Chicago: University of Chicago Press.

Mitchell, S. (2020) 'Through the fractured looking glass', *Philosophy of Science* 87, 771–792.

Mitchell, S., and A.M. Gronenborn (2017) 'After fifty years, why are X-ray crystallographers still in business?', *British Journal for the Philosophy of Science* 68, 703–723.

Mladjenovic, M. (1998) *The Defining Years in Nuclear Physics: 1932–1960s*. Bristol: IOP.

Morgan, M.S., and M. Morrison (eds) (1999) *Models as Mediators: Perspectives on Natural and Social Science*. Cambridge: Cambridge University Press.

Morris, R.J. (1972) 'Lavoisier and the caloric theory', *British Journal for the History of Science* 6, 1–38.

Morrison, M. (2011) 'One phenomenon, many models: inconsistency and complementarity', *Studies in History and Philosophy of Science* 42, 342–351.

Morrison, M. (2015) *Reconstructing Reality: Models, Mathematics, and Simulations*. New York: Oxford University Press.

Morton, J. (1986) 'Developmental contingency modelling: a framework for discussing the processes of change and the consequence of deficiency', in P.L.C van Geert (ed.), *Theory Building in Developmental Psychology*. Amsterdam: North Holland-Elsevier, pp. 141–165.

Morton, J. (2004) *Understanding Developmental Disorders: A Causal Modelling Approach*. New York: Wiley.

Morton J., and U. Frith (1995) 'Causal modelling: a structural approach to developmental psychopathology', in D. Cicchetti and D.J. Cohen (eds), *Developmental Psychopathology, Vol. 1: Theory and Methods*. New York: Wiley, pp. 357–390.

Moss, R.H., J.A. Edmonds, K.A. Hibbard, et al. (2010) 'The next generation of scenarios for climate change research and assessment', *Nature* 463, 747–756.

Mottelson, B.R. (1975) 'Elementary modes of excitation in the nucleus'. *Nobel Lecture*, 11 December, pp. 236–254. https://www.nobelprize.org/uploads/2018/06/mottelson-lecture.pdf.

Moyal, A. (2004) *Platypus: The Extraordinary Story of How a Curious Creature Baffled the World*. Baltimore: Johns Hopkins University Press.

Mulalap, C.Y., T. Frere, E. Huffer, et al. (2020) 'Traditional knowledge and the BBNJ instrument', *Marine Policy*. https://doi.org/10.1016/j.marpol.2020.104103.

Murray, J.S. (2009) 'Blueprinting in the history of cartography', *Cartographic Journal* 46 (3), 257–261.

Nation, K., and M.J. Snowling (1998) 'Individual differences in contextual facilitation: evidence from dyslexia and poor reading comprehension', *Child Development* 69, 996–1011.

National Institute of Child Health and Development (2007) 'Learning disabilities: what are learning disabilities?' https://www.nichd.nih.gov/health/topics/learningdisabilities.

Navarro, J. (2012) *A History of the Electron: J.J. and G.P. Thomson*. Cambridge: Cambridge University Press.

Needham J. (1969/1972) *The Grand Titration: Science and Society in East and West*. London: George Allen & Unwin Ltd.

Needham J. (1970) 'The Chinese contribution to the development of the mariner's compass', in *Clerks and Craftsmen in China and the West: Lectures and Addresses on the History of Science and Technology*. Cambridge: Cambridge University Press, pp. 239–249.

Needham, P. (2000) 'What is water?', *Analysis* 60, 13–21.

Needham, P. (2011) 'Microessentialism: what is the argument?', *Noûs* 45, 1–21.

Neimark, B. (2012) 'Green grabbing at the "pharm" gate: rosy periwinkle production in southern Madagascar', *Journal of Peasant Studies* 39, 423–445.

Nersessian, N. (2010) *Creating Scientific Concepts*. Cambridge, MA: MIT Press.

Neurath, O. (1944) *Foundations of the Social Sciences: International Encyclopedia of Unified Science*, Vol. II, No. 1 (eds O. Neurath, R. Carnap, and C. Morris). Chicago: University of Chicago Press.

Newman, G.E., and J. Knobe (2019) 'The essence of essentialism', *Mind and Language* 34, 585–605.

Newton, I. (1674/1962) 'De aere et aethere', in A.R. Hall and M.B. Hall (eds), *Unpublished Scientific Papers of Isaac Newton*. New York: Cambridge University Press, pp. 214–228.

Nicholson, D.J., and J. Dupré (eds) (2018) *Everything Flows: Towards a Processual Philosophy of Biology*. Oxford: Oxford University Press.

Noakes, R. (2005) 'Ethers, religion and politics in late-Victorian physics: beyond the Wynne thesis', *History of Science* 43, 415–455.

Noakes, R. (2014) 'Industrial research at the Eastern Telegraph Company, 1872–1929', *British Journal for the History of Science* 47, 119–146.

Nolan, D. (2017) 'Naturalised modal epistemology', in R. Fischer and F. Leon (eds), *Modal Epistemology after Rationalism*. Dordrecht: Springer, pp. 7–28.

O'Connor, C. (2019) 'Games and kinds', *British Journal for the Philosophy of Science* 70, 719–745.

Okasha, S. (2002) 'Darwinian metaphysics: species and the question of essentialism', *Synthese* 131, 191–213.

Olulade, O.A., E.M. Napoliello, and G.F. Eden (2013) 'Abnormal visual motion processing is not a cause of dyslexia', *Neuron* 79, 180–190.

O'Malley, M. (2014) *Philosophy of Microbiology*. Cambridge: Cambridge University Press.

O'Neill, O. (2013) 'Science, reasons and normativity', *European Review* 21 (S1), S94–S99.

O'Neill, O. (2018) 'Scientific inquiry and normative reasoning', in *From Principles to Practice: Normativity and Judgement in Ethics and Politics*. Cambridge: Cambridge University Press, pp. 40–52.

Ophir, A., and S. Shapin (1991) 'The place of knowledge: a methodological survey', *Science in Context* 4, 3–21.

Oreskes, N. (2019) *Why Trust Science?* Princeton: Princeton University Press.

Ostergard, R.L., M. Tubin, and J. Altman (2001) 'Stealing from the past: globalization, strategic formation and the use of indigenous intellectual property in the biotechnology industry', *Third World Quarterly* 22, 643–656.

Ostriker, J.P., and P.J.E. Peebles (1973) 'A numerical study of the stability of flattened galaxies: or, can cold galaxies survive?', *The Astrophysical Journal* 186, 467–480.

Paneth, F. (1937) 'The chemical composition of the atmosphere', *Quarterly Journal of the Royal Meteorological Society* 63, 433–438.

Panofsky, E. (1934) 'Jan van Eyck's Arnolfini Portrait', *Burlington Magazine for Connoisseurs* 64, 117–119, 122–127.

Panofsky, E. (1991) *Perspective as Symbolic Form*. Cambridge, MA: MIT Press.

Park, J.H., and L. March (2002) 'The Shampay House of 1919: authorship and ownership', *Journal of the Society of Architectural Historians* 61, 470–479.

Parker, W. (2009) 'II—Confirmation and adequacy-for-purpose in climate modelling', *Aristotelian Society Supplementary Volume* 83 (1), 233–249.

Parker, W. (2011) 'When climate models agree: the significance of robust model predictions', *Philosophy of Science* 78, 579–600.

Patton, L. (2015) 'Methodological realism and modal resourcefulness: out of the web and into the mine', *Synthese* 192, 3443–3463.

Paul, L.A. (2006) 'In defense of essentialism', *Philosophical Perspectives* 20, 333–372.

Paulesu, E., U. Frith, M. Snowling, et al. (1996) 'Is developmental dyslexia a disconnection syndrome? Evidence from PET scanning', *Brain* 199, 143–157.

Paulesu, E., J.F. Démonet, F. Fazio, et al. (2001) 'Dyslexia: cultural diversity and biological unity', *Science* 291 (5511), 2165–2167.

Pauliukaite, R., et al. (2017) 'Theodor von Grotthuss' contribution to electrochemistry', *Electrochimica Acta* 236, 28–32.

Peebles, P.J.E. (2017) 'How the nonbaryonic dark matter theory grew', *Nature Astronomy* 1, 57.

Pérez-Balam, J., J.J.G. Quezada-Euán, R. Alfaro-Bates, et al. (2012) 'The contribution of honey bees, flies and wasps to avocado (*Persea americana*) pollination in southern Mexico', *Journal of Pollination Ecology* 8, 42–47.

Pero, F., and M. Suárez (2016) 'Varieties of misrepresentation and homomorphism', *European Journal for Philosophy of Science* 6, 71–90.

Peschard, I. (2012) 'Modeling and experimenting', in P. Humphreys and C. Imbert (eds), *Models, Simulations and Representations*. London: Routledge, pp. 42–61.

Planck, M. (1906/1913) *The Theory of Heat Radiation* (trans. M. Masius; Second Edition). New York: Dover Publications.

Plutynski, A. (2018) *Explaining Cancer: Finding Order in Disorder*. Oxford: Oxford University Press.

Pollack H.N., S.P. Huang, and P.Y. Shen (1998) 'Climate change record in subsurface temperatures: a global perspective', *Science* 282 (5387), 279–281.

Popper, K. (1978) 'Three worlds'. The Tanner Lecture on Human Values. Talk delivered at the University of Michigan. https://tannerlectures.utah.edu/_documents/a-to-z/p/popper80.pdf.

Portides, D. (2005) 'Scientific models and the semantic view of scientific theories', *Philosophy of Science* 72, 1287–1298.

Portides, D. (2011) 'Seeking representations of phenomena: phenomenological models', *Studies in History and Philosophy of Science A* 42, 334–341.

Potochnik, A. (2017) *Idealization and the Aims of Science*. Chicago: University of Chicago Press.

Potochnik, A. (2021) 'Our world isn't organized into levels', in D.S. Brooks, J. DiFrisco, and W.C. Wimsatt (eds), *Levels of Organization in the Biological Sciences*. Cambridge, MA: MIT Press.

Potochnik, A., and B. McGill (2012) 'The limitations of hierarchical organization', *Philosophy of Science* 79, 120–140.

Powell, H.J. (1919) 'Glass-making before and during the war', *Journal of the Royal Society of Arts* 67, 485–495.

Price, H. (2007) 'Causal perspectivalism', in H. Price and R. Corry (eds), *Causation, Physics and the Constitution of Reality: Russell's Republic Revisited*. Oxford: Oxford University Press, pp. 250–292.

Price, H., with S. Blackburn, R. Brandom, P. Horwich, and M. Williams (2013) *Expressivism, Pragmatism and Representationalism*. Cambridge: Cambridge University Press.

Psillos, S. (1996) 'Scientific realism and the "pessimistic induction"', *Philosophy of Science* 63, S306–S314.

Psillos, S. (1999) *Scientific Realism: How Science Tracks Truth*. London: Routledge.

Psillos, S. (2012) 'Causal descriptivism and the reference of theoretical terms', in A. Raftopoulos and P. Machamer (eds), *Perception, Realism, and the Problem of Reference*. Cambridge: Cambridge University Press, pp. 212–238.

Putnam, H. (1975) 'The meaning of "meaning"', in *Mind, Language and Reality: Philosophical Papers*, Vol. 2. Cambridge: Cambridge University Press, pp. 215–271.

Putnam, H. (1983) 'Reference and truth', in *Realism and Reason: Philosophical Papers*, Vol. 3. Cambridge: Cambridge University Press, pp. 69–86.

Putnam, H. (1990) *Realism with a Human Face*. Cambridge, MA: Harvard University Press.

Putnam, H. (1996) 'Irrealism and deconstruction', in P.J. McCormick (ed.), *Starmaking: Realism, Anti-Realism, and Irrealism*. Cambridge, MA: MIT Press, pp. 179–202.

Quezada-Euán, J.J.G. (2018) *Stingless Bees of Mexico: Biology, Management and Conservation of an Ancient Heritage*. Dordrecht: Springer.

Quine, W.V.O. (1968) 'Ontological relativity', *Journal of Philosophy* 65, 185–212.

Quine, W.V.O. (1969) 'Natural kinds', in *Ontological Relativity and Other Essays*. New York: Columbia University Press, pp. 114–138.

Rafferty, N. (2013/16) 'Pollination ecology', *Oxford Bibliographies Online*. https://www.oxfordbibliographies.com/view/document/obo-9780199830060/obo-9780199830060-0121.xml.

Rainwater, J. (1950) 'Nuclear energy level argument for a spheroidal nuclear model', *Physical Review* 79, 432.

Rainwater, J. (1975) 'Background for the spheroidal nuclear model proposal', *Nobel Lecture*, 11 December, pp. 259–270. https://www.nobelprize.org/uploads/2018/06/rainwater-lecture.pdf.

Ramsay, W. (1895) 'Helium, a gaseous constituent of certain minerals. Part I', *Proceedings of the Royal Society of London* 58, 81–89.

Ramsay, W. (1904) 'Radium and its mysteries', *Scientific American* 90, 9.

Ramsay, W. (1908) 'Percentage of the inactive gases in the atmosphere: a correction to previous calculations', *Proceedings of the Royal Society of London A* 80, 599.

Ramsey, F.P. (1929) 'General propositions and causality', in *The Foundations of Mathematics and Other Logical Essays* (ed. R.B. Braithwaite, with a preface by G.E. Moore). London: Kegan Paul, Trench, Trubner, & Co., 1931, pp. 237–255.

Ramus, F. (2001) 'Outstanding questions about phonological processing in dyslexia', *Dyslexia* 7, 197–216.

Ramus, F. (2003) 'Developmental dyslexia: specific phonological deficit or general sensorimotor dysfunction?', *Current Opinion in Neurobiology* 13, 212–218.

Rashed, R. (1994) 'Fibonacci e la matematica araba', in P. Toubert and A. Paravicini Bagliani (eds), *Federico II e le scienze*. Palermo: Sellerio editore, pp. 324–337.

Rayleigh, Lord (1879) 'On the stability, or instability, of certain fluid motions', *Proceedings of the London Mathematical Society* s1-11, 57–72.

Redish, A. (2006) 'Recent contributions to the history of monetary and international financial systems: a review essay', *European Review of Economic History* 10, 231–248.

Reefe, T.Q. (1977) 'Lukasa: a Luba memory device', *African Arts* 10, 48–50 and 88.

Reydon, T. (2016) 'From a zooming-in model to a co-creation model: towards a more dynamic account of classification and kinds', in C. Kendig (ed.), *Natural Kinds and Classification in Scientific Practice*. Abingdon: Routledge, pp. 59–73.

Rice, C. (2015) 'Moving beyond causes: optimality models and scientific explanation', *Noûs* 49, 589–615.

Rice, C. (2018) 'Idealized models, holistic distortions, and universality', *Synthese* 195, 2795–2819.

Rice, C. (2019) 'Universality and the problem of inconsistent models', in M. Massimi and C.D. McCoy (eds), *Understanding Perspectivism: Scientific Challenges and Methodological Prospects*. New York: Routledge, pp. 85–108.

Rice, C. (2021) *Leveraging Distortions*. Cambridge, MA: MIT Press.

Richlan, F., M. Kronbichler, and H. Wimmer (2009) 'Functional abnormalities in the dyslexic brain: a quantitative meta-analysis of neuroimagining studies', *Human Brain Mapping* 30, 3299–3308.

Riess, A.G., L.M. Macri, S.L. Hoffmann, et al. (2016) 'A 2.4% determination of the local value of the Hubble constant', *The Astrophysical Journal* 826, 1–31.

Robbins, B., and P.L. Horta (eds) (2017) *Cosmopolitanisms. With an Afterword by K.A. Appiah*. New York: New York University Press.

Roberts, M.N. (1998) 'The naming game: ideologies of Luba artistic identity', *African Arts* 31, 56–73 and 90–92.

Roberts, M.N., and A.F. Roberts (1996) 'Luba art and the making of history', *African Arts* 29, 22–35 and 101–103.

Roca-Royes, S. (2011) 'Conceivability and de re modal knowledge', *Noûs* 45, 22–49.

Rohwer, Y., and C. Rice (2013) 'Hypothetical pattern idealizations and explanatory models', *Philosophy of Science* 80, 334–355.

Rorty, R. (1979) *Philosophy and the Mirror of Nature*. Princeton: Princeton University Press.

Rosbaud, P. (1961) 'Victor Moritz Goldschmidt (1888-1947)', in E. Farber (ed.), *Great Chemists*. New York: Interscience Publishers, pp. 361–369.

Rosen, G. (2007) 'The case against epistemic relativism: reflections on Chapter 6 of *Fear of Knowledge*', *Episteme* 4, 10–29.

Rosenberg, J. (2002) *Thinking about Knowing*. Oxford: Oxford University Press.

Rossi, B. 2015. *From Slavery to Aid: Politics, Labour, and Ecology in the Nigerien Sahel, 1800–2000*. Cambridge: Cambridge University Press.

Rovelli, C. (1996) 'Relational quantum mechanics', *International Journal of Theoretical Physics* 35, 1637–1678.

Rovelli, C. (2017) 'Is time's arrow perspectival?', in K. Chamcham, J. Silk, J.D. Barrow, and S. Saunders (eds), *The Philosophy of Cosmology*. Cambridge: Cambridge University Press, pp. 285–296.

Rovelli, C. (2018) *The Order of Time* (trans. E. Segre and S. Carnell). London: Allen Lane.

Rovelli, C. (2021) *Helgoland* (trans. E. Segre and S. Carnell). London: Allen Lane.

Rubin, V.C., W.K. Ford, and N. Thonnard (1980) 'Rotational properties of 21 SC galaxies with a large range of luminosities and radii', *The Astrophysical Journal* 238, 471–487.

Rueger, A. (2005) 'Perspectival models and theory unification', *British Journal for the Philosophy of Science* 56, 579–594.

Ruetsche, L. (2011) *Interpreting Quantum Theories: The Art of the Possible*. Oxford: Oxford University Press.

Rumford, B. Count of (1798) 'An inquiry concerning the source of the heat which is excited by friction', *Philosophical Transactions of the Royal Society of London* 88, 80–102.

Ruphy, S. (2010) 'Are stellar kinds natural kinds? A challenging newcomer in the monism/pluralism and realism/antirealism debate', *Philosophy of Science* 77, 1109–1120.

Ruphy, S. (2011) 'From Hacking's plurality of styles of scientific reasoning to "foliated pluralism": a philosophically robust form of ontologico-methodological pluralism', *Philosophy of Science* 78, 1212–1223.

Ruphy, S. (2016) *Scientific Pluralism Reconsidered: A New Approach to the (Dis)unity of Science*. Pittsburgh: University of Pittsburgh Press.

Russo, F. (forthcoming) *Techno-Scientific Practices: An Informational Approach*. Rowman & Littlefield International.

Rutherford, E., F.W. Aston, J. Chadwick, et al. (1929) 'Discussion on the structure of atomic nuclei', *Proceedings of the Royal Society of London A* 123, 373–390.

Rutter, M., and W. Yule (1975) 'The concept of specific reading retardation', *Journal of Child Psychology and Psychiatry* 16, 181–197.

Ryle, G. (1949/2000) *The Concept of Mind*. London: Penguin Books.

Salis, F., and R. Frigg (2020) 'Capturing the scientific imagination', in A. Levy and P. Godfrey-Smith (eds), *The Scientific Imagination: Philosophical and Psychological Perspective*. Oxford: Oxford University Press, pp. 17–50.

Santos, B. de Sousa (2002) *Towards a New Legal Common Sense: Law, Globalization and Emancipation*. London: Butterworths LexisNexis.

Santos, B. de Sousa, and P. Meneses (eds) (2020) *Knowledges Born in the Struggle: Constructing the Epistemologies of the Global South*. Abingdon: Routledge.

Sarton, G. (1927) *Introduction to the History of Science*, Vol. 1. Washington, DC: Carnegie Institution of Washington; Baltimore: Williams and Wilkins.

Schaffer, S. (1989) 'Glass works: Newton's prisms and the uses of experiment', in D. Gooding, T. Pinch, and S. Schaffer (eds), *The Uses of Experiment: Studies in the Natural Sciences*. Cambridge: Cambridge University Press, pp. 67–104.

Schaffer, S. (1997) 'Experimenters' techniques, dyers' hands and the electric planetarium', *Isis* 88, 456–483.

Schaffer, S. (2004) 'A science whose business is bursting: soap bubbles as commodities in classical physics', in L. Daston (ed.), *Things That Talk: Object Lessons from Art and Science*. New York: Zone Books, pp. 147–194.

Schickore, J., and F. Steinle (eds) (2006) *Revisiting Discovery and Justification*. Dordrecht: Springer.

Schliesser, E. (2021) *Newton's Metaphysics: Essays*. Oxford: Oxford University Press.

Schmidl, P.G. (1996) 'Two early Arabic sources on the magnetic compass', *Journal of Arabic and Islamic Studies* 1, 81–132.

Schmidt, T. (1937) 'Über die magnetischen Momente der Atomkerne', *Zeitschrift für Physik* 106, 358–361.

Schüler, J.J., and T. Schmidt (1935) 'Über Abweichungen des Atomkerns von der Kugelsymmetrie', *Zeitschrift für Physik* 94, 457–468.

Schwalm, C.R., S. Glendon, and P.B. Duffy (2020a) 'RCP 8.5 tracks cumulative CO_2 emissions', *Proceedings of the National Academy of Sciences* 117, 19656–19657.

Schwalm, C.R., S. Glendon, and P.B. Duffy (2020b) 'RCP8.5 is neither problematic nor misleading', *Proceedings of the National Academy of Sciences* 117, 27793–27794.

Schwarz A., A. Gozzi, and A. Bifone (2008) 'Community structure and modularity in networks of correlated brain activity', *Magnetic Resonance Imaging* 26, 914–920.

Searle, J. (1980) '"Las Meninas" and the paradoxes of pictorial representation', *Critical Inquiry* 6, 477–488.

Seidel, L. (1993) *Jan Van Eyck's Arnolfini Portrait: Stories of an Icon*. Cambridge: Cambridge University Press.

Shapin, S. (1998) 'Placing the view from nowhere: historical and sociological problems in the location of science', *Transactions of the Institute of British Geographers* 23, 5–12.

Shaywitz, S.E., J.M. Fletcher, J.M. Holahan, et al. (1999) 'Persistence of dyslexia: the Connecticut Longitudinal Study at adolescence', *Pediatrics* 104, 1351–1359.

Shipley T.F., and J. Zacks (eds) (2008) *Understanding Events: From Perception to Action*. New York: Oxford University Press.

Shiva, V. (2016) *Biopiracy: The Plunder of Nature and Knowledge*. Berkeley: North Atlantic Books.

Siegel, D.M. (1981) 'Thomson, Maxwell and the universal ether in Victorian physics', in G.N. Cantor and M.J.S. Hodge (eds), *Conceptions of Ether: Studies in the History of Ether Theories, 1740–1900*. Cambridge: Cambridge University Press, pp. 239–268.

Siegel, L.S. (1992) 'An evaluation of the discrepancy definition of dyslexia', *Journal of Learning Disabilities* 25, 618–629.

Sjölin Wirling, Y., and Grüne-Yanoff, T. (2021) 'The epistemology of modal modeling', *Philosophy Compass*, 16, e12775.

Sjölin Wirling, Y., and Grüne-Yanoff, T. (forthcoming) 'Epistemic and objective possibility in science', *British Journal for the Philosophy of Science*. https://doi.org/10.1086/716925.

Slater, M. (2015) 'Natural kindness', *British Journal for the Philosophy of Science* 66, 375–411.

Smith, D.E. (1974) 'Women's perspective as a radical critique of sociology', *Sociological Inquiry* 44, 7–14.

Smith, D.E. (1997) 'Comments on Hekman's "Truth and method"', *Signs* 22, 392–398.

Smith, G.E. (2001) 'J.J. Thomson and the electron, 1897–1899', in J.Z. Buchwald and A. Warwick (eds), *Histories of the Electron*. Cambridge, MA: MIT Press, pp. 21–76.

Smith, J.A. (1992) 'Precursors to Peregrinus: the early history of magnetism and the mariner's compass in Europe', *Journal of Medieval History* 18, 21–74.

Snowling, M. (2019) *Dyslexia: A Very Short Introduction*. Oxford: Oxford University Press.

Snyder, J., and T. Cohen (1980) 'Reflexions on "Las Meninas": paradox lost', *Critical Inquiry* 7, 429–447.

Solomon, M. (2001) *Social Empiricism*. Cambridge, MA: MIT Press.

Solomon, M. (2015) *Making Medical Knowledge*. Oxford: Oxford University Press.

Sosa, E. (1991) *Knowledge in Perspective: Selected Essays in Epistemology*. Cambridge: Cambridge University Press.

Sosa, E. (2004) 'Replies', in J. Greco (ed.), *Ernest Sosa and His Critics*. Oxford: Blackwell, pp. 273–325.

Spangler S., A.D. Wilkins, B.J. Bachman, et al. (2014) 'Automated hypothesis generation based on mining scientific literature', *KDD '14*, 1877–1886. http://dx.doi.org/10.1145/2623330.2623667.

Spiegel, E.A. (1987) 'Chaos—a mixed metaphor for turbulence', *Proceedings of the Royal Society A* 413, 87–95.

Spinoza, B. de (2016) *The Collected Works of Spinoza*, Vol. II (ed and trans. E. Curley). Princeton: Princeton University Press.

Spohn, W. (2018) 'How the modalities come into the world', *Erkenntnis* 83, 89–112.

Stainforth D.A., M.R. Allen, E.R. Tredger, and L.A. Smith (2007a) 'Confidence, uncertainty and decision-support relevance in climate predictions', *Philosophical Transactions of the Royal Society A* 365, 2145–2161.

Stainforth D.A., T.E. Downing, R. Washington, A. López, and M. New (2007b) 'Issues in the interpretation of climate model ensembles to inform decisions', *Philosophical Transactions of the Royal Society A* 365, 2163–2177.

Stalnaker, R. (1968) 'A theory of conditionals', in N. Rescher (ed.), *Studies in Logical Theory*. Oxford: Blackwell, pp. 98–112.

Stalnaker, R. (2012) *Mere Possibilities: Metaphysical Foundations of Modal Semantics*. Princeton: Princeton University Press.

Stanford, P.K. (2003a) 'No refuge for realism: selective confirmation and the history of science', *Philosophy of Science* 70, 913–925.

Stanford, P.K. (2003b) 'Pyrrhic victories for scientific realism', *Journal of Philosophy* 11, 551–572.

Stanford, P.K. (2006) *Exceeding Our Grasp: Science, History and the Problem of Unconceived Alternatives*. Oxford: Oxford University Press.

Stanford, P.K. (2015) '"Atoms exist" is probably true, and other facts that should not comfort scientific realists', *Journal of Philosophy* 112, 397–416.

Stanford, P.K., and P. Kitcher (2000) 'Refining the causal theory of reference for natural kind terms', *Philosophical Studies* 97, 97–127.

Stanovich, K.E. (2005) 'The future of a mistake: will discrepancy measurement continue to make the learning disabilities field a pseudoscience?', *Learning Disability Quarterly* 28, 103–106.

Steigman, G. (2007) 'Primordial nucleosynthesis in the precision cosmology era', *Annual Review of Nuclear and Particle Science* 57, 463–491.

Steinle, F. (2002) 'Entering new fields: exploratory uses of experimentation', *Philosophy of Science* 64, S65–S74.

Steinle, F. (2005/2016) *Exploratory Experiments: Ampère, Faraday and the Origin of Electrodynamics* (trans. A. Levine). Pennsylvania: University of Pittsburgh Press.

Stengers, I. (2010/11) *Cosmopolitics I/II*. Minneapolis: University of Minnesota Press.

Sterelny, K. (1996) 'Explanatory pluralism in evolutionary biology', *Biology and Philosophy* 11, 193–214.

Stocker, T.F., D. Qin, G.K. Plattner, et al. (eds) (2013) *Climate Change 2013: The Physical Science Basis. Contribution of Working Group I to the Fifth Assessment Report of the Intergovernmental Panel on Climate Change*. Cambridge University Press, Cambridge, United Kingdom and New York, NY, USA. https://www.ipcc.ch/site/assets/uploads/2018/03/WG1AR5_SummaryVolume_FINAL.pdf.

Stone, R. (1992) 'The Biodiversity Treaty: Pandora's box or fair deal?', *Science* 256, 1624.

Stoney, G.J. (1874/1894) 'Of the "electron," or atom of electricity', *The London, Edinburgh, and Dublin Philosophical Magazine and Journal of Science* 5 (38), 418–420.

Strathern, M. (1996) 'Cutting the network', *Journal of the Royal Anthropological Institute* 2, 517–535.

Strawson, P.F. (1959) *Individuals*. London: Methuen.

Stroud, B. (2020) 'Knowledge from a human point of view', in A. Crețu and M. Massimi (eds), *Knowledge from a Human Point of View*. Dordrecht: Springer, pp. 141–148.

Stuewer, R. (1994) 'The origin of the liquid drop model and the interpretation of nuclear fission', *Perspectives on Science* 2, 76–129.

Suárez, M. (2004) 'An inferential conception of scientific representation', *Philosophy of Science* 71, 767–779.

Suárez, M. (ed.) (2009) *Fictions in Science: Philosophical Essays on Modeling and Idealization*. London: Routledge.

Suárez, M. (2015a) 'Deflationary representation, inference, and practice', *Studies in History and Philosophy of Science* 49, 36–47.

Suárez, M. (2015b) 'Representation in science', in P. Humphreys (ed.), *The Oxford Handbook of Philosophy of Science*. Oxford: Oxford University Press, pp. 440–459.

Suess, H.E. (1947a) 'Über kosmische Kernhäufigkeiten: I. Mitteilung. Einige Häufigkeitsregeln und ihre Anwendung bei der Abschätzung der Haufigkeitswerte für die mittelschweren und schweren Elemente', *Zeitschrift für Naturforschung* 2A, 311–321.

Suess, H.E. (1947b) 'Über kosmische Kernhäufigkeiten: II. Mitteilung. Einzelheiten in der Häufigkeitsverteilung der mittelschweren und schweren Kerne', *Zeitschrift für Naturforschung* 2A, 604–608.

Suess, H.E. (1988) "V.M. Goldschmidt and the origin of the elements', *Applied Geochemistry* 3, 385–391.

Sugden, R. (2009) 'Credible worlds, capacities and mechanisms', *Erkenntnis* 70, 3–27.

Sullivan, P.F., R.R. Pattison, A.H. Brownlee, S.M.P. Cahoon, and T.N. Hollingsworth (2016) 'Effect of tree-ring detrending method on apparent growth trend of black and white spruce in interior Alaska', *Environmental Research Letters* 11, 114007–114012.

Sunder, M. (2007) 'The invention of traditional knowledge', *Law and Contemporary Problems* 70, 97–124.

Sunder, M. (2018) 'Intellectual property in experience', *Michigan Law Review* 117, 197–258.

Tahko, T.E. (2015) 'Natural kind essentialism revisited', *Mind* 124 (495), 795–822.

Tahko, T.E. (2020) 'Where do you get your protein? Or: biochemical realization', *British Journal for the Philosophy of Science* 71, 799–825.

TallBear, K. (2014) 'Standing with and speaking as faith: a feminist-indigenous approach to inquiry', *Journal of Research Practice* 10 (2), N17. http://jrp.icaap.org/index.php/jrp/article/view/405/371.

TallBear, K. (2016) 'Dear indigenous studies, it's not me, it's you. Why I left and what needs to change', in A. Moreton-Robinson (ed.), *Critical Indigenous Studies: Engagements in First World Locations*. Tucson: University of Arizona Press, pp. 69–82.

Tan, A., et al., PandaX-II Collaboration (2016) 'Dark matter results from first 98.7 days of data from the PandaX-II experiment', *Physical Review Letters* 117, 121303(7).

Taylor, C. (1992) 'The politics of recognition', in A. Gutmann (ed.), *Multiculturalism: Examining the Politics of Recognition*. Princeton: Princeton University Press, pp. 25–73.

Teller, P. (2010) '"Saving the phenomena" today', *Philosophy of Science* 77, 815–826.

Teller, P. (2011) 'Two models of truth', *Analysis* 71, 465–472.

Teller, P. (2019) 'What is perspectivism, and does it count as realism?', in M. Massimi and C.D. McCoy (eds), *Understanding Perspectivism: Scientific Challenges and Methodological Prospects*. New York: Routledge, pp. 49–64.

Tenenbaum J.B., C. Kemp, T.L. Griffiths, and N.D. Goodman (2011) 'How to grow a mind: statistics, structure, and abstraction', *Science* 331, 1279–1285.

Textor, M. 2020. 'States of affairs', *Stanford Encyclopedia of Philosophy* (Summer 2020 Edition), E.N. Zalta (ed.). https://plato.stanford.edu/archives/win2016/entries/states-of-affairs/.

Thomson, J.J. (1891) 'XXI. On the illustrations of the properties of the electric field by means of tubes of electrostatic induction', *The London, Edinburgh, and Dublin Philosophical Magazine and Journal of Science* 31 (190), 149–171.

Thomson, J.J. (1893) *Notes on Recent Researchers in Electricity and Magnetism*. Oxford: Clarendon Press.

Thomson, J.J. (1897) 'Cathode rays', *Philosophical Magazine* 44, 293–316.

Thomson, J.J. (1904) *Electricity and Matter: Silliman Lectures*. New York: Charles Scribner's Sons.

Thomson, J.J. (1906) 'Carriers of negative electricity'. *Nobel Lecture*, 11 December, pp. 145–153. https://www.nobelprize.org/prizes/physics/1906/thomson/lecture/.

Thorne, K.S. (2019) 'John Archibald Wheeler: a biographical memoir'. https://arxiv.org/abs/1901.06623v1.

Tobia, K.P., G.E. Newman, and J. Knobe (2019) 'Water is and is not H2O', *Mind and Language* 35, 183–208.

Tobin, E. (2010) 'Microstructuralism and macromolecules: the case of moonlighting proteins', *Foundations of Chemistry* 12, 41–54.

UN (1950) *These Rights and Freedoms*. Lake Success, NY: United Nations, Department of Public Information.

UN (1992) Convention on Biological Diversity, 5 June (1760 U.N.T.S.79). https://treaties.un.org/doc/Treaties/1992/06/19920605%2008-44%20PM/Ch_XXVII_08p.pdf.

UN (2014) Human Rights Council, Twenty-Seventh Session, Joint Written Statement, 28 August (A/HRC/27/NGO/81).

UN (2019a) Traditional Knowledge: Generation, Transmission and Protection. United Nations Economic and Social Council, Permanent Forum on Indigenous Issues, Eighteenth Session, New York, 22 April–3 May (E/C.19/2019/5).

UN (2019b) Implementation of the Convention on Biological Diversity and Its Contribution to Sustainable Development, United Nations General Assembly, Second Committee, Seventy-Fourth Session, 21 November (A/C.2/74/L.66).

van Fraassen, B. (1980) *The Scientific Image*. Oxford: Clarendon.

van Fraassen, B. (2002) *The Empirical Stance*. New Haven: Yale University Press.

van Fraassen, B. (2008) *Scientific Representation: Paradoxes of Perspective*. Oxford: Oxford University Press.

van Inwagen, P. (1998) 'Modal epistemology', *Philosophical Studies* 92, 67–84.

Vellutino, F.R., D.M. Scanlon, E.R. Sipay, et al. (1996) 'Cognitive profiles of difficult to re-mediate and readily remediated poor readers: early intervention as a vehicle for distin-guishing between cognitive and experiential deficits as basic causes of specific reading disability', *Journal of Educational Psychology* 88, 601–638.

Verde, L., T. Treu, and A.G. Riess (2019) 'Tensions between the early and the late universe', *Nature Astronomy* 3, 891–895.

Verreault-Julien, P. (2019) 'How could models possibly provide how-possibly explanations?', *Studies in History and Philosophy of Science* 73, 22–33.

Vickers, P. (2013) 'A confrontation of convergent realism', *Philosophy of Science* 80, 189–211.

Vickers, P. (2017) 'Understanding the selective realist defence against the PMI', *Synthese* 194, 3221–3232.

Viveiros de Castro, E. (1998a) 'Cosmological deixis and Amerindian perspectivism', *Journal of the Royal Anthropological Institute* (N.S.) 4, 469–488.

Viveiros de Castro, E. (1998b) *Cosmological Perspectivism in Amazonia and Elsewhere*. Chicago: Hau Books.

Vogel, J. (2000) 'Reliabilism levelled', *Journal of Philosophy* 97, 602–623.

Vogel, J. (2008) 'Epistemic bootstrapping', *Journal of Philosophy* 105, 518–539.

Waldron, J. (1992) 'Minority cultures and the cosmopolitan alternative', *University of Michigan Journal of Law Reform* 25, 751–793.

Waldron, J. (2006) 'Cosmopolitan norms', in R. Post (ed.), *Another Cosmopolitanism: The Berkeley Tanner Lectures*. Oxford: Oxford University Press, pp. 83–98.

Watson, M.C. (2011) 'Cosmopolitics and the subaltern: problematizing Latour's idea of the commons', *Theory, Culture & Society* 28, 55–79.

Way, K. (1939) 'The liquid-drop model and nuclear moments', *Physical Review* 55, 963–965.

Weber, A. (2018) 'Renegotiating debt: chemical governance and money in the early nineteenth-century Dutch empire', in L.L. Roberts and S. Werrett (eds), *Compound Histories: Materials, Governance and Production, 1760-1840*. Leiden/Boston: Brill, pp. 205–225.

Weisberg, M. (2006) 'Robustness analysis', *Philosophy of Science* 73, 730–742.

Weisberg, M. (2007) 'Three kinds of idealization', *Journal of Philosophy* 104, 639–659.

Weisberg, M. (2013) *Simulation and Similarity*. New York: Oxford University Press.

Weiskopf, D. (2020) 'Representing and coordinating ethnobiological knowledge', *Studies in History and Philosophy of Science C* 84, 101328.

Werrett, S. (2019) *Thrifty Science: Making the Most of Materials in the History of Experiment*. Chicago: University of Chicago Press.

White, S.D.M., J.F. Navarro, A.E. Evrard, and C. Frenk (1993) 'The baryon content of galaxy clusters: a challenge to cosmological orthodoxy', *Nature* 366, 429–433.

Wiggins, D. (1980) *Sameness and Substance*. Cambridge, MA: Harvard University Press.

Wiggins, D. (2001) *Sameness and Substance Renewed*. Cambridge: Cambridge University Press.

Willett, C. (ed.) (1998) *Theorizing Multiculturalism: A Guide to the Current Debate*. Oxford: Blackwell.

Williams, L. (1964) 'Laticiferous plants of economic importance V. Resources of gutta-percha-*Palaquium* species', *Economic Botany* 18, 5–26.

Williamson, T. (2016) 'Modal science', *Canadian Journal of Philosophy* 46, 453–492.

Williamson, T. (2020) *Suppose and Tell: The Semantics and Heuristics of Conditionals.* Oxford: Oxford University Press.

Wilson, A. (2020) *The Nature of Contingency: Quantum Physics as Modal Realism.* Oxford: Oxford University Press.

Wilson, D.B. (1971) 'The thought of late Victorian physicists: Oliver Lodge's ethereal body', *Victorian Studies* 15, 29–45.

Wimsatt, W.C. (1981) 'Robustness, reliability, and overdetermination', in M.B. Brewer and B.E. Collins (eds), *Scientific Inquiry and the Social Sciences.* San Francisco: Jossey-Bass, pp. 124–163.

Winsberg, E. (2018) *Philosophy and Climate Science.* Cambridge: Cambridge University Press.

Winterbottom, A. (2016) *Hybrid Knowledge in the Early East India Company World.* Basingstoke: Palgrave Macmillan.

Woodward, J. (1989) 'Data and phenomena', *Synthese* 79, 393–472.

Woodward, J. (1998) 'Data, phenomena, and reliability', *Philosophy of Science* 67, S163–S179.

Woodward, J. (2003) *Making Things Happen: A Theory of Causal Explanation.* Oxford: Oxford University Press.

Woodward, J. (2007) 'Causation with a human face', in H. Price and R. Corry (eds), *Causation, Physics and the Constitution of Reality: Russell's Republic Revisited.* Oxford: Oxford University Press, pp. 66–105.

Woodward, J. (2021) *Causation with a Human Face.* New York: Oxford University Press.

Worley, S. (2003) 'Conceivability, possibility, and physicalism', *Analysis* 63, 15–23.

Wray, K.B. (2011) *Kuhn's Evolutionary Social Epistemology.* Cambridge: Cambridge University Press.

Wylie, A. (1997) 'Good science, bad science, or science as usual? Feminist critiques of science', in L.D. Hager (ed.), *Women in Human Evolution.* London: Routledge, pp. 29–55.

Wylie, A. (2003) 'Why standpoint matters', in R. Figueroa and S. Harding (eds), *Science and Other Cultures: Issues in Philosophies of Science and Technology.* New York: Routledge, pp. 26–48.

Wylie, A. (2015) 'A plurality of pluralisms: collaborative practice in archaeology', in F. Padovani, A. Richardson, and J.Y. Tsou (eds), *Objectivity in Science: New Perspectives from Science and Technology Studies.* Dordrecht: Springer, pp. 189–210.

Wylie, A. (2020) George Sarton Memorial Lecture 'The Indigenous/Science project: collaborative practice as witnessing'. https://hssonline.org/about/honors/george-sarton-memorial-lecture/.

Xu, F., and J.B. Tenenbaum (2007) 'Word learning as Bayesian inference', *Psychological Review* 114, 245–272.

Yablo, S. (1993) 'Is conceivability a guide to possibility?', *Philosophy and Phenomenological Research* 53, 1–42.

Young, I.M. (1997) 'Unruly categories: a critique of Nancy Fraser's dual systems theory', *New Left Review* 222, 147–160.

Yule, W., M. Rutter, M. Berger, and J. Thompson (1974) 'Over- and under-achievement in reading: distribution in the general population', *British Journal of Educational Psychology* 44, 1–12.

Zinke, J., M. Pfeiffer, O. Timm, W.C. Dullo, and G.R. Davies (2005) 'Atmosphere–ocean dynamics in the Western Indian Ocean recorded in corals', *Philosophical Transactions of the Royal Society A* 363, 121–142.

Zwicky, F. (1933) 'Die Rotverschiebung von extragalaktischen Nebeln', *Helvetica Physica Acta* 6, 110–127.

Index

For the benefit of digital users, indexed terms that span two pages (e.g., 52–53) may, on occasion, appear on only one of those pages.
Figures are indicated by *f* following the page number